Creep and Shrinkage

Hubert Rüsch
Dieter Jungwirth
Hubert K. Hilsdorf

Creep and Shrinkage

Their Effect on the Behavior of Concrete Structures

With 186 Illustrations

Springer-Verlag
New York Heidelberg Berlin

Hubert Rüsch
formerly of
Technische Universität
Munich
Federal Republic of Germany

Dieter Jungwirth
Dyckerhoff & Widmann AG
Munich
Federal Republic of Germany

Hubert K. Hilsdorf
Institut für Baustofftechnologie
Universität Karlsruhe
Karlsruhe
Federal Republic of Germany

Library of Congress Cataloging in Publication Data

Rüsch, Hubert, 1903–1979
Creep and shrinkage.

Bibliography: p.
Includes index.
1. Concrete—Creep. 2. Concrete—Expansion and contraction. 3. Reinforced concrete construction.
I. Jungwirth, Dieter. II. Hilsdorf, Hubert K.
III. Title.
TA440.R79 624.1′8341 82-737
AACR2

Softcover reprint of the hardcover 1st edition 1983

Typeset by Composition House Ltd., Salisbury, England.

9 8 7 6 5 4 3 2 1

ISBN-13: 978-1-4612-5426-3 e-ISBN-13: 978-1-4612-5424-9
DOI: 10.1007/978-1-4612-5424-9

Dedication

Hubert Rüsch, the senior author of this volume, who died in 1979, was one of the great international figures in the field of reinforced concrete. He combined outstanding professional achievements in design, research and teaching with a warm outgoing personality. Aided by his knowledge of five languages, this combination made him a most effective leader in international organizations and led to many personal-professional friendships in a variety of countries.

As a practicing engineer, both in Germany and in Argentina, he was responsible for the design and, often, construction of exceptional concrete structures such as large shell roofs, domes, aircraft hangars, industrial plants, harbor structures, and precast and prestressed concrete systems. His distinguished academic career began in 1948 when he accepted a professorship in concrete construction at his Alma Mater, the Technische Universität München. In his additional related capacity as director of the Materials Testing Institute, which he greatly enlarged and modernized, he conducted and supervised research on a large scale. This resulted in important basic contributions in such areas as the structural performance of concrete, creep, shrinkage, crack formation, inelastic flexural theory, concrete quality control, limit design, and safety theory based on probabilistic reasoning, to mention only the most important ones.

His far-reaching international impact was based not only on these engineering achievements, but also very much on his multilinguality and, particularly, on the ease and warmth with which he established and maintained close personal relationships. When the European Concrete Committee (CEB) began its work in 1953, he immediately became one of its leading members as a Founding Vice President, later its President, and finally its Honorary President. In these capacities, he helped guide the CEB in the creation of its internationally important Model Code, and also established close liaison between the CEB and the American Concrete Institute.

Among his international connections, those with the United States were particularly close. For one year he was Distinguished Visiting Professor at Cornell

University, and later held a similar appointment at the University of Texas in Austin. At both institutions his interaction with colleagues and students was immediate, close, productive, and often long lasting.

His multifarious activities and his 130 publications, including books translated into several languages, brought him many honors and distinctions in Germany and elsewhere. Of these, only those bestowed on him in the United States will be mentioned here: He was elected Foreign Associate of the National Academy of Engineering, an Honorary Member of the American Concrete Institute, which also conferred on him its Wason Medal and Alfred Lindau Award, and he received the Longstreth Medal of the Benjamin Franklin Institute in Philadelphia.

This great man enjoyed life in many of its aspects, in his closeness to his family, in his love for nature, music, art, history, travel, skiing, and mountain climbing, experiences shared on many occasions with this writer.

He will be long and fondly remembered.

Cornell University, Ithaca, NY GEORGE WINTER

Preface

This book is based on *Reinforced Concrete—Prestressed Concrete, Volume 2, Accounting for the Effects of Creep and Shrinkage on the Behavior of Structural Systems* by Hubert Rüsch and Dieter Jungwirth, which appeared in German in 1976. Even then, it was Hubert Rüsch's fervent wish to have his thoughts on the deformations of concrete translated into English in order to reach a wider audience. His earlier efforts to contribute a study to the Series of Monographs of the American Institute had unfortunately not succeeded.

Despite a serious illness, Hubert Rüsch undertook, with his characteristic prudence and thoroughness, the preparatory work for the translation and related revision of his book. Unfortunately fate did not grant him the satisfaction of seeing his work completed. Hubert Rüsch died on October 17, 1979.

In writing this book, Hubert Rüsch drew on his many years of devoted study of the creep problem. These investigations go back to 1934. His awareness of the plastic deformation of concrete under sustained load, which had been reported to him on the occasion of an American sojourn, led him to discover the causes of a major building collapse. At his urging, Professor A. Hummel published, in 1935, a critical survey of the test results then available on concrete creep.

Soon after, he began his work on a German code for prestressed concrete, the first to be completed anywhere in the world. They were set down in DIN 4227 which, however, could only be put into effect after World War II. This code became a model for similar guidelines in other parts of the world and contributed decisively to the headlong growth of the use of structural prestressed concrete. The code contains a section written by Hubert Rüsch, which is one of the first attempts to predict creep in concrete.

In the course of his activities as a researcher and university professor, Hubert Rüsch intensively studied the effect of time on the strength and deformation of concrete. He initiated and directed numerous experimental and theoretical investigations into this problem. In 1958, he organized a RILEM colloquium on "The influence of time upon strength and deformation of concrete," which took place at his institute in Munich and made a significant contribution to the clarification of many open questions on the subject of creep. His pioneering work on the time-related behavior of concrete was assembled in the publication *Researches Toward a General Flexural Theory for Structural Concrete* which in 1962 was honored with the Wason Medal of the American Concrete Institute.

In his last years, he worked on the principle of superposition which was enunciated by his friend and colleague, Douglas McHenry. As in all his work, he strove here to determine the underlying principles without losing sight of the design engineer's practical needs.

What did he and his co-authors want to achieve with the present book?

Decades have passed since Woolson discovered concrete creep in 1905, since Davis and Glanville started systematic tests in 1930 to elucidate this behavior, and since 1937 when Dischinger published the pertinent differential equations. There has been a stream of publications on this subject over the years. Most of these have focused on the rheological properties of concrete. The work of O. Wagner (1958) and A. Neville's book (1970) provided a summary and critical review of these experimental findings. However, mathematical treatment of creep and shrinkage problems in structural engineering developed more slowly. In addition, very few countries have incorporated the solutions proposed by a number of authors into general design practice. It is hoped that the present volume will lead to a faster dissemination of present knowledge. It has been written to serve the builder who wants immediate advice on his practical problems as well as those readers who need to know not only "how" but also "why".

November 1982

DIETER JUNGWIRTH
HUBERT HILSDORF

Table of Contents

List of Symbols

Upper Case Letters

A	cross-sectional area; aging effect
B	support B
C	creep fiber according to Busemann; column of a frame; cement; coefficient
E	modulus of elasticity
F	force; concentrated load
H	horizontal thrust; humidity
I	moment of inertia
L	concentrated life load
M	moment
N	normal force
$N_I N_{II}$	equivalent normal forces
S	action effect; center of gravity
T	half life; temperature
U, W	structural member U coupled with different structural member W
V	shear force; volume
W	water; section modulus
X	statically indeterminate action effects
Z	tensile force

Lower Case Letters

a	deflection
b	width
c	spring constant
d	diameter; differential; uniformly distributed dead load; effective depth

d_c	thickness of concrete cover
e	2.718; eccentricity
f	normal stress
f'_c	specified compressive strength of concrete
f'_{cm}	average compressive strength of concrete
h	total depth of beam; thickness of slab; diameter of concrete member
k	coefficient with proper dimension
l	length of column; span; length of member or element
m	unit moment
n	unit normal force; modular ratio E_e/E_c
$q = d + w$	total unit load
r	radius; radius of gyration
$1/r$	curvature
s	spacing of reinforcing bars; unit snow load; standard deviation
t	time
u	circumference; camber
v	shear stress; unit shear force; volume concentration
w	unit live load; crack width; unit weight
y	lateral deflection; distance of center of reinforcement to centroid of concrete section

Lower Case Greek Letters

α	stiffness ratio; degree of hydration; moment coefficient; angular change of prestressing tendon; dimensionless coefficient
β	dimensionless coefficient; angle; unintended angular change of prestressing tendon
γ	additional safety with regard to buckling; total angular change of prestressing tendon; safety factor; creep factor
δ	dimensionless coefficient; displacement
ξ	dimensionless coefficient
ε	strain
ρ	relaxation coefficient; A_s/A_c = reinforcement ratio
λ	time coefficient
η	coefficient
μ	friction coefficient
φ	creep coefficient
$\varkappa$	dimensionless coefficient

Lower Case Greek Subscripts

γ	safety against buckling

Upper Case Subscripts

B	at support B
C	cracking
D	deformation
DIN	according to German industrial specifications
E	according to Euler
F	at midspan
L	concentrated life load
S	soil
T	temperature
U, W	structural member U coupled with different structural member W

Lower Case Subscripts

a	aggregate
adm	admissable
b	basic; bottom; buckling
c	creep; concrete
ct	contribution of tensile strength of concrete
ce	cement
cs	creep and shrinkage
d	delayed elastic; dead load
dr	drying
e	elastic
eff	effective
f	flow
i	initial
id	idealized
i	1, 2, 3, 4...
m	average value
p	prestress; prestressing tendon
r	action effects or stresses causing cracking; sudden restraint
rel	relaxation
req	required
s	secant modulus; steel reinforcement; shrinkage
sus	sustained load
t	top; tangent; time effect; at time t; tension
tot	total
u	ultimate
v	shear
w	temporary effect
y	yield

Special Subscripts

0	initial value; basic value; basic system; time $t = 0$
1, 2, 3	particular quantities in sequence 1, 2, 3

h	hour
d	days
∞	at time infinity
Δ	finite difference
$\Delta\Delta$	difference of Δ
$\sum$	sum
I, II	particular creep fibers

Superscripts

e	elastic, i.e. neglecting creep effects
I	uncracked sections (concrete)
II	cracked sections (concrete)
0	open system
∞	transition to closed system
$28d$	28 days
	compression

Further symbols are defined in the text as needed

Introduction

When calculating the effects of creep and shrinkage on the behavior of concrete structures, three problem areas are of particular importance:

the knowledge of all possible effects and thus the practical problems requiring a creep and shrinkage analysis [1].

the development of realistic expedients for estimating the coefficient of creep and shrinkage;

information on reliable and simple analytical methods to calculate the effects.

The number of publications on this subject has reached almost avalanche proportions over the last two decades. In particular, this is true for test reports dealing with the study of the various influences on creep deformation. Until 1970, 758 publications are reported [2], [3], and [4]. It shows that since Wagner's [5] summarizing work, published in 1958, which lists 208 references, the amount of knowledge has increased rapidly.

The following explanations are an attempt to provide a condensed review of the present state of knowledge. The three problem areas mentioned above are dealt with in three largely independent sections and explained in the supplements.

As a rule, the influence of steel creep (see Part B, Section 5) can be neglected. Part C, Section 4.3.3 is an exception.

In general, the SI units (International System of Units) are used in the book. In order to facilitate use, the American units are often given in parentheses.

Whenever an equation contains dimensional factors, a second figure is given in parentheses to be used in calculations with American units. If conversions are required, the following conversion factors have to be used:

To convert from:	To:	Multiply by:
m	ft	3.28
m	in.	39.37
m^2	ft^2	10.76
mm^2	in^2	0.00155
N	lbs	0.225
$N/mm^2 = MN/m^2$	psi	145
kg/m^3	lbs/ft^3	0.062

References

[1] Rüsch, H.: Fundamentals for computing the effects of concrete creep, International Association for Bridge Construction and Structural Engineering, Symposium, Madrid, Final Report, 1970.

[2] Shrinkage and creep in concrete, ACI Bibliography No. 7 (1905–1966), American Concrete Institute Committee 209, 1967.

[3] List of additional references to creep and volume changes of concrete (published during period 1901–1964), American Concrete Institute Committee 20, 1967.

[4] Shrinkage and creep in concrete: ACI Bibliograph No, (1966–1970) American Concrete Institute Committee 209, 1970.

[5] Wagner, O.: "Creep in plain concrete," *Ger. Committee Reinforced Concr.* No. 131 (1958).

Part A

How Do Creep and Shrinkage Affect the Behavior of Structures?

For the reader in a hurry:

This part was conceived because even very experienced engineers occasionally are uncertain whether creep and shrinkage can influence action effects[1] or the stress distribution. Section 1 describes the various types of stresses, and Sections 2 and 3 investigate the effect of creep and/or shrinkage on these various types of stresses. Section 4 then summarizes the conclusion of these considerations, breaking them down by type of effect on plain concrete, reinforced concrete, prestressed concrete and composite structures. Thus, if you want a fast answer to your questions, start your search in Section 4; you may then, if you desire, find justification for the answer in Sections 2 and 3, based on the terminology provided in Section 1.

1. Differentiation Among Types of Stresses and of Action Effects According to Their Causes

In order to facilitate an understanding of the effect that creep and shrinkage have on a given structure, it is useful to break down the stresses and action effects[1] occurring in a reinforced concrete structure into the following groups:

Load-induced stresses and action effects result from externally applied loads. Their most significant feature is that they are necessary to satisfy conditions of equilibrium. In statically indeterminate structural systems they must, in addition, satisfy conditions of compatibility.

The generally accepted designation "load" strictly speaking, encompasses only stresses that are caused by loads (such as dead weight or snow) but not those which are caused by forces (such as deceleration or wind pressure). But since the consequences are the same, force induced stresses are included among load induced stresses, as well.

[1] In this book, the term "action effect" is used as a synonym for the various forces acting in a particular section, i.e., moments, shear forces, or normal forces.

Imposed stresses and imposed action effects result from a constraint which, in a statically indeterminate structure, acts between that structure and its supports. This constraint may be caused by a deformation of the structure itself (such as the linear shrinkage or linear temperature gradient) or it may be an external influence (such as the differential settlement of supports or the consequence of the construction process). The constraint is often created intentionally in order to obtain a favorable effect on the stress distribution (Fig. 1.1). This may be achieved, for example, through specific alteration of the construction process.

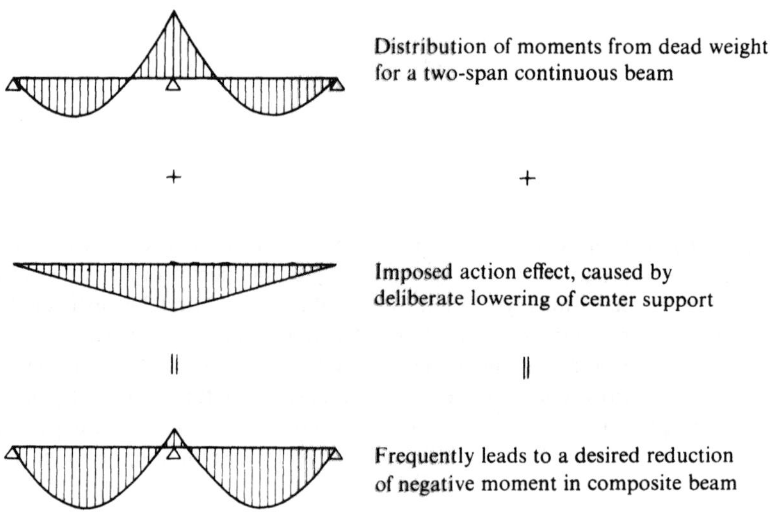

Figure 1.1. Consequences of imposed action effects.

Imposed stresses may be classified as action effects or as reactions, whose role it is to re-establish the conditions of compatibility violated by the imposed deformation. This is why action effects due to imposed deformations are proportional to the stiffness of the structural system. Thus, they decrease as the deformability of the structure grows under the effect of creep or cracking.

Internal stresses occur when the strains created in the various sections of a structure are not mutually compatible (such as nonlinear shrinkage or nonlinear temperature-related strains). The internal stress distribution then has the function of re-establishing compatibility through related strains. Under Hooke's Law, $\sigma = \varepsilon E$, the internal stresses are proportional to the modulus of elasticity. These stresses balance each other within an overall cross section, i.e., their sum equals zero. Neither action effects nor reactions are created.

Naturally, internal stresses can also be created by intentionally imposed strains, such as a pretensioning of tendons embedded in concrete. This type of prestressing is discussed in more detail in Section 3.4.

2. The Effects of Shrinkage

Shrinkage is defined as the reduction in volume of an unloaded concrete at constant temperature. Its primary cause is the loss of water during a drying process. The magnitude of this deformation is described by the shrinkage strain ε_s. The inverse process is called swelling, but it is of little significance in actual practice.

The shrinkage process starts at the surface that is exposed to drying, and it gradually penetrates into the concrete. The corresponding contractions would cause a warping of the cross section, which is usually impossible because of compatibility requirements. Normally, plane sections must remain plane. Shrinkage is then divided into three parts (Fig. 2.1): the mean contraction, ε_s; the part that

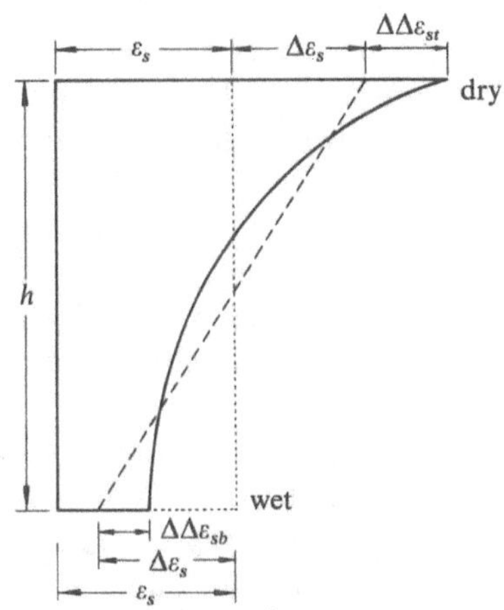

Figure 2.1. Separation of total shrinkage into shrinkage components.

causes the warping, $\Delta\varepsilon_s$; and the part which causes internal stresses, $\Delta\Delta\varepsilon_s$. Generally, neither unimpeded contraction nor unimpeded warping can take place in a statically indeterminate structure. Therefore, a constraint is created which imposes action effects (for example, in a frame). In statically determinate structures, these shrinkage components cause only deformations and do not create any stresses. The $\Delta\Delta\varepsilon_s$ component creates internal stresses in all structures.

Since shrinkage is a gradual process, the corresponding stresses act over extended time periods and are reduced by creep. This is why the question when to take shrinkage stresses into account will be handled together with the problem of creep (cf., Sections 3.2–3.5).

In determining the reduction in shrinkage stresses due to creep, it is best to proceed by first calculating the shrinkage stresses and corresponding action effects disregarding creep. The values thus obtained may be used for the creep calculations (cf., Part C).

3. The Effects of Creep

The creep characteristics of construction materials are defined by the creep coefficient $\varphi = \varepsilon_c/\varepsilon_e$ (cf., Section 4).

3.1. The Effect of Creep on Load-Dependent Action Effects

3.1.1. Statically Determinate Structures

In statically determinate structures, the action effects created by the applied loads are determined exclusively by the conditions of equilibrium. Thus, creep can have an effect only in statically indeterminate structures.

3.1.2. Statically Indeterminate Structures

When the entire structure is made of one material with uniform creep characteristics, as in the case of plain concrete structures, creep cannot affect action effects. This principle applies to reinforced concrete, as well, if the reinforcement distribution is approximately uniform and symmetrical. This is readily understood since one can consider the effects of creep as equivalent to a gradual decrease in the modulus of elasticity.

Load-dependent action effects are affected by creep when elements of the structure are made of concrete of different ages or different compositions, (e.g., normal concrete and lightweight concrete). Generally, the action effects will then be redistributed from the elements with a higher creep rate to the elements with a lower creep rate.

The interaction of structural elements made of two different materials, such as a concrete arch with a steel tie, accentuates the difference in creep deformation. Since the effects of the creep of steel are generally disregarded, the distribution of the action effects may change noticeably in the course of time.

In reinforced concrete two different materials interact; but, as already stated, this is of consequence only if the reinforcement ratio varies or if the reinforcement is asymmetrical. With normally small reinforcement ratios, these effects may be disregarded. All the more so, since in determining action effects, one considers the much more significant effect of cracked sections to be negligible.

When prestressing is applied—and this is also true for partial prestressing—the reinforcement ratio is usually smaller than in reinforced concrete. So in this case also, one is therefore justified in disregarding the effect of creep on the action effects developed by applied loads.

However, such simplification is no longer applicable when composite construction is involved. The proportion of steel within the total cross-sectional area is too large and, generally, its distribution is too non-uniform over the length of a girder.

3.1.3. Theory of Second Order

To be sure, the foregoing considerations apply only as long as the action effects can be calculated on the basis of the theory of first order. When moments have to be determined using the theory of second order, as is the case for buckling problems, taking into account the deformation of the member, the effect of creep can no longer be disregarded. This is also true of statically determinate structural elements.

3.2. The Effect of Creep on Imposed Action Effects

It has been mentioned that action effects due to imposed deformations are proportional to stiffness. Creep reduces this stiffness in all concrete structures. Consequently, action effects due to imposed deformations will decrease in the course of time in all structures, whether they be made of plain, reinforced, or prestressed concrete, and also if they are composite systems.

This decrease in the imposed stresses can be both advantageous and harmful. It is almost always desirable when the constraint is due to natural causes, such as shrinkage, temperature differentials, or settlement of the supports. When, for example, the soil settles farther under the most traveled lane of a concrete highway than under the adjacent lane with less traffic, the concrete should be able to adapt itself like a soft rug to the trough caused by the settlement. In such a case, one would want a concrete with high creep. By contrast, such a decrease is undesirable when the constraint is created in order to affect advantageously the stress distribution. This is the case, for instance, in composite bridges, in prestressed shells, or in continuous girders. For these, the desirable concrete would be one with low creep characteristics.

3.3. The Effect of Creep on Internal Stresses

As stated earlier, the role of internal stresses is to compensate for the strains developed in any given cross section. In general, this requires that plane sections remain plane.

Creep can affect internal stresses in two ways. It can reduce existing internal stresses to the extent that they are of a continuous nature, and it can create its own new internal stresses when materials with different creep characteristics interact in a given cross section. This is shown in the following examples.

Internal stresses are caused primarily by non-uniform volume changes e.g. due to the heat of hydration during hardening of the concrete or, later, to the shrinkage as it proceeds from the surface inward (cf., Section 2). In dimensioning reinforcement, one may almost always disregard these internal stresses. In general, they are risky only at the moment of their inception as they are reduced rapidly and markedly by creep. One must primarily see to it that such internal stresses are kept small enough and that they develop slowly. That is why the requirement that fresh concrete be protected from rapid drying and cooling is of particular significance. It is the only way to prevent that the race between tensile stress and tensile strength (Fig. 3.1) results in the development of cracks.

Frequently, sustained internal stresses are created on prupose. An example of this is the state of internal stress caused by prestressing, whose beneficial effect decreases under the action of creep and shrinkage. Due to the encompassing importance of this problem, it is treated in detail in Section 3.4.

However, there are more harmful internal stresses: those caused by solar radiation. In old concrete that has lost a large part of its deformability, these

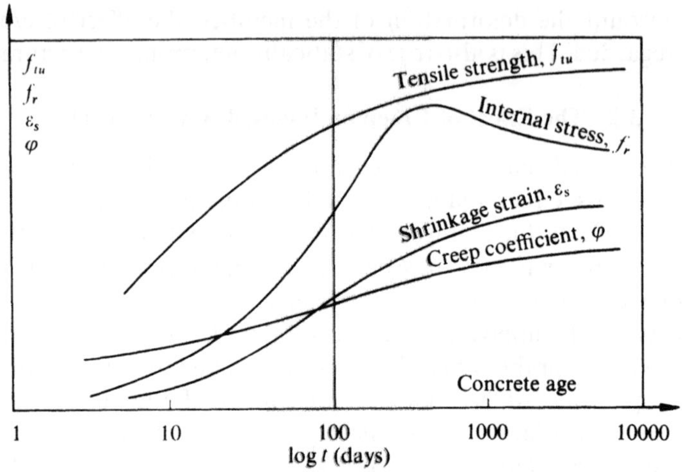

Figure 3.1. Example of development of internal shrinkage stresses. They pass through a maximum due to the effect of creep. As long as the tensile stress does at no time surpass the tensile strength no cracks occur.

stresses may become very high, and they cannot be reduced by creep because thermal stresses develop rapidly and are of short duration. These internal stresses are only rarely taken into account in the analysis. However, surface reinforcement is often provided to prevent the formation of large cracks.

Internal stresses created by creep itself occur primarily in composite materials. One example would be a reinforced concrete column under permanent load. Creep causes contraction of concrete but not of steel. Since plane sections must remain plane, the stresses decrease in the concrete over the course of time, as the stresses increase in the steel. The same redistribution is produced by shrinkage; these internal stresses are, in turn, reduced by creep. As a result of this redistribution caused by creep and shrinkage, compression reinforcement in under-reinforced columns may reach the yield point even under working loads. In order to prevent the columns from buckling, an adequate concrete cover and an appropriate use of lateral ties are required.

In the dimensioning of structural reinforced concrete elements, these internal stresses are never taken into account. They may, however, require structural precautions, such as a close spacing of stirrups in members that will be subjected to high compressive stresses. In prestressed concrete, the level of tensile stresses under working loads is of particular significance. In this connection, one must take into consideration the above-mentioned redistribution of stress. However, this cannot have a measurable effect on load carrying capacity, since the strains caused by the internal stresses are much too small, compared to the failure strain of the materials involved.

However, one cannot forego a verification of the internal stresses created by creep when the steel is placed unsymmetrically and its contribution to the load carrying capacity is very large. This applies, for instance, to composite girders where the creep-induced stress redistribution between concrete and steel may have an appreciable effect on behavior under working loads. The same is true of the redistribution of shrinkage stresses.

3.4. The Effect of Creep on Prestressing

In a statically determinate structure with bonded tendons, initial pretensioning will induce internal stresses in the cross sections of a structural element, since the integral of the steel stresses in each cross section is equal to the integral of the concrete stresses. In the case of unbonded prestressing, on the other hand, it is doubtful whether one is justified in speaking of an internal stress state, since at any section of the structure two distinct cross sections exist: the concrete cross section and the prestressed tendon, which are, however, interrelated. The integral of the stresses acting in an individual section is then no longer zero. The stresses can be grouped into action effects. Thus, this implies a state of constraint if the two cross sections are considered separately. For the total cross section, however, an internal stress state exists similar to pretensioning with bonded tendons.

The compression stresses induced in the concrete by prestressing lead to a reduction of the prestressing force under the influence of the creep of concrete. The shortening due to shrinkage also reduces the prestressing force. When prestressing is applied by means of elastic tendons, the reduction depends on the compression stresses permanently acting along the tendon and on the stiffness of the tendons. It may range from 5 % to 30 % of the initial prestress. If the prestressing is created by forces acting from rigid abutments, the reduction will be considerably greater.

Owing to creep and shrinkage, the reduction of the internal stresses induced by prestressing naturally is proportional to the prestressing force. The same is true of imposed constraints in unbonded prestressing.

However, prestressing can also cause action effects of a different type. They result from the fact that the structure is deformed by the state of stress imposed upon it. Such deformation is unimpeded only in the case of statically determinate systems. In statically indeterminate structures, they are usually not compatible with the support conditions. Since these action effects are caused by the prestressing force which acts upon the concrete cross section as though it were an external load, this type of action effects can be reduced by creep only in the same proportion as the prestressing force that induces them. Thus, they do not behave like imposed action effects, but rather like load-induced action effects. This is why the expression "redistribution action effects" has been adopted.

3.5. The Effect of Creep on Deflections of Structural Elements

The breakdown into various types of stress described above is also useful in evaluating the increased deflections caused by creep.

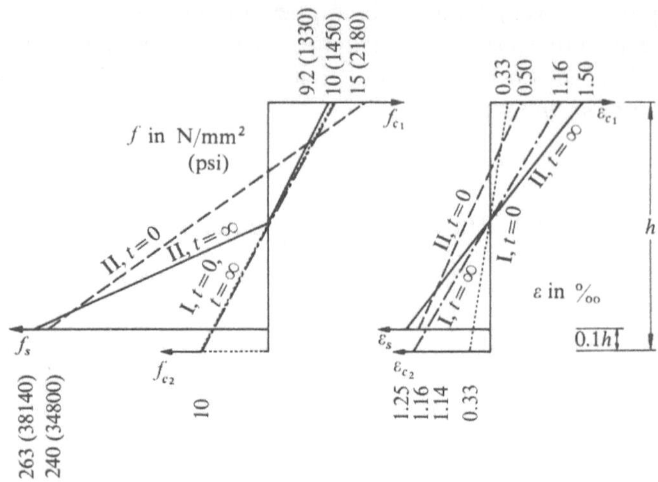

Figure 3.2. Comparison of stresses and strains in uncracked (I) and cracked (II) sections, with and without the influence of creep.

3.5.1. *Load-Induced Stresses*

In plain concrete, the deflections caused by load-induced stresses are increased by an amount proportional to the creep coefficient. This is approximately true also for the uncracked region of reinforced concrete members. In the cracked region, on the other hand, the process is considerably more complex. First of all, even before creep occurs, the deflection is increased considerably by crack formation, as opposed to the case of uncracked sections, since the compression stresses increase in the concrete while, on the tension side, only the steel takes up stresses. This is illustrated by the example given in Fig. 3.2. While the concrete of the compression zone creeps, the neutral axis is lowered. This reduces the compression stresses in the concrete, and creep is slowed down. Therefore, the increase in deflection produced by creep is smaller in uncracked sections than in cracked sections. Nevertheless, total deflection is increased because of the contribution of crack formation, as clearly shown in Table 3.1.

Under load-induced stresses, prestressed concrete behavior is very similar to that of plain concrete. What has been said about reinforced concrete and prestressed concrete applies essentially to composite structures as well.

3.5.2. *Imposed and Internal Stresses*

The deformations which are caused by imposed action effects and internal stresses are generally not affected by creep. However, if the imposed deformation acts on the concrete structure by means of an elastic member, such constraint will exert an effect similar to that of a sustained load. Additional creep deformations are created. Prestressing is a typical example of this process.

Table 3.1. Supplement to Fig. 3.2.

State	Time	ε_{c1}	ε_s; ε_{c2}	$h:r = \dfrac{\varepsilon_{c1} + \varepsilon_{c2}}{0.9h} \cdot h$	Relative Curvature $h:r$ at time $t = 0$ in percent: Uncracked (I)	Cracked (II)
Uncracked	$t = 0$	0.33	0.33	0.66	100	36
	$t = \infty$; $\varphi = 2.5$	1.16	1.16	2.31	350	127
Cracked	$t = 0$	0.50	1.14	1.83	277	100
	$t = \infty$; $\varphi = 2.5$	1.50	1.25	3.06	464	167

Increase in curvature in percent of elastic, uncracked state (I).
Uncracked concrete:
- $t = 0$: 100–100 = 0%
- creep; $t = \infty$: 350–100 = 250%.

Cracked concrete:
- increase due to cracking: 277–100 = 177%
- creep; $t = \infty$: 464–277 = 187%
- cracking and creep: 464–100 = 364%

Increase in curvature in percent of elastic, cracked state (II).
Cracked concrete:
- creep; $t = \infty$: 167–100 = 67%.

4. Summary of the Effects of Creep and Shrinkage

If the creep deformation is expressed in terms of the ratio $\varphi = \varepsilon_c/\varepsilon_e$, i.e., the creep strains under constant stress ε_c and the elastic strain ε_e, while shrinkage is expressed in terms of the shrinkage strain ε_s, the foregoing explanations about the effect of creep and shrinkage under working loads can be summarized as follows. Estimates of these effects are given for greater clarity.

In plain concrete:

Load-induced action effects and stresses which may be calculated by first-order theory do not change under the effect of creep. If second-order theory is applied, one must take into account the effect of the deformations described in detail below. This may be significant, e.g., for long-spanned arches.

Imposed action effects and stresses are reduced through creep by a factor of about $1/2\varphi$, when they are generated rapidly; when they develop only gradually—as in shrinkage—this reduction factor is increased by a constant factor of about 0.2.

Essentially the same applies for internal stresses, e.g., imposed stresses. However, they may be disregarded if they can be kept small by providing expansion joints at sufficiently close intervals and by insuring appropriate curing of the concrete.

If an analysis is made under first-order theory, the deformation caused by sustained load-induced action effects increases to a $(1 + \varphi)$-fold value over the course of time. If second-order theory is used, this increased deformation must be accounted for. Applying the latter theory, the deformations induced by imposed action effects may change, since the effect of creep also raises the action effects. One may always disregard the contribution of internal stresses to deformation.

In reinforced concrete:

Within the range of first-order theory, load-induced action effects are not affected by creep. Thus, changes in load-induced action effects need be considered only when resistance to buckling is verified under second-order theory, it being understood that one may limit this to very slender columns with a high fraction of sustained loads.

When simplified calculations of uncracked sections are carried out to determine the imposed action effects, because only isolated cracks are expected, one must apply the rules that govern plain concrete. The imposed action effects calculated for cracked sections are considerably reduced by crack formation. Creep causes further reduction. Under the assumption of uncracked sections, a factor of roughly $1/(1 + 0.3\varphi)$ may be used for a rapidly developing constraint. In the case of a constraint caused by shrinkage or by foundation settlement in heavy soil, the factor becomes about $1/(1 + 0.2\varphi)$. If the constraint is applied under the assumption of cracked sections, the respective reduction factors are

$$\frac{1}{(1 + 0.1\varphi)} \quad \text{and} \quad \frac{1}{(1 + 0.08\varphi)}.$$

As a rule, internal stresses may be disregarded in reinforced concrete. If they are of a sustained nature, they are considerably reduced by creep. When non-uniform temperature distribution has a significant effect, it is compensated by reinforcement close to the surface.

Owing to the redistribution of internal stresses, deformations created by sustained load-induced stresses for cracked sections are increased over the course of time by a factor of only $(1 + 0.3\varphi)$. Creep sensitivity is so much lower than in plain concrete that the effect of imposed action effects on deformation may be disregarded, and this applies to internal stresses as well.

In partially prestressed concrete:

Load-induced action effects remain unchanged, provided there is no need to apply second-order theory in the analysis.

Under imposed action effects caused by a constraint of given magnitude, reinforced concrete will behave like plain concrete. The redistribution of action effects induced by prestressing decreases owing to the effect of creep at about the same rate as the mean value of the prestressing force by which it is created. Shrinkage effects are similar.

Internal stresses caused by a non-uniform strain distribution (e.g., shrinkage and temperature) are reduced if applied over extended periods. Internal stress conditions created by prestressing are reduced by a factor of approx. $1/2\varphi$

through creep when the prestressing acts against rigid supports. When the prestressing is created by means of elastic tendons, reduction of the prestressing force by creep and shrinkage—save in exceptional cases—may range from 8 % to 20 %. The higher value applies to structures where a high ratio of live load to dead load creates a large creep-inducing stress, as well as to the use of low-quality prestressing steel.

Deformations created by sustained load-induced stresses, in the case of full prestressing, increase by a factor of approximately $(1 + \varphi)$. As the degree of prestressing is lowered, one approaches the values applicable to reinforced concrete. However, this effect is fully felt only in the range of partial prestressing. The same factors apply to the deformation caused by prestressing that counteracts the load-induced stresses. Yet in order to take into account the fact that the prestressing force is reduced by the effect of creep and shrinkage (cf., internal stresses), the analysis should be based on the mean value of the prestressing force which acts during the creep process.

In composite constructions:

Composite structures differ from reinforced or prestressed concrete merely by their higher percentage of steel and its non-uniform distribution. Thus differences are only gradual.

The effect of creep on load-induced action effects is too large to be disregarded.
There is less reduction in the imposed action effects.
The same is true of internal stresses.
Deformations do not increase as much.

Part B

Behavior of Materials

For the reader in a hurry:

Section 1 and parts of Section 3 in Part B which deal primarily with rheological problems are essentially aimed at those readers who are engaged in research or teaching. Sections 2, 3.1, 4, and 5 are of particular interest to the designer, since they discuss methods for the prediction of concrete and steel deformation. Sections 6 and 7 then round out the picture.

Specifically, the individual sections cover the following subject matter:

Section 1. A description of the causes of deformation.
Section 2. The development of compressive and tensile strength of concrete.
Section 3. Proposals for an improved estimate of the modulus of elasticity and of the modulus of deformation of concrete, as well as data on Poisson's ratio. Test results on the magnitude of the various components of creep and shrinkage.
Section 4. A description of two methods for predicting concrete deformations.
Section 5. Deformation characteristics of reinforcing and prestressing steels.
Section 6. A comparison between some deformations measured in actual structures and values predicted in accordance with Section 4.
Section 7. Prospects for future developments.

1. General Rheological Description of the Deformation Process in Concrete

The realization that concrete not only shrinks but also creeps when subjected to sustained stresses is a relatively recent insight. Early in this century, Woolson [1] and others (cf., [2]) had recognized the creep phenomenon in principle and had begun the first investigations. However, it was not until 1930 that a systematic study of this property was undertaken through the work of Davis [3] and of Glanville [4]. Thereafter, interest grew very rapidly. Very soon it became barely possible to retain an overview of the innumerable publications on the subject. By 1946, the American

Concrete Institute Committee 209 was entrusted with organizing studies on "volume changes and plastic flow in concrete" and to insure a critical review of the findings. In 1958, Wagner [5] published his analysis of most of the studies published up to that time. This provided the first overview. In the more than 20 years since then, countless further test reports have been published. They can be found, for example, in the ACI bibliographies ([6] and [7]). Beyond that, there developed a knowledge of the underlying physical principles which can offer the key to the multiplicity of observed phenomena. It therefore became necessary to assemble and review critically the insights and findings accumulated to date. A notable contribution along these lines was Neville's book [2]. In the following exposition, there is also a rough overview of the new insights and their effect on the suggestions regarding the evaluation of the magnitude of creep and shrinkage. This includes particular attention to the needs of the engineer faced with practical design problems.

1.1. Creep

1.1.1. Concept

In rheology, the study of the mechanical behavior of deformable bodies, a distinction is often made between creep and relaxation experiments, which examine, respectively, the course of strain under sustained stress and the stresses that correspond to a strain that is kept constant. The problems that interest the engineer fall mostly between these two cases. He sees creep as the property of a material to develop time-dependent deformations under the effect of stresses, and he assumes that the two rheological concepts cited above have a common cause.

In the following, creep in concrete is described both in terms of the creep strain ε_c (Fig. 1.1) and, in the usual manner, by the creep coefficient $\varphi = \varepsilon_c/\varepsilon_e$. All statements refer exclusively to the behavior of normal concrete under working loads, i.e. under stresses that do not exceed approx. $0.5\,f'_c$. Tensile creep is not examined since it is of less significance and since there are not enough test results to document differing opinions.

The analytical description of the magnitude of creep, as well as the understanding of the physical processes that cause creep in concrete, are made considerably simpler by breaking down creep into several components. Such components are delayed elastic strain and flow, or basic creep and drying creep. The following sections examine the magnitude of these creep components

1.1.2. Separation of Creep into Delayed Elastic Strain and Flow

It was realized early on that the concept "creep" lacks precise definition because it encompasses not only irreversible but also reversible strain components. For if a test piece that has been subjected to a sustained load is relieved of the load, one observes a noticeable strain recovery (cf., Fig. 1.1).

In this connection, one used to speak of reversible creep—and some do to this day. In 1935, Hummel [8] was probably the first to suspect that this involves delayed elasticity. Almost all materials—though in greatly varying degree—exhibit the property that part of the reversible strain appears with some delay. In most steels, this strain component is actually so small that almost their entire creep strain under high stresses is of a permanent nature. In some rocks, on the other hand, strains caused by sustained loading are 80% reversible and form about 5% of the immediate elastic strain [10]. In many polyphase materials, the delayed elastic strain is considerably greater. This includes concrete, in which the predominantly viscous hydrated cement paste interacts with predominantly elastic aggregates. In that case, delayed elasticity reaches a level of about 40% of the instantaneous elastic strain (cf., Section 2.1.2).

The development in time of the delayed elastic strain component also varies considerably among the various materials. However, as a rule, one can assume that the higher the fraction of the elastic strain, the slower the recovery. The half-life ranges from a few hours to several weeks for concrete.

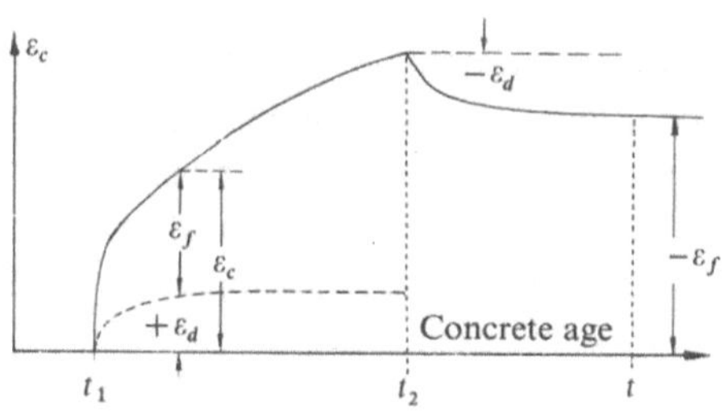

Figure 1.1. Schematic presentation of a creep experiment which starts at a concrete age t_1 and which ends at a concrete age t_2. Creep, ε_c, is separated into delayed elasticity, ε_d, and into flow ε_f.

Delayed elastic strains can be observed only after the load has been removed, but, to the extent that it encompasses only elastic strains, the consensus is that it is of the same magnitude under load (Fig. 1.1).

Consequently we have the equation

$$\varepsilon_c = \varepsilon_d + \varepsilon_f. \tag{1.1}$$

(creep = delayed elastic deformation + flow)

This breakdown is particularly significant when the creep-inducing stress decreases considerably within the period of observation—either because the sustained load changes or because creep reduces the stresses created by some constraint (relaxation). In such cases, not only the elastic strains but also the delayed elastic strains counteract the stress reduction resulting from the flow.

The following example, in which it was assumed that the sustained loads act only for a limited time, shows the error that is committed by not taking into account the reversibility of creep:

Strain component	With breakdown	Without breakdown
When load is applied	$+\varepsilon_e$	$+\varepsilon_e$
Under sustained load	$+\varepsilon_d + \varepsilon_f$	$+\varepsilon_c$
When load is removed	$-\varepsilon_e$	$-\varepsilon_e$
After removal of load	$-\varepsilon_d$	
Total	$+\varepsilon_f$	$+\varepsilon_c - \varepsilon_d + \varepsilon_f$

1.1.3. *Differentiating Between Rapid Initial Strain and Flow*

The observation that a considerable portion of flow occurs on the first day after the sustained load is applied has received little attention in the past, despite the fact that studies have pointed it out on several occasions ([11] and [12]). This "rapid initial strain" is particularly pronounced in young concrete (Fig. 1.2). Up to now, it has been ascribed to the delayed elastic strain [12]. In fact, it is mostly irreversible.

Since the rapid initial strain differs clearly from the other flow components both by the strain rate and by its pronounced dependence on loading age, i.e., the age of

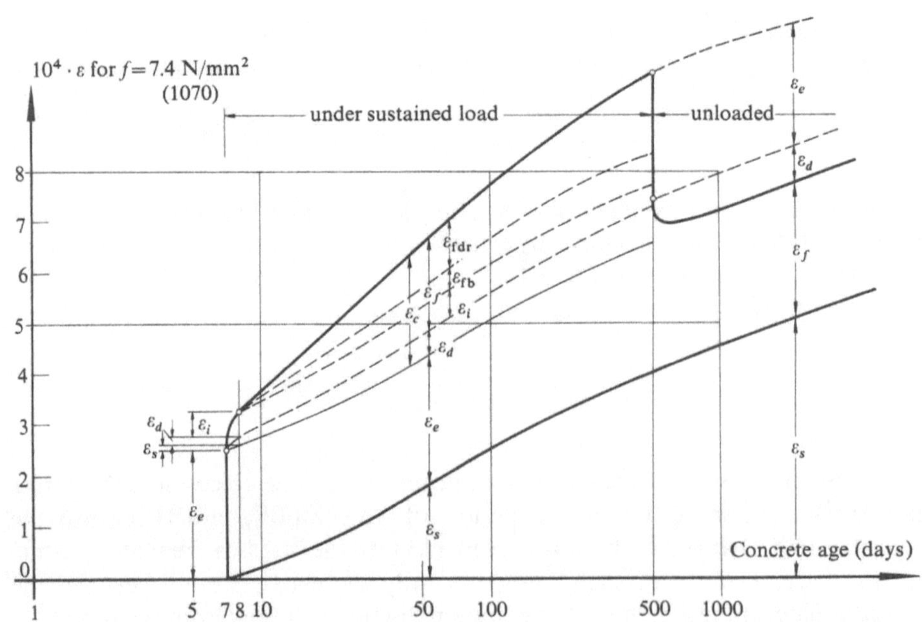

Figure 1.2. Measured values of creep and separation into strain components.

the concrete when load is first applied, it is appropriate to deal with it separately when considering the rheological model representations. We have as a result:

$$\varepsilon_c = \varepsilon_d + \varepsilon_i + \bar{\varepsilon}_f. \tag{1.2}$$

(creep = delayed elastic strain + rapid initial strain + remaining flow)

1.1.4. Distinction between Basic Flow and Drying Flow

The very significant proposal to distinguish clearly between phenomena induced by distinct causes, namely, basic creep and drying creep, was first alluded to by Neville [13] in 1955, introduced by Hansen [14] in the interpretation of creep, and fully validated by Ali and Kesler [15] by 1964. In this context, basic creep is understood to be the creep ε_{cb} that occurs under sealed conditions, i.e., no moisture exchange takes place with the ambient atmosphere, while drying creep, ε_{cdr}, designates the increase in creep strain resulting from the drying process. The predominant cause of basic creep is viewed as the property of delayed elasticity as well as the viscous deformation of the hydrated cement gel. Drying creep is ascribed to an accelerated movement of water molecules in the pore system of the hydrated cement paste caused by the external load.

If, in accordance with Sections 1.1.2 and 1.1.3, the strain components ε_d and ε_i are first separated, the remaining flow, $\bar{\varepsilon}_f$, can only be broken down into the two components, basic flow ε_{fb} and drying flow ε_{fdr}:

$$\bar{\varepsilon}_f = \varepsilon_{fb} + \varepsilon_{fdr}. \tag{1.3}$$

(remaining flow = basic flow + drying flow)

From the causes presented for these two strain components, one can conclude that basic flow depends primarily on cement paste content and moisture content, while drying flow is closely related to shrinkage and is therefore mainly affected by the moisture gradient and the dimensions of the member. Like shrinkage, drying flow can be partially reversed by wetting. This characteristic, which is of little interest in practice, was demonstrated by tests [16] in which the test pieces where placed under water after the load had been removed.

1.2. Shrinkage

Shrinkage designates the time-dependent strains which concrete suffers at constant temperature without any external strains. Since it is essentially a consequence of the drying process, the dimensions of the structural element have a great influence on the shrinkage strain, ε_s. This influence has often been underestimated. Thick elements lose water very slowly. Actual shrinkage starts only after the capillary water has evaporated and gel water is lost. This again shows the close relationship to drying flow which also takes place in gel water.

1.3. Influence of Temperature on Creep and Shrinkage

Creep and shrinkage strains increase with temperature since the deformability of the hydrated cement gel is greater and drying is accelerated. As the temperature rises, its influence becomes increasingly pronounced. Thus, creep shows only a slight dependence on temperature for concrete protected against drying at temperatures of 0 °C to 20 °C, (32 °F to 68 °F) whereas a temperature rise from 20 °C(68 °F) to 100 °C (212 °F) increases creep by a factor of 2 to 3.

When concrete is allowed to dry at the higher temperature, there occurs an initially intensified drying flow; but later the flow rate drops markedly since dry concrete flows very little. Naturally, this process is highly affected by the thickness of the structural member.

In special cases, such as in reactor technology, one can therefore not disregard the effect of temperature on creep. Unfortunately, within the framework of this book, we must forego the study of such special cases. Under normal conditions, one may assume that a mean temperature of 0° C (32 °F) hardly changes creep characteristics with respect to 20 °C, (68 °F), but that a mean temperature of 40 °C (104 °F) increases creep by about 25%, on the average.

1.4. Summary

In the foregoing explanations, creep strain, ε_c has been broken down into several components which not only differ in their rheological cause, but also behave differently with respect to external influences. These conclusions are summarized in the following chart (which includes shrinkage) and are illustrated by an example in Fig. 1.2.

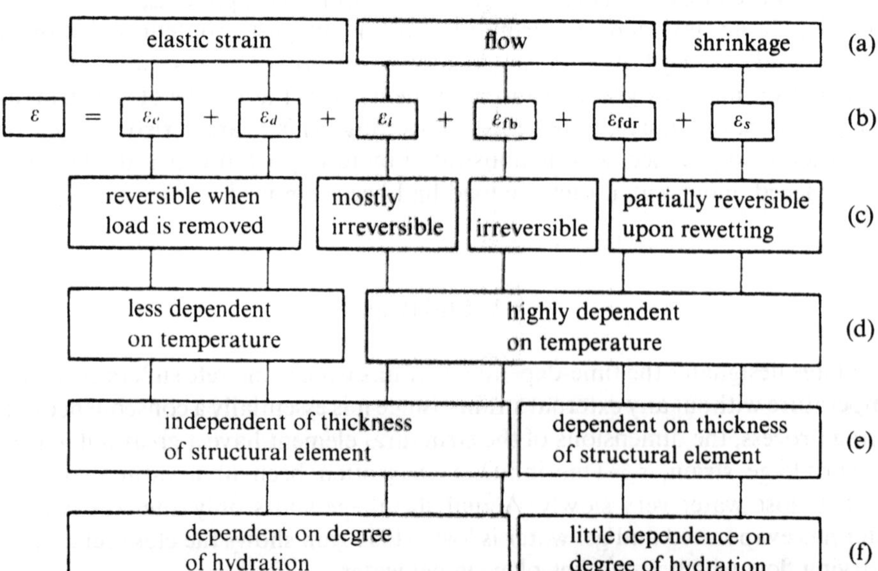

2. Development of the Strength of Concrete

2.1. Effect of Time on the Compressive Strength of Concrete

The rate of hydration, and consequently the time-related process of strength development of concrete, depends on the type of cement used. In Fig. 2.1,[1] the coefficient $\beta_t = f'_{ct}/f'_{c28}$ for concrete made of normal cement (Type I), of rapid

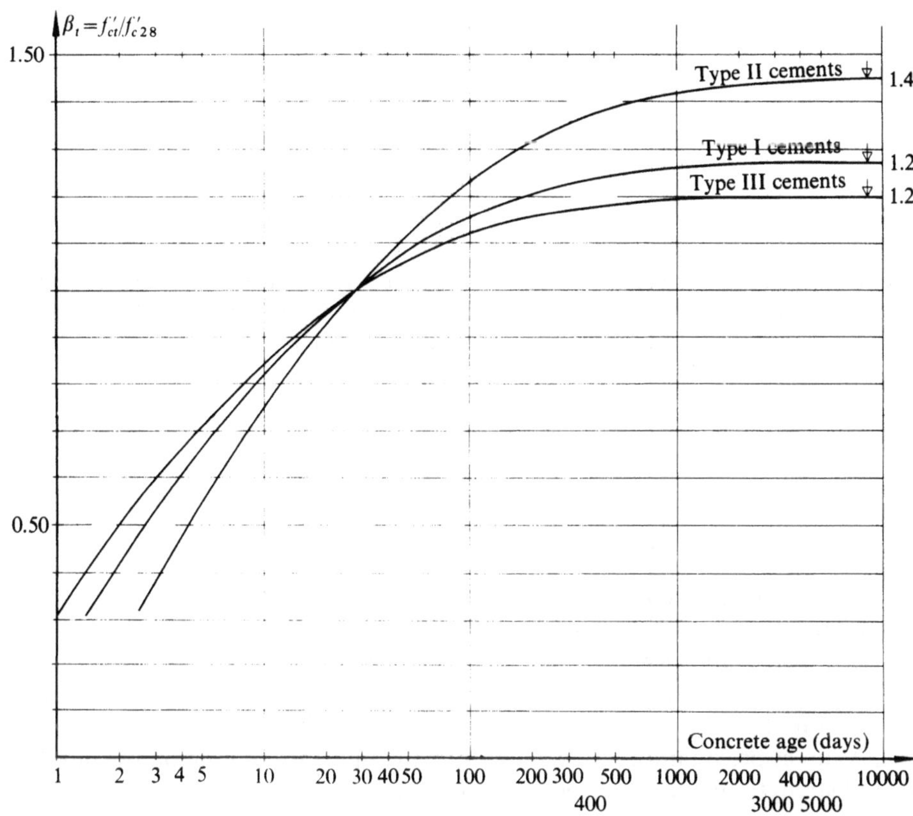

Figure 2.1. Coefficient $\beta_t = f'_c/f'_{c28}$ to describe the effect of concrete age on concrete compressive strength.

[1] The relations given in Fig. 2.1 are based on experiments on concrete made with German cements. Here, as well as in the subsequent sections, it has been assumed that the properties of the following German and US cements are similar:

Germany (DIN 1164)	USA (ASTM C 150)	
Z25; Z35L; Z45L	Type II; IV	slow
Z35F: Z45F	Type I	normal
Z55	Type III	rapid

cement (Type III), and of slow cement (Type II) is plotted against the age of the concrete. These curves apply to concretes that were moist cured up to an age of 7 days at approximately 20 °C (68 °F) and were then allowed to dry. If the storage temperature is above 20 °C, hydration is accelerated, and at lower temperatures, hydration of concrete is correspondingly slower. The effect of temperature can be accounted for by substituting for the actual age an effective age given by Eq. (2.1) [38]:

$$t = \sum \frac{T\,°\mathrm{C} + 10}{30} \Delta t \tag{2.1}$$

or

$$t = \sum \frac{T°(\mathrm{F}) + 14}{82} \Delta t.$$

In Eq. (2.1), Δt is the number of calendar days at a nearly constant concrete temperature of T °C (T °F). For $T = 20$ °C (68 °F), the effective age and the actual age are identical.

The influence of the type of cement can also be taken into account by adjusting the age of the concrete. This is particularly advantageous in order to take into consideration the influence of the type of cement on concrete creep (discussed later). To this effect, we set

$$\begin{aligned} t &= \beta_{ce} \sum \frac{T\,°\mathrm{C} + 10}{30} \Delta t, \\ t &= \beta_{ce} \sum \frac{T\,°\mathrm{F} + 14}{82} \Delta t. \end{aligned} \tag{2.2}$$

In this formula, β_{ce} equals 2.0 for a Type I cement; β_{ce} equals 1.0 for a Type II cement; and β_{ce} equals 3.0 for a Type III cement.

Inserting the appropriate value for the coefficient β_{ce}, the development of strength with time for the various type of cement can be described by a single curve shown in Fig. 2.2. This relationship was determined on the basis of a large number of tests. As an approximation, we can describe the influence of time on the development of strength by means of the following equation:

$$\frac{f'_{ct}}{f'_{c\infty}} = 1 - 1.4\sqrt{t_{\mathrm{eff}}}. \tag{2.3}$$

2.2. The Tensile Strength of Concrete

The following equation expresses the relationship between the tensile strength of concrete and its compressive strength:

$$f_{tu28} = C\sqrt[3]{(f'_{c28})^2}. \tag{2.4}$$

Using the coefficient represented in Fig. 2.1, we then obtain the following equation for the influence of time on the tensile strength:

$$f_{tut} = C\sqrt[3]{(\beta_t f'_{c28})^2}. \tag{2.5}$$

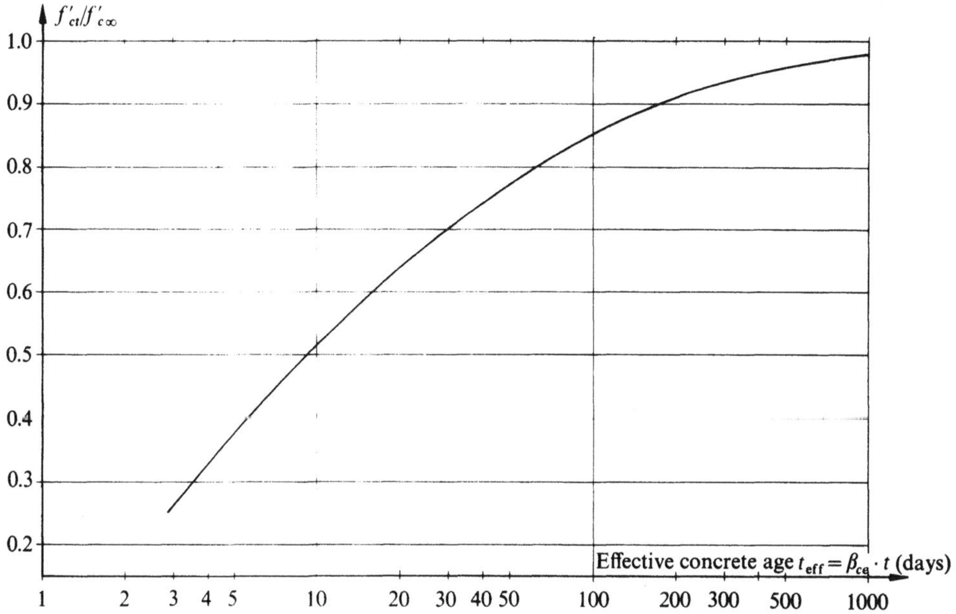

Figure 2.2. Relation between relative concrete strength and effective concrete age.

Values for C which are taken from a literature survey (17) are given in Table 2.1. They are valid for members that are 100 mm (4″) thick. For members with a thickness of 400 mm (16″), one obtains values which are roughly 25 percent lower (cf., [17] for details)

Table 2.1

Type of strength	Lower limit $C_{5\%}$ (N/mm^2)	(psi)	Mean value C_m (N/mm^2)	(psi)	Upper limit $C_{95\%}$ (N/mm^2)	(psi)
Concentric tensile strength	0.19	0.97	0.27	1.40	0.35	1.83
Flexural strength	0.39	2.05	0.57	2.64	0.62	3.24

The factors C given in Table 2.1 are valid only for laboratory tests where the experimental scatter is limited. Under practical conditions, one must expect considerably larger scatter, which is also dependent on the quality of the concrete used. In Table 2.2 values are proposed for the factor C under practical conditions. They are taken from [18].

Table 2.2

Type of strength	Range of compressive strength (N/mm²)	(psi)	Lower limit $C_{5\%}$ (N/mm²)	(psi)	Mean value C_m (N/mm²)	(psi)	Upper limit $C_{95\%}$ (N/mm²)	(psi)
Concentric	10/20	2000/3000	0.23	1.21	0.41	2.13	0.59	3.05
tensile	20/30	3000/4500	0.22	0.13	0.36	1.86	0.50	2.59
strength	30/60	4500/9000	0.21	0.11	0.33	1.70	0.44	2.29
Flexural	10/20	2000/3000	0.47	2.43	0.77	3.99	1.06	5.50
strength	20/30	3000/4500	0.45	2.34	0.67	3.51	0.90	4.67
	30/60	4500/9000	0.44	2.29	0.62	3.21	0.79	4.13

3. Data on the Magnitude of Strain

3.1. Modulus of Elasticity

3.1.1. Significance

As a rule, it is not necessary to know the modulus of elasticity of concrete E_c with precision in order to determine the stresses and action effects caused by loads. It is enough to use the approximations according to Section 3.1.4. However, it is another matter when imposed or internal stresses occur, for instance, in the case of prestressing. This applies particularly when the concrete is strained at a very early age, since the data given in Section 3.1.4 are valid for 28-day-old concrete. In addition, the type of aggregates and the rate of loading often have an effect that can no longer be disregarded, even when considering only the domain of normal concrete. More precise information on E_c is also of importance for an estimate of deflections.

The formulas derived below are summarized in Section 3.1.8.

3.1.2. General View

Widely differing suggestions regarding the dependence of E_c on strength of the concrete may be found in the literature. This is due in part to the fact that up to now no standard test method has been agreed on for the determination of E_c. As shown by Fig. 3.1, which is not to scale, there are palpable differences between the initial tangent modulus E_t, the secant modulus E_s, and the unloading modulus, E_e, which are even magnified when one distinguishes between the values obtained for a first loading and repeated loading. The order of magnitude of the differences is shown by the example presented below. It is valid for 28-day-old concrete loaded in the range of working stress, and it does not include the permanent strains induced by repeated loads. In old concrete, the differences are reduced, while in very young concrete, they are larger. The secant modulus measured during loading is the one least affected by permanent strains and is, therefore, most frequently regarded as the governing

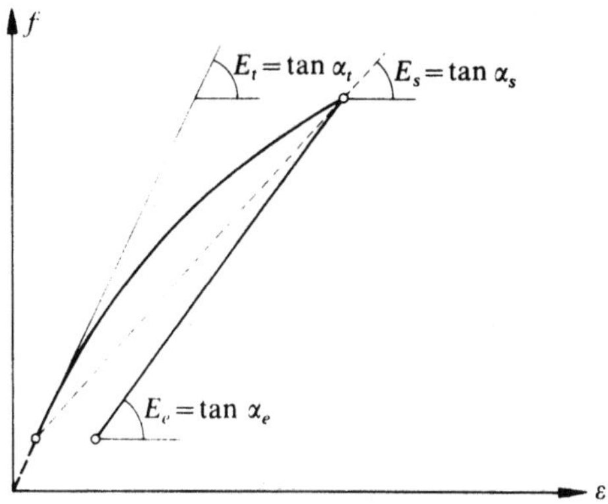

Figure 3.1. Schematic presentation of strains upon initial loading.

parameter. Assigning to it a value of 100 percent, one obtains the following relationship:

	Tangent modulus	Secant modulus	Unloading modulus
At first loading	112%	91%	106%
After 10 load repetitions	-	100%	101%

Another cause of differences among test results is the lack of an agreed standard loading and unloading rate. Further causes are the effects of minute surface cracks in the specimens which may depend on the manufacturing and the storage procedures of the concrete.

For the engineer faced with practical problems, the problem of estimating the modulus of elasticity is significant since he normally makes only a rough differentiation between short- and long-term loads. He assumes that the effect of loads of short duration is defined with sufficient precision by values for E_c provided in Fig. 3.1, and that time-dependent deformations need be taken into account only in the case of sustained loads. However, this is too great a simplification of the actual process. The concept "short duration" comprises loading periods that range from approximately a few seconds to several days, and this does not consider whether such loads are applied only once or repeatedly. In the following, therefore, an attempt is made to present reasonably realistic moduli.

3.1.3. *Definitions*

The base value of the modulus of elasticity E_0.

The base value of the modulus of elasticity is defined as the secant modulus for a load up to $f_c'/3$, reached within approximately 1 min. This type of loading is nearly equivalent to the case where a rapidly applied stress acts for about 30 sec.

The Modulus of Deformation, E_D

In practice, even changes in stress which are generated at a considerably slower rate are often considered of "short duration" and the corresponding strain is then determined by means of a standardized value of E_c. However, in this process, one disregards the effect of creep which develops very rapidly during the initial time after loading. Up to now, allowance for creep was considered only if the load was applied for more than one day. However, the deformations developing within the first 24 hr after load application can be determined approximately by means of the "modulus of deformation, E_D."

3.1.4. Determination of the Base Value of the Modulus of Elasticity, E_0

Most suggestions found in the literature for determining E_0 take into account only the dependence of E_0 on the strength of the concrete and, in this way, intend also to compensate for the effect of age of the concrete. It has also been proposed to take into account the effect of aggregates by introducing concrete density as a parameter. In this manner, the hope was that a steady transition from heavyweight to lighweight concrete could be found [19]. Accordingly, the ACI Building Code gives the following relationship between the modulus of elasticity and the compressive strength of concrete:

$$E_0 = w_c^{1.5} \times \underset{(33)}{0.044}\sqrt{f'_c}, \tag{3.1}$$

where w_c is the unit weight of the concrete in kg/m^3 (lbs/ft^3) and f'_c is the compressive strength of the concrete in N/mm^2 (psi).

In our opinion, however, the influence of the modulus of elasticity of aggregate is so great in the case of lightweight concrete that one cannot renounce determining experimentally the modulus of elasticity of the concrete. However, if one limits oneself to normal-weight concrete and to the higher-strength concretes customary in structural engineering as well as to medium- and low-slump concretes, one can obtain satisfactory values with the following formula:

$$E_0 = \underset{(1910)}{4.3}\beta_a w_c \sqrt[3]{f'_c}, \tag{3.2}$$

where β_a is a coefficient that is dependent on the type of aggregates.

Approximate values are $\beta_a \sim 0.7$ for sandstone, $\beta_a \sim 0.9$ for limestone, granite, $\beta_a \sim 1.0$ for quartzite, and $\beta_a \sim 1.1$ for basalt or dense limestone. w_c is the unit weight of the concrete in kg/m^3 (lbs/ft^3) at the moment of change in the stress (for estimated values, cf., Section 3.1.8). f'_c is the compressive strength in N/mm^2 (psi) at the moment of change in the stress.

However, the values for β_a given above are based only on a limited number of experimental data (cf., e.g., [21]). In the future, the β_a values should be ascertained with precision by means of reliable tests.

For very short duration of loading, such as 0.5 sec, E_0 may increase by approximately 10%. The formula used in such cases is

$$E_0 = \underset{(2090)}{4.7}\beta_a w_c \sqrt[3]{f'_c}. \tag{3.3}$$

3.1.5. Determination of the Modulus of Deformation, E_D, for Duration of Loading Up to 24 hr

It has already been shown in Section 1 that the load-dependent strain of the concrete ε_t can be broken down into the instantaneous elastic strain, ε_0, the delayed elastic strain, ε_d, and flow ε_f:

$$\varepsilon_t = \varepsilon_0 + \varepsilon_d + \varepsilon_f.$$

The last two terms depend on the duration of load application. In Sections 3.3.1 and 3.3.2, equations will be presented for estimating ε_d and ε_f during the initial 24 hr after load application. Accordingly, Eq. (3.15) yields the following approximation for the delayed deformation after 24 hr:

$$\varepsilon_d = 0.4\beta_{d24h}\varepsilon_{e28} = 0.4 \times 0.28\varepsilon_{e28} = 0.11\varepsilon_{e28},$$

where the value $\beta_{d24h} = 0.28$ was taken from Fig. 3.4 which shows the time development of ε_d. The flow strain after 24 hr corresponds to the fast initial deformation ε_i defined in Section 1.1.3 and formulated in Section 3.3.2. Its magnitude can be estimated by means of Eq. (3.16)

$$\varepsilon_{f24h} = \varepsilon_i = \varepsilon_0 0.8\left(1 - \frac{f'_{ct}}{f'_{c\infty}}\right).$$

We then get the following strain occurring during the first 24 hr:

$$\varepsilon_t = \varepsilon_0 + 0.11\varepsilon_{e28} + \varepsilon_0 0.8\left(1 - \frac{f'_{ct}}{f'_{c\infty}}\right). \tag{3.4}$$

Since the delayed elastic strain is only a small component, we will only make a very small error by using the approximation $\varepsilon_{e28} = \varepsilon_0$ in Eq. (3.4).

If we want to estimate the deformation at a given moment between application of the load and after 24 hr, we can use the coefficient β_i. It describes the development of time-dependent strains during the first 24 hr; these strains are composed of the delayed elastic strain and of flow. β_i is plotted in Fig. 3.2 against the duration of load application. This relationship, however, is based at this time on only a limited number of tests.

The ratio $f'_{ct}/f'_{c\infty}$ used in Eq. (3.4) represents the development in time of the compressive strength of the concrete. The latter was already discussed in Section 2.1 and is shown in Fig. 2.2. However, we can also use the coefficient β_t in accordance with Fig. 2.1 if we make the following transformation:

$$\frac{f'_{ct}}{f'_{c\infty}} = \beta_t \frac{f'_{c28}}{f'_{c\infty}} = \beta_t \beta_\infty.$$

In accordance with Fig. 2.1, the coefficient β_∞ depends on the type of cement.

For Type I cement,

$$\beta_\infty = \frac{1}{1.27} = 0.785.$$

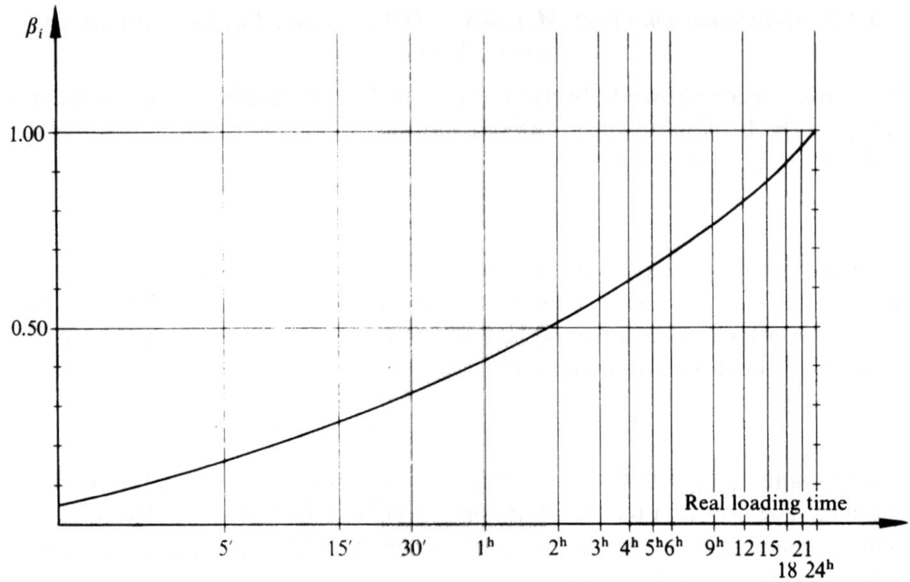

Figure 3.2. Development of creep strain during the first 24 hours after load application.

For Type II cement,

$$\beta_\infty = \frac{1}{1.45} = 0.69.$$

For Type III cement,

$$\beta_\infty = \frac{1}{1.20} = 0.835.$$

Taking these coefficients into account, and after simplifying, Eq. (3.4) yields

$$\varepsilon_t = \varepsilon_0[1 + \beta_t(0.91 - 0.8\beta_t\beta_\infty)]. \tag{3.5}$$

For load application lasting up to 24 hr, the modulus of deformation E_D is

$$E_D = \frac{1}{\varepsilon_t}. \tag{3.6}$$

3.1.6. *Effect of Repeated Loads*

For repeated loads, the component of fast initial strain is dropped from Eq. (3.4). Thus, we have

$$\varepsilon_t = \varepsilon_0(1 + 0.11\beta_i). \tag{3.7}$$

3.1.7. *Effect of Age of Concrete on the Modulus of Elasticity*

The effect of the age of concrete on the modulus of elasticity can readily be estimated by means of Eq. (3.2):

$$\frac{E_t}{E_{28}} = \frac{w_t}{w_{28}}\sqrt[3]{\frac{f'_{ct}}{f'_{c28}}} = \frac{w_t}{w_{28}}\sqrt[3]{\beta_t}.$$

Drying has little influence on unit weight. For $t = 7$ days, $w_t/w_{28} \sim 1.01$, and for $t = 1$ year, $w_t/w_{28} \sim 0.99$. This very slight influence can be disregarded, so that we can write

$$E_t = \beta_e E_{28}, \tag{3.8}$$

where

$$\beta_e = \sqrt[3]{\beta_t}. \tag{3.9}$$

The values for β_e given in Fig. 3.3 are computed in accordance with Fig. 2.1. In Section 14.1, Fig. 14.2 also presents the inverse value

$$\frac{1}{\beta_e} = \beta_\varepsilon.$$

3.1.8. *Summary*

The foregoing shows clearly that one makes do with a rough approximation if one attempts to describe the deformations induced by loads of short duration merely by means of a single value of the modulus of elasticity dependent on the compressive strength of the concrete. Yet this is the method almost always used in actual practice. The fact that it has been satisfactory in the majority of cases can be ascribed to the following reasons:

(a) The structural deformation induced by short-term loads is usually not of great interest since, over the course of time, more pronounced values are brought about by shrinkage, cracking, and creep. A more precise computation of the elastic deformation and, consequently, of E_c may nevertheless be required when an appreciable portion of the load is applied to very young concrete.

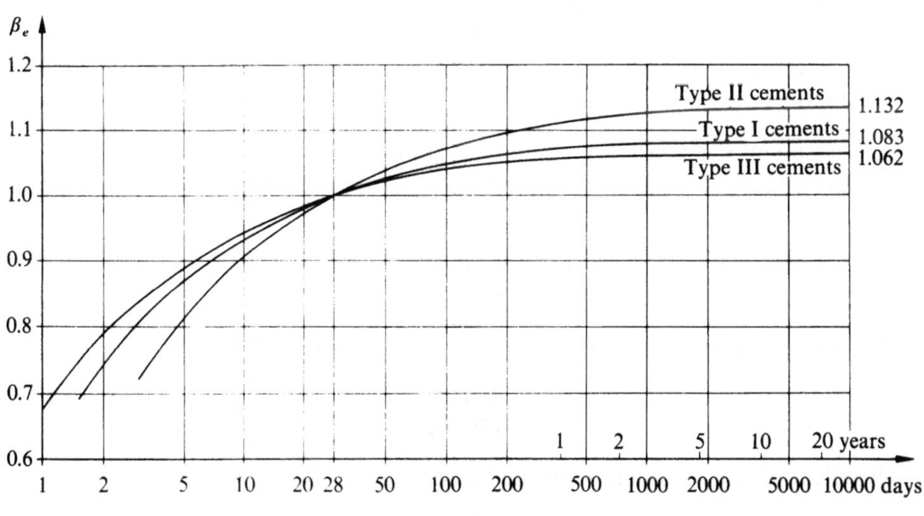

Figure 3.3. Effect of concrete age on modulus of elasticity: $E_t = \beta_e/E_{28}$. When concrete is cast in the winter time the effective concrete age has to be estimated from Eq. (2.1) taking into account curing temperature.

(b) The flow coefficient φ_f and the delayed elastic strain ε_d are relative to the strain occurring at 28 days: $\varphi_f = (\varepsilon_e - \varepsilon_d)/\varepsilon_{e28}$. The creep strain is therefore determined correctly even when the concrete is already subjected to a sustained load at the age of a few days. Knowledge of the dependence between E_c and the age of concrete is thus of no importance for the determination of creep strain.

(c) Similarly, this applies to the influence that the various aggregates have on creep. If among the tests described in [10]—which investigate the influence of the type of aggregate on creep—one considers only those aggregates that are of interest in structural engineering: Rhine river gravel,[1] quartz, granite, and basalt. One notes that Rhine river gravel produces the largest creep strains, and basalt the smallest creep strains. However, the creep coefficient $\varphi_\infty = \varepsilon_{c\infty}/\varepsilon_{e28}$ remains almost unchanged as can be seen in the following table:

Measured variable	Rhine river gravel	Basalt	Ratio
$10^6\ \varepsilon_{e28}$	4.5	2.9	1.55
$10^6\ \varepsilon_{c\infty}$	11.2	6.8	1.65
φ_∞	2.49	2.34	1.06

The coefficient β_a appearing in Eq. (3.2) thus has no noticeable influence on the magnitude of the creep coefficient.

Therefore a more precise determination of the modulus of elasticity is of evident significance in practically only three cases:

In ascertaining the elastic strain when the member is loaded at the age of only a few days.

When one wants to check whether the computational assumptions were effective by comparing the calculated strains with the actual measurements.

In determining the vibration amplitude or the resonant frequency of structures or structural members.

Recapitulation of the Formulas

The basic value E_0 of the modulus of elasticity (duration of load application about 30 sec) is

$$E_0 = 4.3\beta_a w \sqrt[3]{f'_c}. \qquad (3.2)$$

(1910)

The basic value for very rapidly changing processes, such as vibrations is

$$E_0 = 4.7\beta_a w \sqrt[3]{f'_c}. \qquad (3.3)$$

(2090)

The effect of age of concrete on the basic value E_0 is

$$E_t = \beta_e E_{0,28}. \qquad (3.8)$$

[1] Rhine river gravel is composed mostly of quartz with impurities of lime and granite.

The modulus of deformation for first-time loading and duration of loading between 1 min and 24 hr (cf., Section 3.1.5) is

$$E_{Dt} = \frac{E_{0t}}{1 + \beta_i(0.91 - 0.8\beta_t \beta_\infty)}. \tag{3.10}$$

The modulus of deformation for repeated loading after a few days (cf., Section 3.1.6) is

$$E_{D,t} = \frac{E_{0t}}{1 + 0.11\beta_i}. \tag{3.11}$$

In these formulas:

β_a is a coefficient to allow for the type of aggregate:

$$\begin{aligned} \beta_a &= 0.7 \quad \text{(for sandstone)} \\ &= 0.9 \quad \text{(for limestone, granite)} \\ &= 1.0 \quad \text{(for quartzite)} \\ &= 1.1 \quad \text{(for basalt, dense limestone)} \end{aligned}$$

w is the unit weight of concrete in $kg/m^3 (lbs/ft^3)$. If no experimental figures are available, it can be estimated as follows for concrete made of dense limestone,

$$w = 2300 + \frac{f'_{c28}}{0.3}. \tag{3.12}$$

$$\left(143 + \frac{f'_{c28}}{700}\right)$$

For quartzite,

$$w = 2200 + \frac{f'_{c28}}{0.3}. \tag{3.13}$$

$$\left(136 + \frac{f'_{c28}}{700}\right)$$

For sandstone,

$$w = 2180 + \frac{f'_{c28}}{0.3}. \tag{3.14}$$

$$\left(135 + \frac{f'_{c28}}{700}\right)$$

f'_c is the strength of concrete in N/mm^2 (psi).
β_i is a coefficient dependent on the duration of loading in accordance with Fig. 3.2.
β_t is a coefficient dependent on the relative strength of concrete f'_{ct}/f'_{c28} in accordance with Fig. 2.1.

β_∞ is a coefficient dependent on the type of cement.

$$\text{Type I: } \beta_\infty = 0.785;$$

$$\text{Type II: } \beta_\infty = 0.69;$$

$$\text{Type III: } \beta_\infty = 0.835.$$

β_e is a coefficient dependent on the age of concrete in accordance with Fig. 3.3.

Examples showing the application of these formulas are given in Appendix I, Section 1.

3.2. Poisson's Ratio

Poisson's ratio is defined as the ratio between the lateral strain and the axial strain which occurs under uniaxial stress. When the stress on the concrete is within the range of work loads, Poisson's ratio ranges from 0.10 to 0.30, a value of 0.20 generally being used in analysis.

Poisson's ratio depends on the composition of concrete and on the type of aggregate. Since no generally valid mathematical formulas have yet been worked out for these relationships, Poisson's ratio for a given concrete must be determined experimentally if a particularly high accuracy is required. Under normal conditions, however, this is not necessary.

Poisson's ratio depends to a large degree on the magnitude of the load. With stresses above approximately $0.4 f'_c$, it increases rapidly as the load is raised, owing to the development of microcracks, and it can even reach values > 0.5. Poisson's ratio changes very little under a sustained load. This signifies that the creep coefficients are approximately the same for axial and lateral strain.

3.3. Creep

3.3.1. Delayed Elastic Strain Development Over Time

As mentioned earlier, the magnitude of the delayed elastic strain component can be observed only when a load is removed after sustained application. Since this recovery process stretches over several months, it must be assumed that the full value can occur under load only if the sustained load has acted for a sufficiently long duration. It has not been proven that the development over time of the delayed elastic strain is the same under a sustained load and after unloading. In the absence of sufficient experimental data, the correctness of this assumption is the basis of the following.

Figure 3.4 is a curve describing creep recovery after unloading with time. It is based on a large number of tests with varying loading periods of at least 1 year and presents the mean value for these findings. In the process, it was determined that within the range of the concretes normally used in structural engineering, only relatively minor deviations from the average time curve occur even when considering very different loading ages or climatic conditions. One half of the ultimate strain is reached after about 30 days. 90 % of the final value is reached after approximately 1 year. Similar results had been reported by Grasser as early as 1964 [22].

Thus, it can be postulated that the time curve for delayed elastic strains is largely independent of the specimen's shape or age and of the concrete composition within the range that is of interest for structural engineering. The fact that other notions are found in some test reports can be generally explained. For instance, one often finds the statement that elastic recovery has come to a halt after only a few days. This usually involves cases where the sustained load has been applied for only a short time. Very often it is also overlooked that shrinkage counteracts the elastic recovery. This creates the impression of a standstill (cf., Fig. 1.2) as soon as the initial rapidly developing fraction of the recovery has faded.

The Relative Value

The final value of the delayed elastic strain can also be expressed as a fraction of the elastic strain. An analysis of 78 individual tests shows that this fraction ranges from 0.20 to 0.50 and has a mean value of 0.300 [23]. The following are some characteristic results.

Roll [24] investigated the influence of mix proportions on the delayed elastic strain under a creep stress of $f = 0.35f'_c$. His findings were as follows:

Cement content kg/m³ Water-cement ratio	1010 0.31	590 0.44	435 0.49	322 0.53
$\varepsilon_{d\infty}/\varepsilon_{e28}$	0.24	0.28	0.32	0.41

In the same study, values of $\varepsilon_{d\infty}$ were reported for four different mix proportions and for various values of relative sustained stress f/f'_c. If these strains are related to the elastic strains observed at the age of 28 days for the respective degree of loading f/f'_c, it is found that the relative stress level within the range $f/f'_c = 0.20$ to $f/f'_c = 0.65$ has no distinct effect on the ratio $\varepsilon_{d\infty}/\varepsilon_{e28}$. The deviations from the mean are on the order of $\pm 6\%$. Matters are different when—as is customary—for

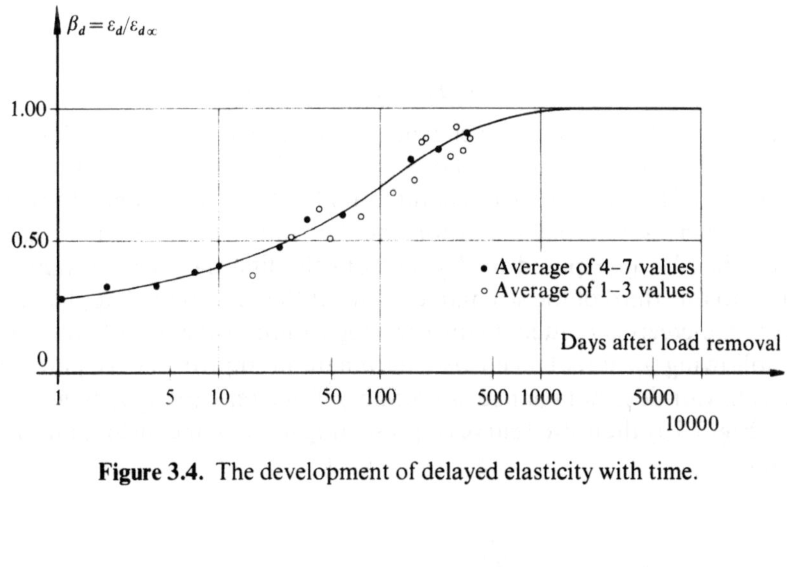

Figure 3.4. The development of delayed elasticity with time.

ε_{e28} the elastic strain under work load, i.e. for $f/f'_c \sim \frac{1}{3}$ is used. The values given in [24] then yield the following results for the average of the four tested mix proportions.

f/f'_c	0.20	0.35	0.50	0.65
$\varepsilon_{d\infty m}/\varepsilon_{e28}$	0.24	0.28	0.31	0.44

In [25], studies on the creep of mass concrete for gravity dams are reported. The ratio $\varepsilon_{d\infty}/\varepsilon_{e28}$ was noticeably smaller for aggregate sizes between 0 and 115 mm than for 0 to 75 mm (3.0 in.) (0.17 < 0.36).

The effects of loading age and type of curing preceeding the test remain relatively small if the load is sustained long enough. In still unpublished tests run in Munich [26] one half of the specimens were water-cured before tests began at ages 1, 3, 7, 28, and 90 days, while the other half was exposed after only 1 day to a constant environment of 65% relative humidity under 20 °C (68 °F). The load was removed after about 800 days. The final values for $\varepsilon_{d\infty}/\varepsilon_{e28}$ had a coefficient of variation of $\pm 6\%$ about the mean value of 0.33.

It is not entirely clear to what degree the type of stress has an effect. In torsion tests described by Ishai [27] the values of the ratio $\varepsilon_{d\infty}/\varepsilon_{e28}$ were observed to be twice as high as in compression tests. However, since in these tests only 50-mm-thick specimens were used, made of mortar containing 80 vol. % cement and since microcracking was observed, these findings, by themselves, do not appear to be conclusive. Unfortunately, since then, no other test results bearing on strain recovery in concrete specimens subjected to torsion have been published.

The tests available and analized to date indicate that the ratio $\varepsilon_d/\varepsilon_{e28}$ is not a constant. However, since no clear dependency of this ratio on specific parameters has been demonstrated, and since the deviations from a constant mean value are limited, we will assume the approximation

$$\varepsilon_d = 0.4\beta_d \varepsilon_{e28}, \tag{3.15}$$

where the coefficient β_d can be obtained from Fig. 3.4.

3.3.2. *The Fast Initial Strain*, ε_i

In Figs. 3.5 and 3.6 the time-dependent strain as reported in [26] for concrete specimens of 200-mm diameter and a compressive strength of approximately 50 N/mm^2 (7100 psi) were plotted separately for loading ages between 1 and 90 days.

All specimens were moist cured for one day and then stored at 65% relative humidity, thus being exposed to drying from the first day. The sustained stress amounted to one third of the strength existing at the time of load application.

In Fig. 3.5, creep is plotted against the logarithm of the age of concrete t. All curves, following an initial steep rise, terminate in approximately parallel lines.

This behavior does not change appreciably if one replaces ε_c by flow ε_f. This is shown in Fig. 3.5 by the dotted curves representing the extremes in loading age, 1 day and 90 days.

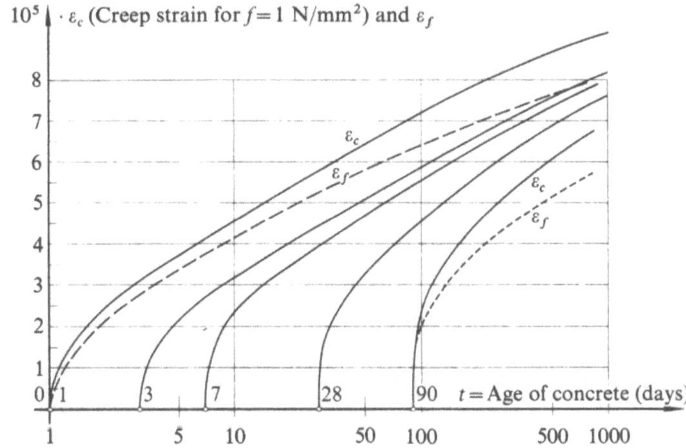

Figure 3.5. Effect of age at loading on creep ε_c (solid lines) and on flow ε_f (broken lines).

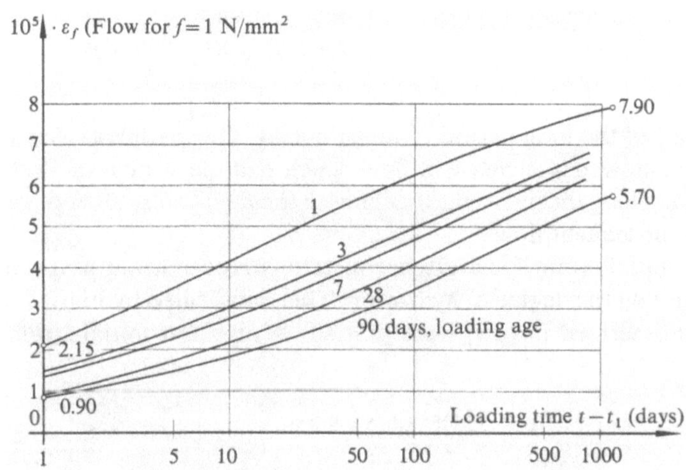

Figure 3.6. Effect of duration of loading and age at loading on flow.

One gets an entirely different impression, however, if one plots along the abscissa the duration of loading $t - t_1$ instead of the age of concrete. In Fig. 3.6, flow ε_f is plotted to illustrate this difference. All the curves then follow a similar course. They differ appriciably only in the initial value and slightly in their slope.

The observation cited in Section 1.1.3 that the magnitude of the fast initial strain depends mainly on the loading age is shown by the following comparison between the flow observed after 3 years of sustained load in specimens which were allowed to dry after 1 day of moist curing. In the following table, the strains are related to a stress $f = 1$ N/mm^2.

Load applied at the age of:	1 day	90 days	Decrease
Fast initial strain	2.15	0.90	1.25
Subsequent flow	5.75	4.80	0.95
Total flow in 3 years	7.90	5.70	2.20

Thus the flow in the test pieces loaded at the age of 90 days had dropped to about 70 % of the flow at a loading age of 1 day. About 60 % of this difference was caused by the fast initial strain and only 40 % by the difference in flow rate after a load duration of more than one day.

The studies described above also investigated the effect of curing. This is why one half of the specimens remained in a fog room. The findings are presented in the following table:

Load applied at the age of:	1 day	90 days	Decrease
Fast initial strain	2.15	0.25	1.90
Strain increase in 3 years	5.75	4.60	1.15
Total flow in 3 years	7.90	4.85	3.05

As a result of the long period of moist curing, the specimens that were loaded after 90 days showed a decrease in flow which had the strongest effect on the fast initial strain. The gain in strength achieved through moist storage reduced only slightly the subsequent flow.

The fast initial strain for a given concrete composition appears to depend in great measure on the degree of hydration. This is indicated by Fig. 3.7 in which the tests results described in [26] were plotted, i.e., the fast initial strain versus the

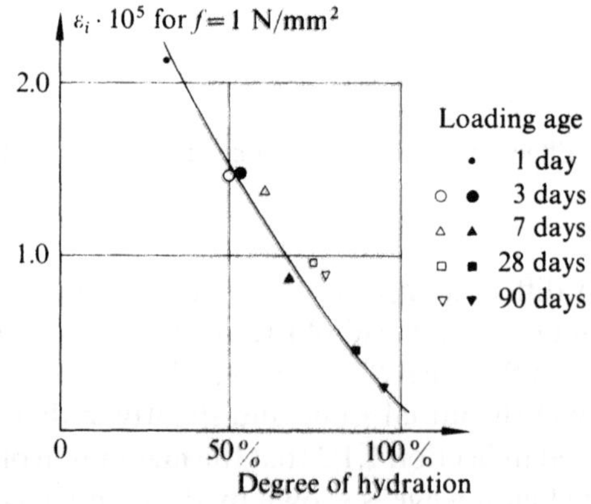

Figure 3.7. Fast initial strain as a function of degree of hydration, observed in experiments on high-strength concrete loaded at different ages [26]. The open symbols represent data for 1 day of moist curing. The filled symbols represent data for moist curing up to load application and subsequent storage at 65 % RH.

corresponding degree of hydration. The degree of hydration existing at the time of loading was computed in accordance with [28]. The diagram leads to the conclusion that the fast initial strain for complete hydration approaches a very small value which is equal to total flow after 1 day of loading, i.e., the slope of the ε_f curve (cf., Fig. 3.6).

In [29], test findings are reported for thin specimens, 70 × 70 × 280 mm, (2.75 × 2.75 × 11 in.) which, after a short period of moist curing, were exposed to an atmosphere with 50% relative humidity, and were then loaded at ages 7, 14, 21, 27, 28, 90, and 180 days. The results show a similar dependency of the fast initial strain on the degree of hydration of concrete. For practical applications, the fast initial strain in relation to the relative compressive strength $f'_{ct}/f'_{c\infty}$ can be estimated as follows:

$$\varepsilon_i = \varepsilon_0 0.8\left(1 - \frac{f'_{ct}}{f'_{c\infty}}\right), \tag{3.16}$$

where ε_0 is the elastic strain defined in Section 3.1.4.

3.3.3. *Development of Flow with Time*

Many theories have been presented to explain the occurrence of flow. Clear overviews can be found in [2], [15], and [30]. Most authors have attempted to ascribe all observed phenomena to a single physical principle. However, since concrete is composed of very diverse basic elements—hydrated cement paste, aggregates, and larger pores—and since hydrated cement paste itself represents a similarly complex system—hydrated cement, unhydrated cement, capillary water, capillary pores, gel water, gel pores—it must be assumed that the contributions to the observed flow strain come from more than one source. For the engineer engaged in practical operations, these theories are only significant to the extent that they are needed to understand the approach selected to estimate the time-dependent strains. This can be achieved by a simplified approach.

The foregoing has shown that the fast initial strain—whose magnitude presumably depends on the degree of hydration and the moisture content of the concrete—is followed by a gradually increasing flow strain characterized by a fairly constant slope.

In this connection, it is an interesting question whether flow is ever going to reach a limit. In the view of several authors (e.g., [31], [32], and [34]), the creep of sealed specimens, which at a later age is primarily due to basic flow, obeys the exponential law $\varepsilon_c = at^n$ which, if plotted on a double logarithmic scale, is represented by a rising straight line. Then, the basic creep—and along with it, after separation of the relatively fast developing component, ε_d, the basic flow as well—will never reach a limit. With the exponents which were given for the plotted test results, one would have to expect that the basic flow might continue to grow appreciably over the time period that is of interest in practice, so that it would approximately double during the time period 5–50 years.

Insofar as this interpretation is based on a graphic representation of the test results, the sensitivity of a double logarithmic scale to small experimental errors is to be taken into account, which automatically excludes extrapolation over longer time periods. This is exemplified by Fig. 3.8 with a test series in which water-cured

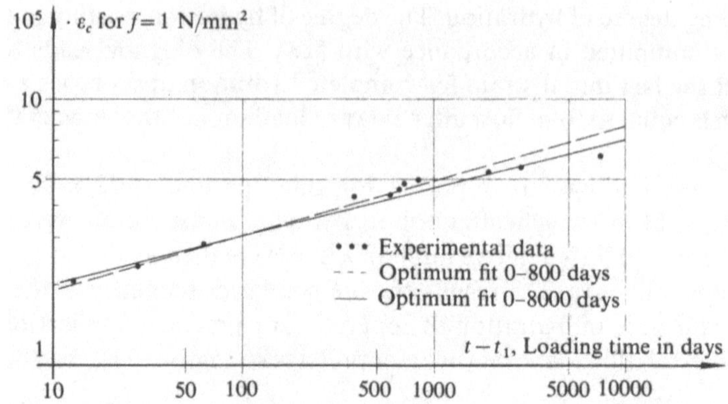

Figure 3.8. When plotted on a double logarithmic scale the creep strains measured on a specimen stored in water can be represented by the broken or by the solid line. However, one should bear in mind that—because of the logarithmic scale—deviations between a data point and the straight line for long durations of loading lead to errors which are ten times as large than for short durations of loading.

test pieces were observed for 20 years [33]. In Fig. 3.8, the creep curves as obtained through the straight-line approximation shown in Fig. 3.7 are plotted in the usual semilogarithmic scale and are compared to the experimental data. Figure 3.8 shows that under a long-time sustained load the exponential law breaks down and that, even in the case of water curing, the experimental results tend toward a limiting value for creep, so that the strain increase between 5 and 50 years may be only 25 %.

However, interesting theoretical considerations were presented in support of the above-cited potential law [32] which, without any doubt, are of great significance for the case of a limited period of sustained loading. Yet they do not take into account the inevitable aging processes. For one, the viscous strain of the cement gel will lead gradually to redistribution of the inner stresses toward stiffer components (such as neighboring aggregates); for another, it must be assumed that the readiness of the gel to deform viscously is reduced over the course of time as a result of the stiffening of the hydration products owing to growing crystallization and polymerization. Both effects explain why not only drying flow, but also basic flow, will tend toward a limiting value over the course of time. (See Fig. 3.9.)

3.3.4. *The Basic Flow* ε_{fb}

The value of ε_{fb} depends on the volume fraction of cement paste and on the moisture content of the concrete. The first parameter fluctuates only within narrow limits in structural concrete, which is considered here exclusively. With regard to moisture, one may assume that, in practice, when the load is first applied, the concrete has not yet lost so much of its initial moisture that one need expect an appreciable effect. In this moist state, the basic flow fraction is greatest and it drops with increasing drying, slowly at first and then at an increased rate, as can be seen in Fig. 3.10—taken from [36]—which shows the example of cement paste which, however, was under load for only 6 days. According to [15], the concrete specimens, which had been dried prior to loading under 50 % relative humidity, reached, after 3 months of

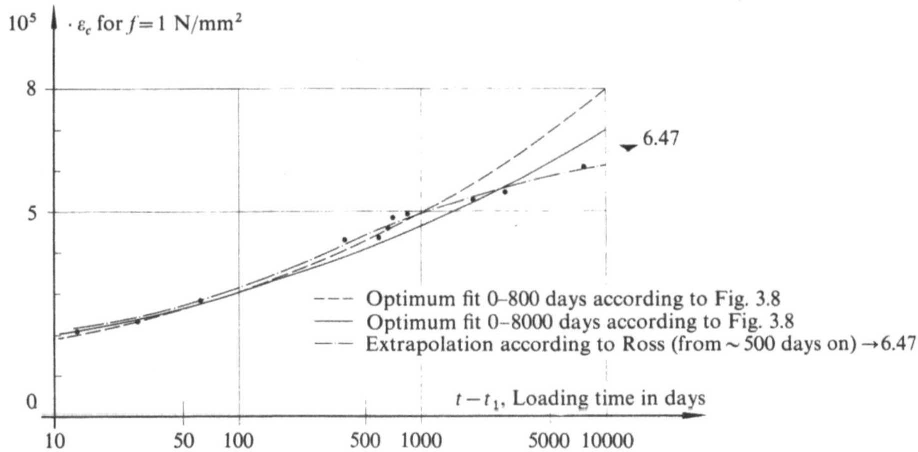

Figure 3.9. Now the data from Fig. 3.8 are plotted on a semi-logarithmic scale together with the parabolic functions described by the straight lines in Fig. 3.8. These functions do no longer fit the data points. However, an extrapolation according to Ross [35], when applied to experiments with sufficiently long duration of loading, represents the data well.

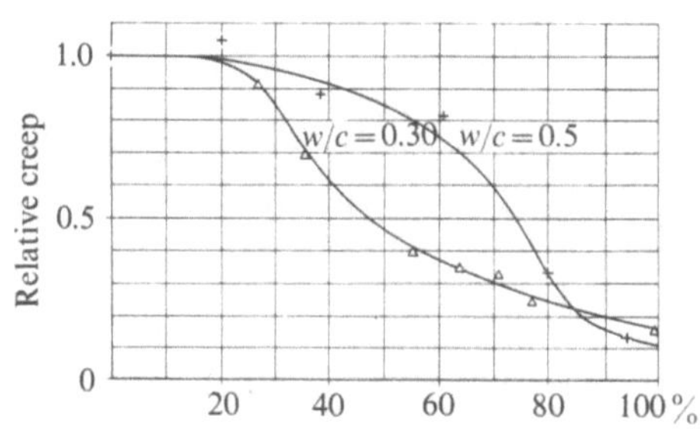

Figure 3.10. Relative creep strain after 6 days of loading for specimens which had been pre-dried prior to loading. Creep is given as a function of the creep of sealed specimens. Rütz [36].

loading, only 40% of the creep strain observed on specimens which had been continuously water cured. According to L'Hermite [16], basic creep after complete predrying is reduced to about 5% of the creep of a moist specimen.

3.3.5. *Drying Flow* ε_{fdr}

In order to determine the magnitude of drying flow, we have at present only limited data at our disposal because tests on the same concrete and at the same age have only rarely been undertaken both under sealed conditions and in a drying atmosphere. Since drying flow depends mainly on the moisture gradient and the dimensions of the test piece, but is also affected by the porosity of concrete, it is difficult to achieve the desired breakdown of flow into two components. Bažant *et al.* introduced possibilities for the numerical analysis of basic creep and drying creep without, however, having accomplished the desired breakdown into basic flow and drying flow ([34] and [37]).

Figures 3.11 and 3.12 show two examples in which the creep strain measured in [33] and [29] could be broken down into ε_d, ε_i, ε_{fb}, and ε_{fdr}. For this purpose, ε_d had to be determined in accordance with the information presented in Section 3.3.1, since no observations were available for ε_d. The table on page 41 shows that drying flow in those cases that occur in practice may far exceed basic flow.

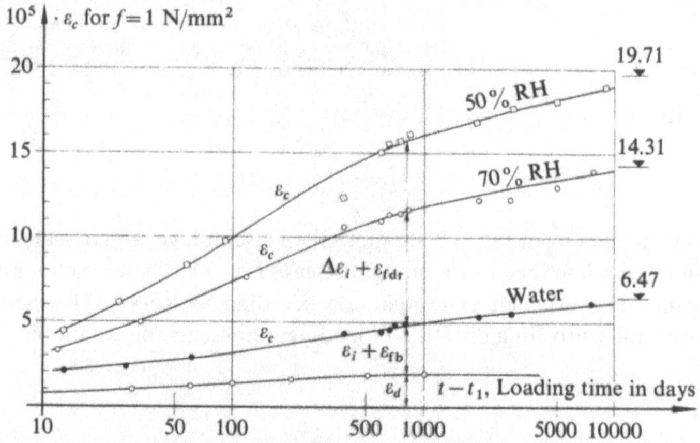

Figure 3.11. Effect of relative humidity on creep of concrete. Cylinders, diameter 100 mm; (3.9 in.) compressive strength approximately 20 N/mm² (2800 psi); age at loading 28 days; duration of loading approximately 20 years. Delayed elasticity has been estimated from other experiments.

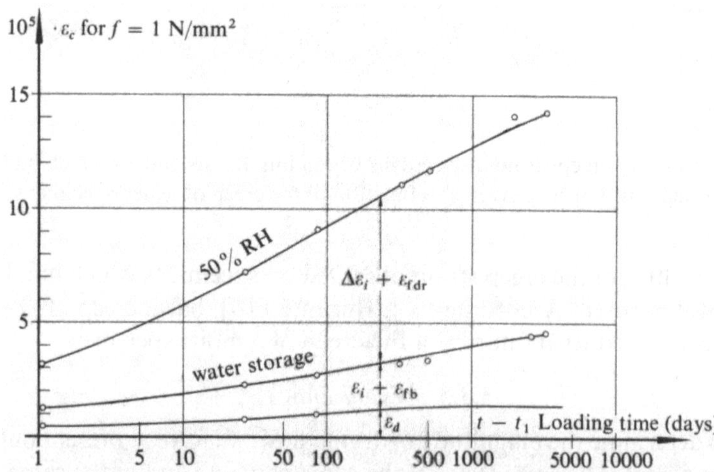

Figure 3.12. These creep curves are taken from [29]. They describe creep of specimens 70 × 70 × 280 mm (2.8 × 2.8 × 11 in.) with a compressive strength of approximately 20 N/mm² (2800 psi). Age at loading 7 days; duration of loading approximately 10 years. Delayed elasticity has been estimated from other experiments.

Author	Loading age (days)	Duration of load, (years)	Relative humidity (%)	Dimensions of specimen (mm)	$10^5 \times$ strain per unit stress $(N/mm^2)^{-1}$			Fraction (%)		
					ε_d	$\varepsilon_i + \varepsilon_{fb}$	$\Delta\varepsilon_i + \varepsilon_{fdr}$	ε_d	$\varepsilon_i + \varepsilon_{fb}$	$\Delta\varepsilon_i + \varepsilon_{fdr}$
Troxell and Raphael [33]	28	20	70	100/350	1.9	4.1	7.7	14	30	56
Troxell and Raphael [33]	28	20	50	100/350	1.9	4.1	12.7	10	22	68
Mamillan [29]	7	10	50	40/40/160	1.3	3.4	9.0	9	25	66
Mamillan [29]	28	10	50	40/40/160	3.4	3.4	7.1	9	29	62
Mamillan [29]	365	10	50	40/40/160	0.8	1.3	3.0	16	26	58

In Section 1.1.4, it was hypothesized that drying flow also follows a course that approximates that of the shrinkage strains. To illustrate this, Fig. 3.14 shows the ratios derived from Figs. 3.11 and 3.13. It is undeniable that there is a close connection between both terms. Total identity cannot be expected since determination of the creep strain by breaking down the total strain into load-dependent creep and load-unrelated shrinkage is itself an approximation justified only by the simplification it achieves.

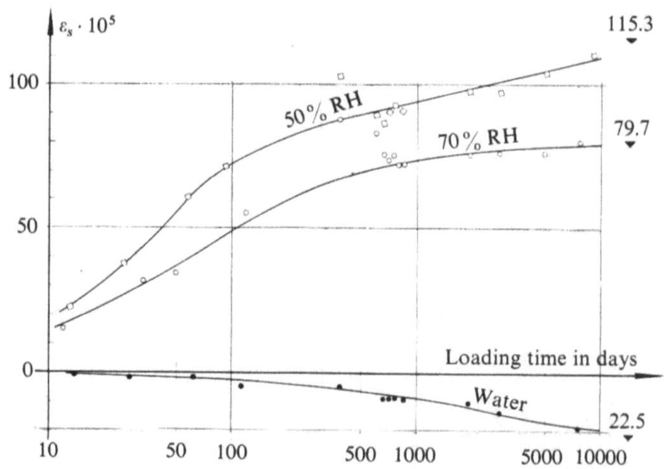

Figure 3.13. Shrinkage of the specimens described in Fig. 3.11 [33].

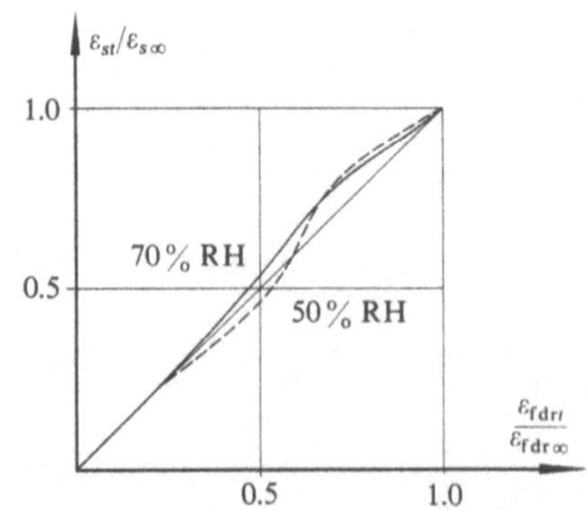

Figure 3.14. Relation between shrinkage ε_{st} and drying flow

$$\varepsilon_{fdr} = \varepsilon_c - \varepsilon_d - \varepsilon_i - \varepsilon_{fb}.$$

Values are given as fractions of the corresponding final values which had been obtained through an extrapolation according to Ross.

3.3.6. Total Flow

It has already been pointed out in Section 3.3.5 that the test results available at this time are not yet sufficient to break down the residual flow $\bar{\varepsilon}_f$ into its components ε_{fb} and ε_{fdr}. We must therefore make do with estimating only the total flow ε_f. This is usually adequate for practical applications. However, it may be advantageous, in the process, to break down ε_f into the components ε_i, fast initial deformation, and $\bar{\varepsilon}_f$, residual flow. The magnitude of flow is a function of numerous parameters. These include, in particular, the degree of hydration and/or the age of concrete, its moisture content and its composition, as well as all the factors that affect the drying of concrete, i.e., the relative humidity of the ambient air and the dimensions of the member. Thus, it is not possible to provide simple relationships as in the case of the strain components ε_e, ε_d, and ε_i. However, Section 4 introduces methods that make it possible to estimate flow.

3.3.7. The Principle of Superposition

For an estimate of creep under varying stress, frequently the principle of superposition is used, which first was applied to concrete by McHenry. He stated the following [56]:

> The strains produced in concrete at any time t by a stress increment applied at any time t_0 are independent of the effects of any stress applied either earlier or later than t_0. The stress increment may be either positive or negative, but stresses which approach the ultimate strength are excluded.

McHenry showed the validity of this principle on the basis of experiments on sealed specimens which, therefore, showed only basic creep. Various investigations conducted later, however, showed that the principle of superposition may give erroneous results, particularly in experiments where changes in load cause a decrease of total concrete strains [57].

This is illustrated in Fig. 3.15 in which the results of experiments according to [26] are shown. Three similar concrete specimens are exposed to three different load histories as shown in Fig. 3.15a:

Experiment A: Age at loading 28 days.
Experiment B: Age at loading 90 days.
Experiment C: Age at loading 28 days, unloaded at an age of 90 days.

The creep strains observed on these three specimens are shown in Fig. 3.15b.

If the principle of superposition were correct, then the creep strain of specimen C after unloading at time $t_2 = 90$ days should follow from the superposition of the creep strain of specimens A and B. A − B = C (see Fig. 3.15a):

$$\varepsilon_c(t_1\, t_2 t) = \varepsilon_c(t_1\, t) - \varepsilon(t_2 t).$$

Figure 3.15b shows that for the above example creep recovery of specimen C is overestimated when applying the principle of superposition. Furthermore, after longer durations of unloading, an increase rather than a decrease of strain would be predicted.

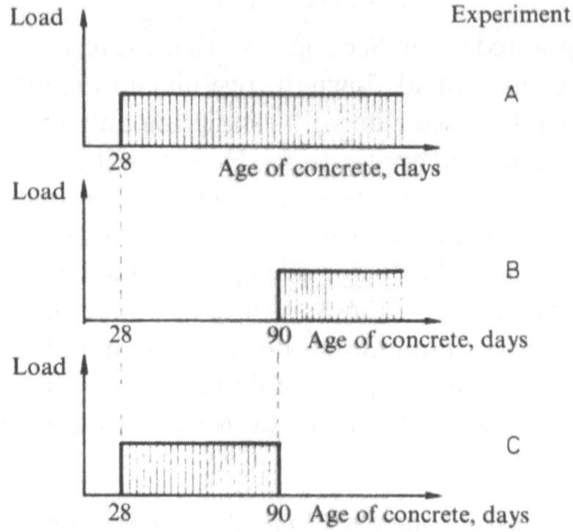

Figure 3.15a. Experimental program to evaluate principle of superposition: A − B = C.

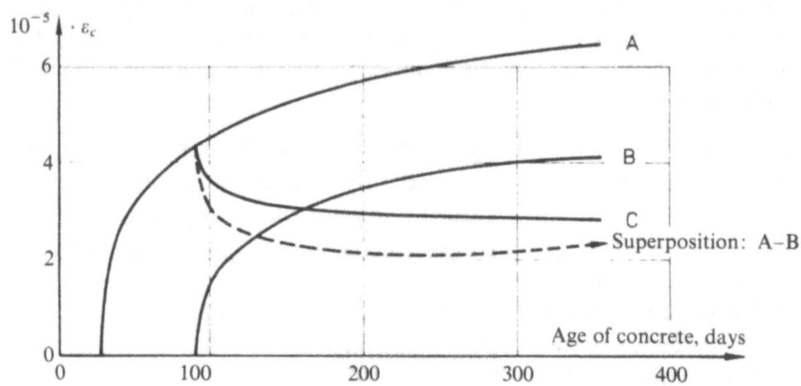

Figure 3.15b. Creep strains of specimens subjected to the load history given in Fig. 3.15a.

A more correct result will be obtained if creep is separated into irreversible flow ε_f and into reversible delayed elasticity ε_d as shown in the preceding sections. Then the creep of specimen C after unloading is for a flow function according to Section 4.3

$$\varepsilon_c(t_1 t_2 t) = \varepsilon_f(t_1 t_2) + \varepsilon_d(t_1 t_2 t) - \varepsilon_d(t_2 t).$$

Such an approach is described in more detail in Section 4.3. Possibilities to apply the principle of superposition considering nonlinearity between stress and strain are not dealt with here because such approaches are too elaborate for most practical applications.

3.4. Shrinkage

It was pointed out in Section 1.2 that shrinkage in concrete is essentially a consequence of the drying process. Verbeck [39] showed that within a certain humidity range in hydrated cement paste, moisture loss and shrinkage are proportional to each other. Consequently, all the parameters that affect the drying of concrete also have an effect on shrinkage [28].

Thus, significant parameters influencing shrinkage in concrete are the water content or the water/cement ratio in the concrete and the relative humidity of the ambient air. As the water/cement ratio increases and the relative humidity drops, shrinkage increases. According to [28], this relationship can approximately be expressed by the equation

$$\varepsilon_{s\infty} = \beta_1 p \sqrt{\frac{100 - \mathrm{H}}{100}}, \tag{3.17}$$

where $\varepsilon_{s\infty}$ is the final shrinkage of concrete, β_1 is a proportionality coefficient, p is the total porosity, which depends on the water/cement ratio and on the age of concrete, and H is the relative humidity in %.

The content and stiffness of the aggregate contained in concrete are also of particular significance for shrinkage. Since aggregates generally shrink very little or not at all, shrinkage in concrete decreases as the aggregate content is raised, approximately according to the following equation [40]:

$$\varepsilon_s \approx \varepsilon_{s0}(1 - v_a)^2, \tag{3.18}$$

where ε_s is the shrinkage of concrete, ε_{s0} is the shrinkage of hydrated cement paste, and v_a is the volume concentration of aggregates.

Finally, shrinkage in concrete clearly decreases as the modulus of elasticity of the aggregate increases because a stiff aggregate strongly restrains the shrinkage of the cement paste.

The influence of specimen size on shrinkage is of great practical significance. Thick members, which can only dry slowly, correspondingly also shrink slowly. As an approximation, the development of shrinkage over time can be expressed by the following equation [28]:

$$\frac{\varepsilon_s}{\varepsilon_{s\infty}} = 1 - \exp\left[-\alpha\left(\frac{t}{\bar{l}^2}\right)^{1/3}\right], \tag{3.19}$$

where t is the duration of drying, $\bar{l} = V/S$ is the relative thickness of specimen, V is the volume of specimen, S is the specimen's surface exposed to drying, and α is the coefficient of diffusion for concrete which can be estimated in accordance with [28].

Thick members not only shrink more slowly than thin ones; for practical applications, one can also assume that they shrink less than thin members, because—under certain circumstances—they reach the limit of shrinkage only after thousands of years. Hansen and Mattock [42] give the following equation for the effect of size of the member on shrinkage:

$$\varepsilon_{s\infty} = \beta_2 l^{-r\bar{l}}, \tag{3.20}$$

where β_2 and r are coefficients that depend on the other parameters that affect shrinkage.

Simplified relationships for the estimation of shrinkage in structural concrete will be presented in Section 4.

3.5. Deformations of Lightweight Concrete

Most of the lightweight concretes used for structural applications are made of lightweight aggregates. Such aggregates have a porosity which is considerably larger than that of conventional aggregates. Thus, their unit weight as well as their modulus of elasticity are significantly smaller. Therefore, most of the deformation characteristics of lightweight concrete differ from the corresponding characteristics of normal-weight concrete.

3.5.1. *Modulus of Elasticity*

As a first approximation, the lower bound value of the modulus of elasticity of concrete E_c may be given as a function of the moduli of the aggregates and of the cement paste, respectively:

$$\frac{1}{E_c} = \frac{1 - v_a}{E_{cc}} + \frac{v_a}{E_a}, \tag{3.21}$$

where E_a is the modulus of elasticity of aggregates, E_{cc} is the modulus of elasticity of the hydrated cement paste, and v_a is the volume concentration of aggregates.

More detailed information regarding this relationship are given in [53]. According to Eq. (3.21), the modulus of elasticity of the concrete decreases as the modulus of the aggregates, E_a, increases. Therefore, the modulus of elasticity of lightweight aggregate concretes is significantly smaller than that of normal-weight concretes of equal strength. Depending on the unit weight of the aggregate, the modulus of elasticity of lightweight concretes may be 30–50 % of the modulus of a normal-weight concrete.

Generally, the moduli of the aggregates are not known. Then the modulus of elasticity of a lightweight concrete may be estimated from its unit weight, e.g., according to Eq. (3.1). Nevertheless, an experimental measurement of E_c is more reliable.

3.5.2. *Creep*

The creep of concrete is also affected by the stiffness of its aggregates. In general, aggregates do not creep. However, a redistribution of stresses may occur under the action of a sustained load, resulting in a stress increase in the aggregates and a stress decrease in the hydrated cement paste. This redistribution of stresses is more pronounced the stiffer the aggregates. Concrete made of stiff aggregates, therefore, creeps less than concrete made of soft aggregates [10]. Consequently, the creep strain of lightweight aggregates is in most instances larger than that of normal-weight aggregates (e.g., [54]).

This, however, does not hold true for the creep coefficient φ. The reduction of the stiffness of the aggregates has a greater effect on elastic strains than on creep, so that the creep coefficient of lightweight aggregate concrete is often only one half of the creep coefficient of normal-weight concrete of equal strength.

3.5.3. Shrinkage

In Section 3.4, we have already stated that aggregates restrain shrinkage of the hydrated cement paste more when their modulus of elasticity is large. Therefore, shrinkage of lightweight concrete is considerably larger than shrinkage of normal-weight concrete. Depending on the stiffness of the lightweight aggregates or on the unit weight of the lightweight aggregate concrete, shrinkage of lightweight concrete may increase by 20–50% of the shrinkage of normal-weight concrete.

4. Prediction Methods

4.1. Overview

Prediction methods are intended to provide design engineers with the means to estimate creep and shrinkage strain for a given concrete rapidly, with sufficient precision and using known parameters.

The methods available today can be divided into three categories in accordance with the required expenditure:

Level 1 methods: These simple procedures permit a rough estimate of the final deformation as a function of a few parameters, such as thickness of member, relative ambient humidity, and age of concrete.

Level 2 methods: On the basis of diagrams or equations, one can estimate the influence of all significant parameters known to the engineer on the magnitude and the time development of deformation in concrete.

Level 3 methods: These are procedures with particularly high precision requirements which are used for special problems in which the correct estimate of creep and shrinkage deformations to be expected is of unusual importance.

In addition, the methods differ in their fundamental principles: some methods present the time-dependent deformations of concrete as product of several coefficients, β, dependent on the principal parameters

$$\varepsilon_c = \varepsilon_{c0}\beta_1\beta_2\beta_3\cdots.$$

In other methods, the deformation of concrete is presented as a sum of various deformation components, e.g.,

$$\varepsilon_c = \varepsilon_d + \varepsilon_f \qquad \text{(I)}$$

or

$$\varepsilon_c = \varepsilon_{\mathrm{cb}} + \varepsilon_{\mathrm{cdr}}. \qquad \text{(II)}$$

Table 4.1 provides an overview of the more recent prediction methods known to us.

Table 4.1. Prediction methods for creep and shrinkage strains.

Method	Reference	Year	Level	Principle
CEB 70	[44]	1970	2	Product
Meyers *et al.*	[45]	1972	1–2	–
Rüsch–Jungwirth	[46]	1976	2	Sum I
BCS	[47]	1977	1–2	–
Branson	[48]	1977	2	Product
CEB 78	[49]	1978	2	Sum I
ACI 209	[50]	1978	2	Product
DIN 4227	[51]	1979	2	Sum I
Bažant–Panula	[34]	1979	3	Sum II
Bažant–Panula (simplified method)	[43]	1979	2–3	Sum II

In the following sections, two methods are presented briefly which we consider to be of particular practical significance.

4.2. The ACI 209 Method

This method was developed by subcommittee II of ACI committee 209 [50]. It was based mainly upon a method proposed by Branson [48].

4.2.1. Creep

The ACI 209 method allows the prediction of a final creep coefficient φ_∞ as well as the prediction of the development of creep over time.

The final creep coefficient may be estimated from the following equation:

$$\varphi_\infty = 2.35\beta_{t_1}\beta_H\beta_d\beta_S\beta_F\beta_{AC}. \tag{4.1}$$

The coefficient β_{t_1} takes into account the age of concrete at load application t_1. It may be estimated from the following expression:

$$\beta_{t_1} = 1.25t_1^{-0.118} \quad \text{for} \quad t_1 \geq 7 \text{ days}. \tag{4.2}$$

The coefficient β_H may be used to estimate the effect of the relative humidity, H, of the surrounding environment:

$$\beta_H = 1.27 - 0.0067H \quad \text{for} \quad H \geq 40\%. \tag{4.3}$$

The effect of the average thickness of the concrete member may be taken into account using the coefficient β_d. Relationships are given for three different ranges of d as follows

(a) $50 \leq d \leq 150$ mm. ($2 \leq d \leq 6$ in)

d	50(2)	75(3)	100(4)	125(5)	150(6)
β_d	1.3	1.17	1.11	1.04	1.00

(4.4)

(b) 150 < d < 380 mm. (6 < d < 15 in)

$$\beta_d = 1.14 - 0.00091d(0.023d) \quad \text{for} \quad t - t_0 \leq 365 \text{ days},$$
$$\beta_d = 1.10 - 0.00067d(0.017d) \quad \text{for} \quad t - t_0 > 365 \text{ days}. \tag{4.5}$$

(c) $d \geq 380$ mm. ($\geq$ 15 in)

$$\beta_d = \tfrac{2}{3}(1 + 1.13e^{-0.0212v/s}), \tag{4.6}$$
$$(1.13e^{-0.54v/s})$$

where v/s is the ratio of volume to surface area of the member.

The coefficient β_S takes into account the consistency of the fresh concrete. The following equation is given to estimate β_S:

$$\beta_S = 0.82 + 0.00264S\,(0.067) \tag{4.7}$$

where S is the slump of fresh concrete in mm.

The coefficient β_F takes into account the content of fine aggregate particles according to the following relationship:

$$\beta_F = 0.88 + 0.0024F, \tag{4.8}$$

where F is the content of fine particles (<4.8 mm) as a percentage of the total aggregate content.

The effect of the air content of the concrete may be taken into account using the coefficient β_{AC}:

$$\beta_{AC} = 0.46 + 0.09\text{AC} \geq 1.0, \tag{4.9}$$

where AC is the air content of fresh concrete in vol. %.

The development of creep over time may be deduced from the following hyperbolic function:

$$\varphi(t, t_1) = \frac{(t - t_1)^{0.6}}{10 + (t - t_1)^{0.6}}\,\varphi_\infty. \tag{4.10}$$

The total load-dependent strain per unit stress may be deduced from the following creep function:

$$\Phi(t, t_1) = \frac{1}{E_c(t_1)} - \frac{\varphi(t, t_1)}{E_c(t_1)}. \tag{4.11}$$

In this equation, the modulus of elasticity of the concrete at the time of load application, E_c, may be estimated from the concrete compressive strength at an age of 28 days, f'_{c28} and from the concrete unit weight, ρ(kg/m^3).

$$E_c(t_1) = 0.04326\sqrt{\rho^3 f'_c(t_1)}, \tag{4.12}$$
$$(33)$$

where

$$f'_c(t_1) = \frac{t_1}{a + bt_1}\,f'_{c28}. \tag{4.13}$$

The constants a and b are functions of the particular type of cement. For normal Type I cement, $a = 4$ and $b = 0.85$; for rapidly hardening Type III cement, $a = 2.8$ and $b = 0.92$.

This method is applicable to both structural normal-weight and lightweight concrete and to concretes made of various types of cement. However, an initial curing period of 7 days is required. An age at load application of $t_1 < 7$ days is excluded.

4.2.2. Shrinkage

The ACI 209 method allows the prediction of a final shrinkage strain, $\varepsilon_{s\infty}$, as well as the prediction of the development of shrinkage over time.

The final shrinkage strain may be estimated from

$$\varepsilon_{s\infty} = 780 \times 10^{-6}(\beta_{CP}\beta_H\beta_d\beta_S\beta_F\beta_{ce}\beta_{AC}). \tag{4.14}$$

The coefficient β_{CP} takes into account the effect of curing periods differing from 7 days.

curing period in days	1	3	7	14	28	90
β_{CP}	1.2	1.1	1.0	0.93	0.86	0.75

(4.15)

The coefficient β_H is used to estimate the effect of relative humidity:

$$\beta_H = 1.40 - 0.010H, \qquad 40 \le H \le 80\%$$

$$\beta_H = 3.00 - 0.030H, \qquad 80 < H \le 100\%. \tag{4.16}$$

The effect of average thickness of the member may be taken into account as follows:

(a) $50 \le d \le 150$ mm. ($2 \le d \le 6$ in)

d	50(2)	75(3)	100(4)	125(5)	150(6)
β_d	1.35	1.25	1.17	1.08	1.00

(4.17)

(b) $150 < d \le 380$ mm ($6 < d < 15$ in).

$$\begin{aligned} \beta_d &= 1.23 - 0.0015d\ (0.038d), \qquad t - t_1 \le 365 \text{ days},\\ \beta_d &= 1.17 - 0.0011d\ (0.029d), \qquad t - t_1 > 365 \text{ days}. \end{aligned} \tag{4.18}$$

(c) $d > 380$ mm (> 15 in).

$$\beta_d = 1.2e^{-0.047v/s}\,(1.2e^{-0.12v/s}), \tag{4.19}$$

where v/s is the volume/surface ratio of the drying member.

The effect of concrete consistency may be estimated from

$$\beta_S = 0.89 + 0.00162S\ (0.041), \tag{4.20}$$

where S = slump of fresh concrete in mm (in).

β_F is used to take into account the content of fine aggregates:

$$\begin{aligned} \beta_F &= 0.30 + 0.0014F, \qquad F \leq 50\%, \\ \beta_F &= 0.90 + 0.002F, \qquad F > 50\%, \end{aligned} \tag{4.21}$$

where F is the content of fine aggregates ($<$4.8 mm) in percent. The coefficient β_{ce} indicates the effect of cement content:

$$\beta_{ce} = 0.75 + 0.00061C\ (0.00036), \tag{4.22}$$

where C is the cement contant in kg/m^3 (lbs/yd^3).

β_{AC} is used to estimate the effect of air content:

$$\beta_{AC} = 0.95 + 0.008A \geq 1.0, \tag{4.23}$$

where A is the air content of fresh concrete in vol. %.

The development of shrinkage over time may be expressed by

$$\varepsilon_{st} = \frac{t}{35 + t}\,\varepsilon_{s\infty}, \tag{4.24}$$

where t is the duration of drying in days.

4.2.3. Lightweight Aggregate Concrete

The modulus of elasticity of lightweight aggregate concrete may be estimated from Eq. (4.12), taking into account the unit weight of the concrete.

The relationships given in Sections 4.2.1 and 4.2.2 for the prediction of shrinkage and creep of normal-weight concrete are also valid for lightweight aggregate concrete.

4.3. The Rüsch–Jungwirth Method

This method is the basis for the subsequent formulations of the CEB-FIP Model Code 1978 [49] and for the German Prestressed Concrete Code DIN 4227-1979 [51].

4.3.1. The Modulus of Elasticity

Equations (3.2) and (3.12)–(3.14) in Section 3.1.8 apply here.

4.3.2. Creep

Creep is expressed by the following equation which is valid for constant stress:

$$\varepsilon_c = \frac{\varphi_t \cdot f}{E_{0,28}}, \tag{4.25}$$

where $E_{0,28}$ is the mean modulus of elasticity at an age of 28 days [cf., Eq. (3.2)].

The creep coefficient φ_t is broken down into the delayed elastic strain and flow. This yields the following equation which expresses the case of a constant sustained load applied between t_1 and t_2. At the moment $t > t_2$, i.e., after the unloading of the specimen, the following equation applies for φ_t:

$$\varphi_t = \varphi_{d0} \cdot \beta_{d(t_2 - t_1)} + \varphi_{f0}(\beta_{ft_1} - \beta_{ft_2}) - \varphi_{d0}\beta_{d(t_2-t_1)}\beta_{d(t_1-t_2)}\beta_{d(t-t_2)}. \quad (4.26)$$

The first term of this equation represents the delayed elastic strain. The second term is the flow which occurs during loading. The third term corresponds to the delayed elastic recovery after unloading at time t_2. (See also App. I, Section 2.3)

In Eq. (4.26):

φ_{d0} is the final value of the delayed elastic strain. According to Eq. (3.15), $\varphi_{d0} \sim 0.4$.

β_d is a coefficient to account for the development over time of the delayed elastic strain in accordance with Fig. 3.4.

φ_{f0} is the base value of flow given in column 3 of Table 4.2.

β_f is a coefficient, in accordance with Fig. 4.1, which expresses the development over time of flow, taking into account the effective thickness of member d_{eff} and effective age t.

t is the effective age of concrete at the particular instance being considered.

t_1, t_2 are the effective ages at the start (t_1) and end (t_2) of the loading period, respectively.

For sudden change in stress the principle of superposition is valid (Section 3.3.7).

The effective thickness of the member is given by the following equation:

$$d_{\text{eff}} = \beta_{\text{eff}} \frac{2A}{u}, \quad (4.27)$$

Table 4.2. Basic flow coefficients and basic shrinkage coefficients for various environmental conditions.

	1	2	3	4	5	6
	Environmental condition to which the member is exposed	Average relative ambient humidity in % (approx.)	Basic flow coefficient[a]	Basic shrinkage strain[c]	Coefficient β_{eff} according to Eq. (4.27)	Adjusted basic shrinkage strain
1	In water		0.8	10×10^{-5}	30	10×10^{-5}
2	In very humid air, e.g., directly above water level	90	1.3	-10×10^{-5}	5.0	-13×10^{-5}
3	Outdoors	70	2.0	-25×10^{-5}	1.5	-32×10^{-5}
4	In dry air, e.g., indoors[b]	40	3.0	-40×10^{-5}	1.0	-46×10^{-5}

[a] Valid for medium slump concrete: for low or for high slump concrete, the values should be raised or reduced by 25%, respectively.

[b] Under normal conditions, indoors the average value of relative humidity is higher.

[c] In CEB/FIP 1978 and in DIN 4227 these values have been increased by 30%.

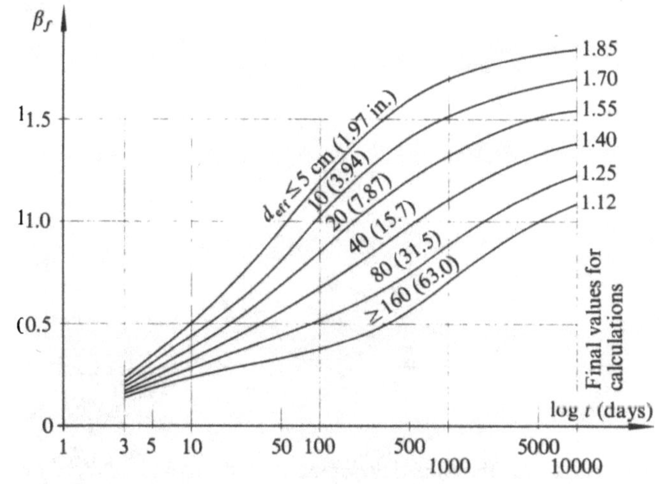

Figure 4.1. Coefficient β_f for the time development of flow. A more detailed presentation of these values is given in Fig. 1.4, Part C.

where A is the cross-sectional area; u is the portion of the cross section exposed to drying, including any internal surfaces; β_{eff} is a coefficient dependent on climatic conditions. It may be taken from column 5 in Table 4.2. It takes into account the influence of relative humidity on the development over time of creep and shrinkage strain.

Equation (2.2) applies to estimate the effective age.

4.3.3. Shrinkage Strain

The following equation applies to shrinkage strain:

$$\varepsilon_{s,t} = \varepsilon_{s0}(\beta_{st} - \beta_{st_1}), \tag{4.28}$$

where ε_{s0} is the base value of shrinkage strain from Table 4.2, column 4. t and t_1 are the effective ages of concrete at the start (t_1) and end (t), respectively, of the time period for which shrinkage effect is to be determined. β_s is the coefficient for the development of shrinkage strain over time, which is a function of the effective thickness of the member, d_{eff}, and the effective age, t or t_1, in accordance with Fig. 4.2.

Equation (4.27) applies to the effective thickness of the member. In determining the effective age in accordance with Eq. (2.2), it must be kept in mind that one should always set $\beta_{ce} = 1$, regardless of the type of cement, since the very slight influence of the type of cement on shrinkage may be disregarded.

The numerical examples given in the following sections have been calculated on the basis of the values given in Table 4.2, column 4. However, more recent evaluations of experimental data showed that the values for basic shrinkage should be increased to the values given in column 6 of Table 4.2 though the examples given in Part C are based on column 4.

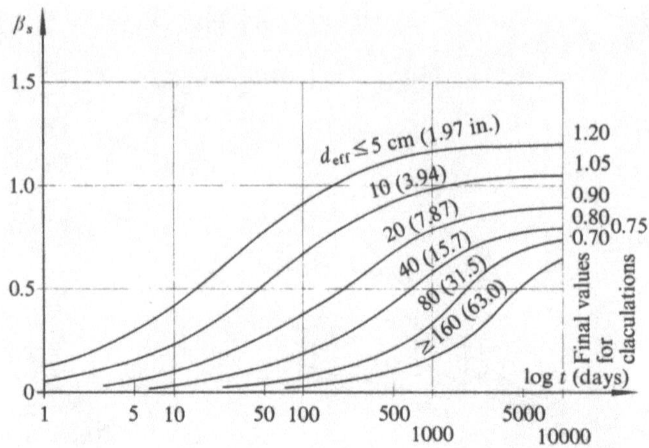

Figure 4.2. Coefficient β_s for the time development of shrinkage. A more detailed presentation is given in Fig. 1.6, Part C.

4.3.4. Lightweight Aggregate Concrete

The Rüsch–Jungwirth method in its original form is valid only for normal-weight concrete. However, it may also be applied to lightweight aggregate concrete if the following corrections are made [55].

The modulus of elasticity $E_{0,28}$ may be taken from the following table:

Unit weight (kg/m³) (lbs/ft³)	1000 62	1200 74	1400 87	1600 99	1800 112	2000 124
Modulus of elasticity (MN/m²) (ksi)	5000 725	8000 1160	11 000 1600	15 000 2175	19 000 2750	23 000 3330

The delayed elastic strain φ_{d0} is the same for lightweight and normal-weight aggregate concrete. The basic value of flow φ_{f0} as given in Table 4.2 has to be reduced as follows in order to obtain the value for lightweight concrete $(\varphi_{f0})_{LW}$

$$(\varphi_{f0})_{LW} = \alpha\varphi_{f0}.$$

The basic value of shrinkage as given in Table 4.2 has to be increased as follows:

$$(\varepsilon_{s0})_{LW} = \beta\varepsilon_{s0}.$$

The coefficients α and β may be taken from the following table:

Compressive strength of lightweight concrete	α	β
$8 < f_c' < 15$ N/mm² (1160) (2175 psi)	1.3	1.5
$15 < f_c' < 55$ N/mm² (2175) (7975 psi)	1.0	1.2

Table 4.3. Final creep coefficients and final shrinkage coefficients for normal weight concrete.

	Moisture conditions			
	humid, outside (RH 75%)		dry, indoors (RH 50%)	
Effective thickness d_{eff}	small[a] <200 mm (8 in.)	large >600 mm (24 in.)	small <200 mm (8 in.)	large >600 mm (24 in.)
Creep coefficients, φ,				
Age at loading				
Young (3–7 days)	2.7	2.1	3.8	2.9
Normal (7–60 days)	2.2	1.9	3.0	2.5
Old (>60 days)	1.4	1.7	1.7	2.0
Shrinkage, ε_s				
Age from which shrinkage has to be taken into account				
Young (3–7 days)	26×10^{-5}	21×10^{-5}	43×10^{-5}	31×10^{-5}
Normal (7–60 days)	23	21	32	30
Old (>60 days)	16	20	19	28

[a] For 200 mm (8 in.) < d_{eff} < 600 mm (24 in.), you may interpolate linearly.

4.3.5. *Simplified Method*

According to the definitions of Section 4.1, the Rüsch–Jungwirth prediction method may be considered a Level 2 procedure. For many practical cases, however, it is sufficient to estimate only the final value of the concrete strains which occur after a very long period of time without taking into account their development over time (Level 1). The final values of creep and shrinkage may be taken from Table 4.3, which is also included in the CEB/FIP model code [49].

4.4. Comparison of the ACI 209 and Rüsch–Jungwirth Methods

A comparison of various prediction methods has been reported in [23]. It was found that the creep and shrinkage values estimated by the two methods differed considerably. Nevertheless, a comparison of the two methods with experimental findings showed consistent reliability. The mean coefficient of variation for the prediction of creep functions measured about 25% for both methods.

The ACI 209 method is based on a product principle and is, therefore, ill suited for the prediction of sudden change in stress and, in particular, of unloading. The Rüsch–Jungwirth method, by contrast, is based on a sumation principle, with a breakdown into delayed elastic strain and flow. Consequently, this method is always better suited to estimate creep after load removal. In the basic approach to the description of flow, this method is based on the assumption that the flow in concrete at varying ages can be described by a single function. This concept was selected because it is particularly suited for application of the computational method described in Part C. To be sure, this type of representation reflects only imperfectly the actual behavior of the material. The accuracy achieved, however, leads to the conclusion that the error caused by assuming one unique flow function is within tolerable limits.

5. Deformation Characteristics of Reinforcing and Prestressing Steels

Like concrete, steel may creep. Since, however, creep in steel is ascribable to other causes than in concrete, e.g., to stress-induced dislocation movements, it becomes noticeable only under very high stresses. Consequently, creep can be completely disregarded in the reinforcing steel used in reinforced concrete construction. In prestressing steel, the manufacturing process has a great influence. In practice, however, we refer to relaxation rather than creep because this behavior is studied experimentally as stress reduction under constant strain. However, in prestressed concrete the stress loss of the prestressing steel can only partially be described by relaxation because of the added effect of concrete creep.

Up until now, there have been no well-founded rheological conceptions which might predict reliably the effect which steel composition and the manufacturing process has on the relaxation process. Relaxation behavior must, therefore, be determined experimentally for each variety of steel. The experience gained to date permits only the following generally valid statements:

(a) In the prestressing steels used at the present time, the modulus of elasticity is modestly affected by the type of tendon; but it is only to a negligible degree dependent on the type of manufacturing process. One estimates for:

Single wires and bars: $E_p = 2.1 \times 10^5$ N/mm^2 (3×10^7 psi).

Strands: $E_p = 2.0 \times 10^5$ N/mm^2 (2.85×10^7).

If cables are used, in exceptional cases, one must, in addition, determine the strain in the cable.

(b) The stress loss occurring in relaxation tests, within the range that is of interest in practice, develops over time in approximate accordance with the following logarithmic law:

$$\Delta f = a + b \log t.$$

As Fig. 5.1 shows, the constants a and b are highly dependent on the degree of prestressing.

(c) The stress loss is also dependent in large measure on the manufacturing process, as Fig. 5.2 illustrates.

(d) In prestressing steels which exhibit high stress loss, it may occur in a relaxation test that when the degree of prestressing, f_{po}/f_{pu}, is markedly raised, the final stress β_{pt}/f_{pu} may even decrease. This is shown in Fig. 5.3, which also yields the percentage of stress reduction.

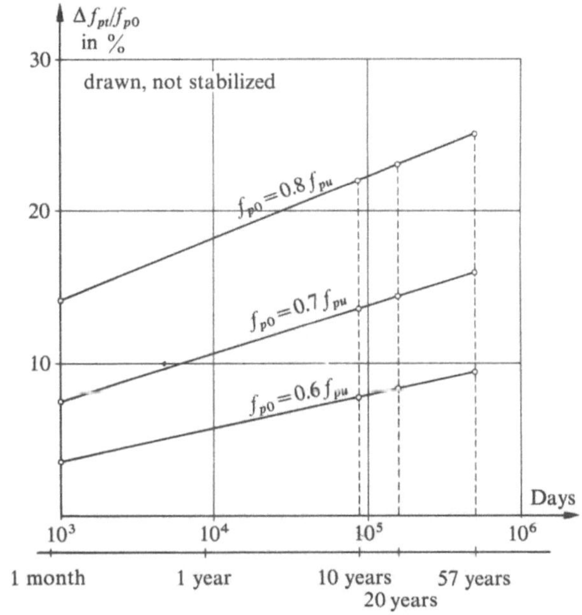

Figure 5.1. The relative loss of stress as observed in a relaxation test at 20°C (68°F) may be described by a logarithmic function. It allows extrapolation to a limited extent.

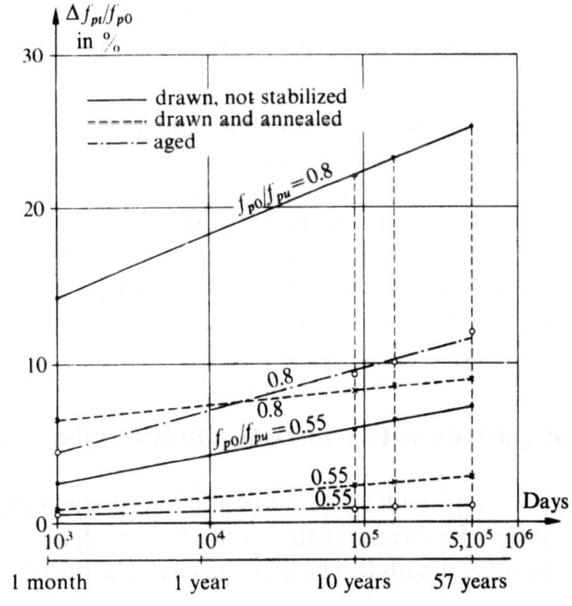

Figure 5.2. The relative values of loss of stress also depend on the type of steel.

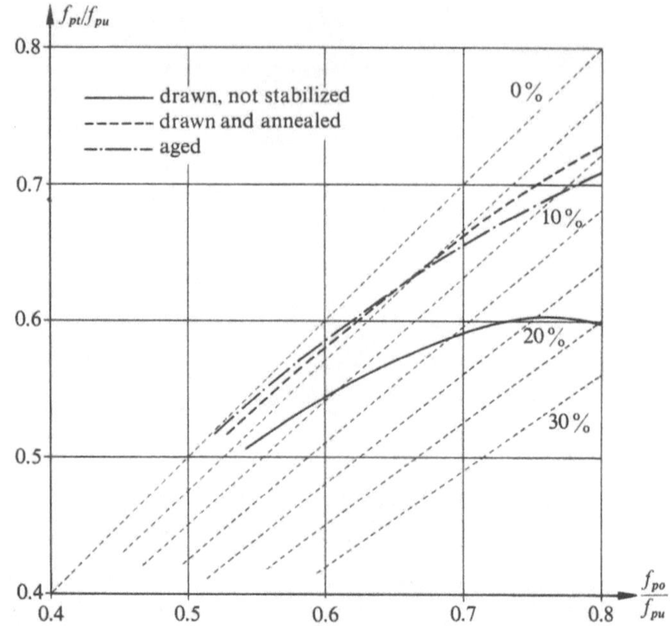

Figure 5.3. Final steel stress of three types of prestressing tendons as a function of initial stress. The heavy lines describe relaxation to be expected after approximately 57 years at room temperature. The dotted lines allow an estimate of the loss of stress in percent.

(e) Temperature also has a considerable influence on the relaxation-induced stress loss. An increase from 20 °C to 40 °C (60 °F to 104 °F) may raise the losses shown in Fig. 5.3 by about $\frac{1}{3}$.

The foregoing indications are a warning against raising the permissible prestressing stress too much, compared to present-day customary values, as is sometimes proposed (e.g., nearly to the yield point).

In light of the differences that exist in the behavior of the various types of steel under sustained load, one must insist on the requirement that only those steels be authorized for prestressing whose characteristics have been determined by thorough testing and are guaranteed by the manufacturer.

6. Comparison with Construction Site Observations

The validity of the data provided for the prediction of creep and shrinkage was investigated by means of measurements made on 15 bridges. The dimensions of these structures and other details can be found in [52]. For brevity, only the average

Table 6.1. Comparison of creep and shrinkage measurements on 15 bridges.

(a) *Mean parameters for the investigated bridges*

$2A/u = 430$ mm
Effective loading age: 28 days
Mean value of assumed relative humidity: 80 %
Mean modulus of elasticity: $E_c = 30\,000$ MN/m^2
Mean creep-inducing stress: $f_{sus} = 4.5$ MN/m^2

(b) *Measured values extrapolated in accordance with Ross* [35]

Mean value of creep coefficient: $\varphi_\infty = 1.6$
Mean value of shrinkage strain: $\varepsilon_{s\infty} = 11 \times 10^{-5}$
Elastic deformation: $\varepsilon_e = f_{sus}/E_c = 4.5/30\,000 = 15 \times 10^{-5}$
Total deformation: $\varepsilon = \varepsilon_e \varphi_\infty + \varepsilon_{s\infty} = (15 \times 10^{-5})\,1.6 + 11 \times 10^{-5} = 35 \times 10^{-5}$ $\triangleq 100\%$

(c) *Estimate according to the Rüsch–Jungwirth method*

$d_{eff} = \beta_{eff} 2A/u = 43(3.25) = 140$ mm
$\varphi_\infty = 1.65(1.15 - 0.32) + 0.4 = 1.75$
$\varepsilon_{s\infty} = 17.5 \times 10^{-5}(0.7 - 0.05) = 11.5 \times 10^{-5}$
$\varepsilon = [15(1.75) + 11.5] \times 10^{-5} = 37.8 \times 10^{-5}$ $\triangleq 108\%$

(d) *According to* CEB-FIP *1970*

$\varphi_\infty = 1.9(1)0.72 = 1.35$
$\varepsilon_{s\infty} = (21 \times 10^{-5})0.55(0.9) = 10.5 \times 10^{-5}$
$\varepsilon = [15(1.35) + 10.5] \times 10^{-5} = 30.8 \times 10^{-5}$ $\triangleq 88\%$

(e) *According to ACI 209*

$\varphi_\infty = 2.35(0.67)0.84(0.73) = 0.97$
$\varepsilon_{s\infty} = (78 \times 10^{-5})0.6(0.43) = 20.1 \times 10^{-5}$
$\varepsilon = [15(0.97) + 20.1] \times 10^{-5} = 34.7 \times 10^{-5}$ $\triangleq 99\%$

results are given here. To this effect, those dimensions and environmental conditions that differed only slightly were averaged, and the final strain values extrapolated in accordance with Ross [35] were adjusted to a mean stress of 4.5 N/mm^2. (640 psi). The resulting data are given in Table 6.1. They were compared to the values computed using the various methods. Reasonable agreement has been obtained with all methods.

To check the development of strains over time, measurements effected over a 20-year period at the Rhine bridge at Worms, as well as those made for 10 years at the Rhine bridge at Bendorf, were incorporated into the β_f curves. Here, too, there was good agreement. (See Fig. 6.1.)

——— Bridge across the river Rhine at Bendorf, $d_{eff} = 60 \cdot 3.25 \geq 160$ cm (63 in.), obs $\varphi = 1.9$ cal $\varphi = 1.85 \cdot \varphi_{f0} = 1.65$

------ Bridge across the river Rhine at Worms; second arch

······· Bridge across the river Rhine at Worms, arch No. 5, $d_{eff} = 53 \cdot 3.25 \geq 160$ cm (6.3 in.) obs $\varphi = 1.6 \cdot$ cal $\varphi = 1.76 \cdot \varphi_{f0} = 1.65$

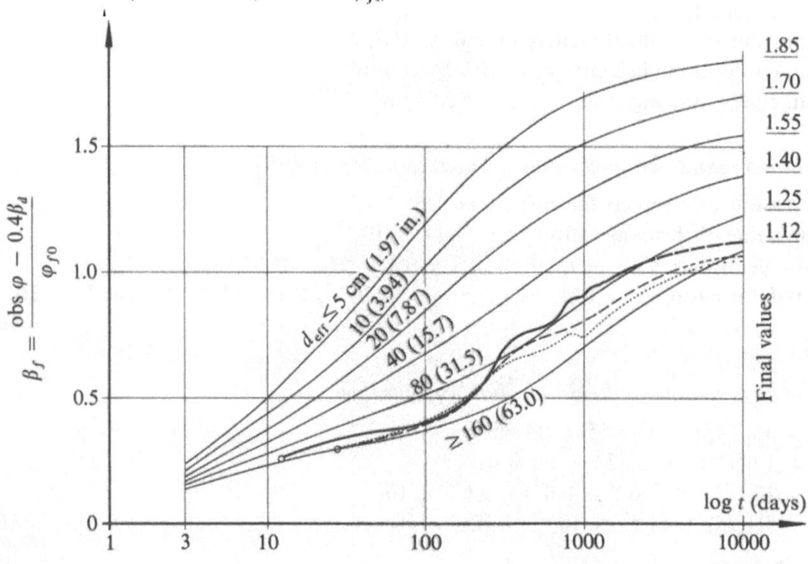

Figure 6.1. For two span bridges across the River Rhine the development of β_f has been estimated from measured values for $\beta_f = (\varphi_{\text{measured}} - 0.4/\beta_d)\varphi_{f0}$. There is good agreement with the relations given in Fig. 4.1.

7. Prospects for Future Developments

Sections 1, 3.3, and 3.4 developed rheological concepts based on the analysis of test results; these concepts point at the possibility that the principles used to predict creep strains can be greatly perfected in the years to come. Since continued development of knowledge in most technological fields is in constant flux, any newly proposed approach—including the one introduced in Section 4.3—can only be an imperfect and temporary solution to existing problems.

If further investigations confirm the validity of the concepts mentioned above, it is conceivable that better documented numerical data might justify the application of the following formula to satisfy high-accuracy requirements in the determination of creep strain:

$$\varphi_t = \varphi_d(t_1)\beta_d(\Delta t) + \varphi_i(t_1, \%\,\text{RH}) + \varphi_f(\%\,\text{RH})\beta_{ft}(\Delta t, d_w, t_1). \tag{7.1}$$

In this formula, the first term, which represents the delayed elastic strain, is expressed in a more general form than in the previous formulations. Recent studies appear to indicate that the coefficient φ_d is not a constant and depends on factors such as the loading age. It is also possible that the coefficient β_d does not depend solely on the time under load. For it is probable, contrary to assumptions made hitherto, that delayed elastic strain develops faster under load than the creep recovery observed after unloading.

The second term in Eq. (7.1) is aimed at representing the fast initial strain discussed in Section 3.3.2. It is a function of the loading age t_1 and probably also of relative humidity (%RH). The influence of loading age can be taken into account by means of the ratio $f'_{ct}/f'_{c\infty}$ in accordance with Eq. (3.16). This formulation was already used in the CEB/FIP prediction method [49].

The third term describes the development of flow over time. It is a function of ambient conditions and of the coefficient β_f. This coefficient takes into account not only development over time under load Δt, but also the influence of the drying rate—a function of the effective thickness d_{eff}—and of the loading age t_1.

Such a proposal would signify a further step toward moving away from the previously used multiplication approach and would thus facilitate adaptation to the test findings without becoming too intricate. The more difficult computation caused by the separate treatment of φ_i can, in all probability, be compensated for by simplifying the determination of φ_f. Separation of the fast initial strain would enhance the validity of assuming a single flow curve, β_{ft}, independent of the loading age.

Nevertheless, it is undeniable that a formulation using flow curves that depend on the loading age generally yields closer agreement with experimental data. This drawback is more than compensated for however, by the simplicity of this method.

Finally—despite the understandable desire to perfect the prediction methods—one should always consider the possibility of simplifying the prediction methods as long as the analytical problem at hand permits the attendant reduction in precision.

References for Part B

[1] Woolson, I. H.: "Some remarkable tests indicating 'Flow' of concrete under pressure," *Engineering News*, **54**, 18, 1905.

[2] Neville, A. M.: *Creep of Concrete: Plain, reinforced and Prestressed*, North Holland, Amsterdam, 1970.

[3] Davis, H. E.: "Flow of concrete under sustained compression stresses," *Proc. ASCE*, Part 3, May 1928.
Davis, R. E. and H. E.: "Flow of concrete under sustained compression stresses," *Proc. ASTM*, **30**, Part 2, 1930.
Davis, R. E. and H. E.: "Flow of concrete under the action of sustained load." J. Am. Concr. Inst. **37**, 1931.
Davis, R. E., Davis, H. E., and Brown, H. E.: "Plastic flow and volume changes of concrete." *Proc. ASTM*, **37**, Part 2, 1937.
Davis, R. E., Davis, H. E., and Hamilton, J. S.: "Plastic flow of concrete under sustained stress." *Proc. ASTM*, Part 2, 1934.

[4] Glanville, W. H.: "The creep or flow of concrete under load." Building Research Technical Paper No. 12. H.M. Stationary Office, 1930.
Glanville, W. H. and Thomas, F. G.: "Further investigations on the creep or flow of concrete under load." In: "Studies in reinforced concrete, Part III," Department of Scientific and Industrial Research. Building Research Technical Paper No. 12, London, 1930, Technical Paper No. 21. London, 1939.

[5] Wagner, O.: "Das Kriechen unbewehrten Betons" (Creep of unreinforced concrete). Deutscher Ausschuss für Stahlbeton, Heft 131, Verlag Wilhelm Ernst and Sohn, Berlin, 1958.

[6] *ACI Committee 209, Annotated Bibliography on Shrinkage and Creep in Concrete—1905–1964*, American Concrete Institute, Detroit, 1967, 102 pp.

[7] Lorman, W. R.: *List of Additional References to Creep and Volume Changes of Concrete 1901–1964*, American Concrete Institute, Detroit, 1967, 58 pp.

[8] Hummel, A.: "Vom Kriechen und Fliessen des erhärteten Betons und seiner praktischen Bedeutung" (Creep and flow of hardened concrete and its practical significance), *Zement*, No. 50/51, 1935.

[9] Kordina, K.: "Influence of time upon strength and deformation of concrete." Final report on RILEM Symposium, 1958, RILEM Bulletin No. 9, 1960.

[10] Rüsch, H., Kordina, K., and Hilsdorf, H. K.: "Der Einfluss des mineralogischen Charakters der Zuschläge auf das Kriechen von Beton" (The influence of the mineralogical character of the aggregates upon creep of concrete), Deutscher Ausschuss für Stahlbeton, Heft 146, Verlag Wilhelm Ernst and Sohn, Berlin, 1962.

[11] Alexandrowski, S. W.: "Design of concrete and reinforced concrete structures with regard to temperature and humidity effects taking into account creep" (in Russian), Moscow, 1966.

[12] Illston, J. M.: "The delayed elastic deformation of concrete as a composite material." International Conference on the Structure of Concrete. Cement and Concrete Association, London, 1968.

[13] Neville, A. M.: "Theories of creep in concrete," *J. Am. Concr. Inst.*, **52**, 1955.

[14] Hansen, T. C.: "Creep of concrete, a Discussion of some fundamental problems," Bulletin No. 33 of the Swedish Cement and Concrete Institute at the Royal Institute of Technology, Stockholm, 1958.

[15] Ali, J., and Kesler, C. E.: "Mechanisms of creep in concrete." *Symposium on Creep of Concrete*, ACI Special Paper No. 9, Detroit, 1964.

[16] L'Hermite, R.: "What do we know about plastic deformation and creep of concrete?" RILEM Bulletin No. 1, Paris, March 1959.

[17] Heilman, H.: "Beziehung zwischen Zug- und Druckfestigkeit des Betons" (Relations between tensile and compressive strength of concrete), Beton, Heft 2, Beton-Verlag, Düsseldorf, 1969.

[18] Rüsch, H.: "Die Ableitung der charakteristischen Werte der Betonzugfestigkeit" (A deduction of the characteristic values of concrete tensile strength), Beton, Heft 2, Beton-Verlag, Düsseldorf, 1975.

[19] Pauw, A.: "Static modulus of elasticity of concrete as affected by density," *J. Am. Concr. Inst.* **57**, 6, 1960.

[20] *ACI 318—Building Code Requirements for Reinforced Concrete*, American Concrete Institute, Detroit, 1977.

[21] Meyer, F.: "Einfluss unterschiedlicher Zuschläge auf die Druckfestigkeit und den *E*-Modul von Beton" (Influence of type of agggregate on the compressive strength and the modulus of elasticity of concrete), Beton, Heft 3, Beton-Verlag, Düsseldorf, 1972.

[22] Grasser, E.: "Der Einfluss der verzögert elastischen Verformung auf das Verhalten der Betontragwerke" (The effect of delayed elasticity on the behavior of concrete structures), Institut für Massivbau, TU München, July, 1964.

[23] H. S. Müller and H. K. Hilsdorf.: "Comparison of prediction methods for creep coefficients of structural concrete with experimental data," *Fundamental Research on Creep and Shrinkage of Concrete*, Sijthoff and Nijhoff Publishers BV, The Hague, 1982.

[24] Roll, F.: "Long-time creep recovery of highly stressed concrete cylinders," Creep of Concrete, ACI Special Publication No. 9, Detroit, 1964.

[25] Polivka, M., Pirtz, D., and Adams, F.: "Studies of creep in mass concrete," Paper 12, On Mass Concrete, ACI Special Publication No. 6, Detroit, 1964.

[26] Hilsdorf, H., Rüsch, H., and Kordina, K.: "Der Einfluss der Lagerungsart von der Belastung sowie von Spannungsänderungen auf das Kriechen von Beton" (The influence of type of curing prior to load application as well as of stress variations on creep of concrete) (unpublished).

[27] Ishai, O.: "Elastic and inelastic behavior of cement mortar in torsion," Creep of Concrete, ACI Special Publication No. 9, Detroit, 1964.

[28] Hilsdorf, H. K.: Austrocknung und Schwinden von Beton im Stahlbetonbau (Drying and shrinkage of concrete and reinforced concrete structures), Berichte aus Forschung und Praxis, Berlin, München, Verlag Ernst and Sohn, 1969. "A method to estimate the water content of concrete shields," *Nuclear Engineering and Design*, **6**, 1967.

[29] Mamillan, M., and Lelan, J.: "Le fluage du béton," Annales de l'Institut Technique du Batiment et des Travaux Publics, No. 246, Paris, June, 1968.

[30] Trost, H.: "Spannungs-Dehnungs-Gesetz eines viskoelastischen Festkörpers wie Beton und Folgerungen für Stabtragwerke aus Stahlbeton und Spannbeton" (Stress-strain law of a viscoelastic solid similar to concrete and its consequences for structures made of reinforced and prestressed concrete), Beton, 16, Heft 6, Beton-Verlag, Düsseldorf, 1966.

[31] Browne, R. D. and Blundell, R.: "The influence of loading age and temperature on the long-time creep behavior of concrete in a sealed, moisture-stable state," *Materials and Structures*, **2**, 1969.

[32] Wittmann, F. and Setzer, M.: "Vergleich einiger Kriechfunktionen mit Versuchsergebnissen" (Comparison of creep functions with test results), *Cement and Concrete Research, Vol. 1*, Pergamon Press, New York, 1971.

[33] Troxell, G. E., Raphael, I. M., and Davis, R. E.: "Long-time creep and shrinkage tests of plain and reinforced concrete," *ASTM Proc.*, **58**, 1958.

[34] Bazant, Z. P. and Panula, L. P.: "Practical prediction of time dependent deformations of concrete," *Materials and Structures*, **11**, No. 65, 1978; **11**, No. 66, 1978; **12**, No. 69, 1979.

[35] Ross, A. D.: "Concrete creep data," *The Structural Engineer*, 15, No. 8, 1937.

[36] Rütz, W.: "Das Kriechen des Zementsteins im Beton und seine Beeinflussung durch gleichzeitiges Schwinden" (Creep of hydrated cement paste in concrete as influenced by simultaneous shrinkage), Deutscher Ausschuss für Stahlbeton, Heft 183, Verlag Ernst and Sohn, Berlin, 1966.

[37] Bazant, P. and Najjar, L. J.: Drying of concrete as a nonlinear diffusion problem, *Cement and Concrete Research, Vol. 1*, Pergamon Press, New York, 1971.

[38] Saule, A. G. A.: "Principles underlying the steam curing of concrete at atmospheric pressure," *Magazine of Concrete Research*, No. 6, 1951.

[39] Verbeck, G. J. and Helmuth, R. H.: "Structure and Physical Properties of Cement Paste," Proceedings of the Fifth International Symposium on the Chemistry of Cement, Part III, Tokyo, 1968.

[40] Pickett, G.: "Effect of aggregate on shrinkage of concrete and a hypothesis concerning shrinkage," *ACI Journal*, **52**, 1956.

[41] Hansen, T. C. and Nielsen, K. E. C.: "Influence of aggregate properties on concrete shrinkage," *J. Am. Concr. Inst.*, **63**, 1966.

[42] Hansen, T. C. and Mattock, A. H.: "Influence of size and shape of member on the shrinkage and creep of concrete," *J. Am. Concr. Inst.*, **63**, 1966.

[43] Bazant, Z. P., Panuala, L.: "Simplified prediction of concrete creep and shrinkage from strength and mix," Structural Eng. Report No. 78-10/640, Department of Civil Engineering, Technological Institute, Northwestern University, Evanston, IL, 1978.

[44] CEB/FIP International recommendation for the design and construction of concrete structures, Paris–London, 1970.

[45] Mayers, B. and Branson, D. E.: "Design aid for predicting creep and shrinkage properties of concrete," *J. Am. Concr. Inst.*, **69**, 1972.

[46] Rüsch, H. and Jungwirth, D.: *Stahlbeton—Spannbeton* (Reinforced Concrete—Prestressed Concrete), Band 2, Werner-Verlag, Düsseldorf, 1976.

[47] " A simple design method for predicting the elastic modulus and creep of structural concrete," British Concrete Society, London, 1977.

[48] Branson, D. E.: *Deformation of Concrete Structures*, McGraw-Hill, New York, 1977.

[49] CEB/FIP Model Code, Comité Euro-International du Béton, Paris–London–Berlin, 1978.

[50] American Concrete Institute Committee 209, Subcommittee II, "Prediction of creep, shrinkage and temperature effects, 2," Draft, Detroit, October 1978.

[51] DIN 4227, Deutsches Institut für Normung, Spannbeton, Richtlinien für Bemessung und Ausführung" (Specifications for design and construction of prestressed concrete), Berlin, 1979.

[52] Jungwirth, D.: "Erfassung der Grösse und des zeitlichen Verlaufes der Kriech- und Schwind-verformungen von Breton—Auswertung der Messungen von Bauwerken unter Einbezug von Laborergebnissen" (Estimate of magnitude and development with time of creep and shrinkage strains of concrete—Evaluation of measurements on structures taken into account laboratory experiments), IVBH Symposium, Vorbericht, Madrid, 1970.

[53] Hansen, T. C.: "Influences of aggregates and voids on modulus of elasticity of concrete, cement mortar and cement paste," *J. Am. Concr. Inst.*, **62**, 1965.

[54] Shideler, J. J.: "Lightweight aggregate concrete for structural use," *J. Am. Concr. Inst.*, **54**, 1957.

[55] DIN 4219, "Leichtbeton und Stahlleichtbeton mit geschlossenem Gefüge, Teil 1 und 2," (Lightweight concrete and reinforced lightweight concrete with a dense structure, part 1 and 2), Berlin, 1979.

[56] McHenry, D. A.: "A new aspect of creep in concrete and its application to design," *ASTM Proc.*, **43**, 1943.

[57] Ross, A. D.: "Creep of concrete under variable stress," *J. Am. Concr. Inst.*, **54**, 1958.

Part C

Examples

For the reader in a hurry:

Generally speaking, Part C can be used by design engineers even if they have not read the foregoing chapters. To this effect, Section 1 presents a clear review of the methods discussed in Part **B** which make it possible to estimate the magnitude of creep and shrinkage. The same was done in Section 2 with regard to the mathematical tools which are discussed in more detail in Appendix II.

The following table provides a summary of the problems treated. It is intended to facilitate the search for the appropriate section.

Section	Problem	Effect			Type of construction						Structural System determin. D indeterm. I		Creep				Numerical method					
		Loads	Constraint	Shrinkage	Plain concrete	Reinforced concrete	Bonded prestressed concrete	Unbonded prestressed concrete	Partially prestressed concrete	Composite construction	External	Internal	Taking into account ε_d	Uniform concrete	2 types of concrete	Creep of reinforcement	Expanded Dischinger	Trost	Busemann	Iteration with $\alpha_{i,u}$	Finite differences (step by step method)	f_m-Method
3.1	Tied arch	x		x	x						1I	D		x			x					
3.2	Prestressing without bond	x		x				x			D	1I		x			x					
4.2	Reinforced concrete column	x				x					D	1I		x			x					
4.3.1	Prestressed concrete beam	x	x	x			x					1I		x			x					
4.3.2	Prestressed concrete beam	x	x	x			x					1I		x								x
4.3.3	Prestressed concrete beam	x	x	x			x					1I		x		x						x
4.4	Partial prestressing								x													
5	Steel composite section	x								x		2I		x					x			
6.2	Two-strand prestressing	x	x	x			x					2I		x						x		
6.3.1	Two-strand prestressing	x	x	x			x					2I		x							x	

7.4	Two-span continuous prestressed concrete girder	x	x	x			x	1I	1I		x						x
8.3.1	Three-span continuous girder on elastic supports		x		x			2I			x						x
8.3.2	Three-span continuous girder on elastic supports		x		x			2I			x			x			
9.2	Rapid settlement of foundation		x		x			1I			x		x				
9.3.2	Slow settlement of foundation, affinity with ε_c		x		x			1I			x		x				
9.3.3	Slow settlement of foundation (Dimel)		x		x			1I			x			x			
9.3.4	Slow settlement of foundation (Trost and Mainz)		x		x			1I			x						x
9.3.6	Slow settlement of foundation, new auxiliary values		x		x	x		1I		x							x
9.4	Coupling of precast members of same age	x	x		x			1I			x		x				
10.2	Contraction joint			x	x						x						
10.3	Impeded shrinkage deformation			x	x			1I			x		x				
10.3.5	Danger of shrinkage cracking			x	x			1I			x		x				
10.4	Shrinkage not affinitive to creep			x	x					x							x
10.5	Influence of shrinkage reinforcement			x		x		1I			x						x
11.1.2	Two-span continuous girder made of different types of concrete	x	x		x			1I				x	x				
11.1.3	Frame made of different types of concrete	x	x		x			1I				x	x				
11.2.2	Cross-section of different concrete types	x	x	x	x				2I			x			x		
11.2.3	Cross-section of different concrete types (approximation)	x	x		x				1I			x	x				
12.1	Two-span continuous girder of different concrete types	x	x		x			1I		x		x				x	
14.3.1.1	Deflection calculation for prestressed concrete beam	x	x				x				x						
14.3.1.2	Deflection calculation for prestressed concrete bridge deck	x	x				x				x						
14.3.1.3	Curvature of a cracked reinforced concrete section	x				x			1I		x						x
14.3.1.4	Prefabricated reinforced concrete girder	x		x		x					x						
14.3.2.2	Influence of change of structural system on camber	x	x		x			1I			x						
14.3.2.3	Camber in free cantilever construction	x	x	x			x				x						

Introduction

Anyone who must familiarize himself thoroughly with the problems of creep will appreciate the following collection of examples taken from actual practice. However, the specialist, too, wants to determine which method of computation will give him the fastest usable solution to his particular problem.

Since climatic conditions are never totally predictable, there is an inherent uncertainty in the input data φ and ε_s. It would, therefore, not be sensible to strive for "precise computation methods." Whenever the influence of creep and shrinkage is of decisive importance, one is better advised to estimate the potential scatter for φ and ε_s (ca. $\pm 20\%$) and then, in the light of these limiting values, examine possible consequences by using simple calculation techniques.

1. Brief Compilation of Formulations for an Estimate of Instantaneous and Time-Dependent Strains

In the following the Rüsch–Jungwirk method to estimate concrete strains as given in the German prestressed concrete code DIN 4227 is presented in more detail because the examples to demonstrate the effect of creep on structural behavior given in the subsequent chapters are based on this method. However, it should be pointed out that other methods such as the one proposed by ACI committee 209 are equally suited for the numerical methods presented in this book.

1.1 The Modulus of Elasticity of Concrete (cf., Part B, Sections 3.1 and 4)

For 28-day-old concrete, the values of E for concrete of different strength are given in Table 1.1. (f'_{c28} is the 5% defect of cube strength after 28 days in N/mm².)

The age dependence of E can be estimated by means of the coefficient β_e plotted in Fig. 1.1. (Fig. 14.2 in Section 14.1 gives the inverse value $1/\beta_e$):

$$E_t = E_{28}\beta_e. \tag{1.1}$$

However, this is only significant for an estimate of immediately occurring deformations. The creep coefficient is always related to the modulus of elasticity of 28-day-old concrete. Thus,

$$\sum \varepsilon = \frac{f}{E_t} + \frac{f}{E_{28}}\varphi + \varepsilon_s. \tag{1.2}$$

Table 1.1

Strength of concrete f'_{c28} (N/mm²)	10 (1434)	15 (2151)	25 (3585)	35 (5019)	45 (6453)	55 (7887)	(psi)
$10^{-3}\,E_{28}$ (N/mm²)	22 (3155)	26 (3728)	30 (4302)	34 (4876)	37 (5306)	39 (5593)	(psi)

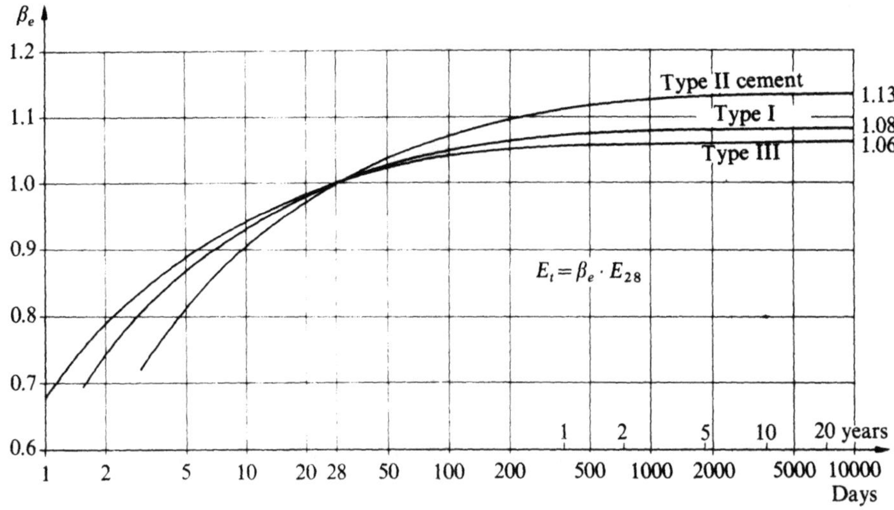

Figure 1.1. Effect of concrete age on modulus of elasticity $E_t = \beta_e E_{28}$.

In exceptional cases, when one needs to know more exactly the immediately occurring deformation, the procedure in Part B, Section 3.1.4 can be followed.

1.2. The Creep Coefficient (cf., Fig. 1.2; cf., Part B, Sections 3.3 and 4)

Figure 1.2. Description of Eq. (1.3).

$$\varphi = 0.4\beta_{d(t_2-t_1)} + \varphi_{fo}(\beta_{ft_2} - \beta_{ft_1}) - 0.4\beta_{d(t_2-t_1)}\beta_{d(t-t_2)}, \quad (1.3)$$

where

β_d is the coefficient for the time development of the delayed elastic strain in accordance with Fig. 1.3.

φ_{fo} is the base value for flow in accordance with Table 1.2.

β_f is the coefficient for the flow-time relationship in accordance with Fig. 1.4. Figure 1.5 facilitates the determination of the difference $\beta_{ft_2} - \beta_{ft_1}$.

t_f is the effective age of concrete for flow

$$t_f = \beta_{ce} \cdot \beta_T \cdot t. \quad (1.4)$$

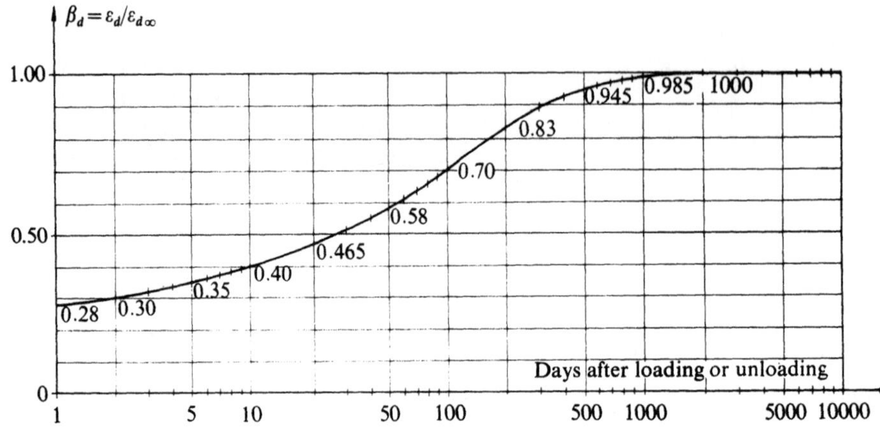

Figure 1.3. Coefficient β_d for the development of delayed elasticity with time (see also Table 1.3).

Table 1.2. Basic flow coefficient, basic shrinkage strain and coefficient β_{eff} for various environmental conditions. (The values in bold face are taken from CEB/FIP 1978 or from DIN 4227; the intermediate numbers were interpolated graphically.)

Location of members	Average relative ambient humidity (%)	Values for medium slump concrete[a]		
		Basic flow coefficient φ_{f0}	Basic shrinkage strain ε_{s0}[b]	Coefficient β_{eff} for effective thickness
In water	100	0.8	+10	30
	95	1.1	0	10
In very humid air, e.g., immediately above water level	90	1.3	−10	5
	80	1.7	−20	2.4
Outdoors	70	2.0	−25	1.5
	60	2.4	−30	1.2
	50	2.7	−35	1.0
In dry air, e.g., indoors	40	3.0	−40	1.0

[a] For low or high slump concrete the values for φ_{f0} and ε_{s0} must be respectively reduced and increased by 25%.

[b] For corrected values see B4.2.

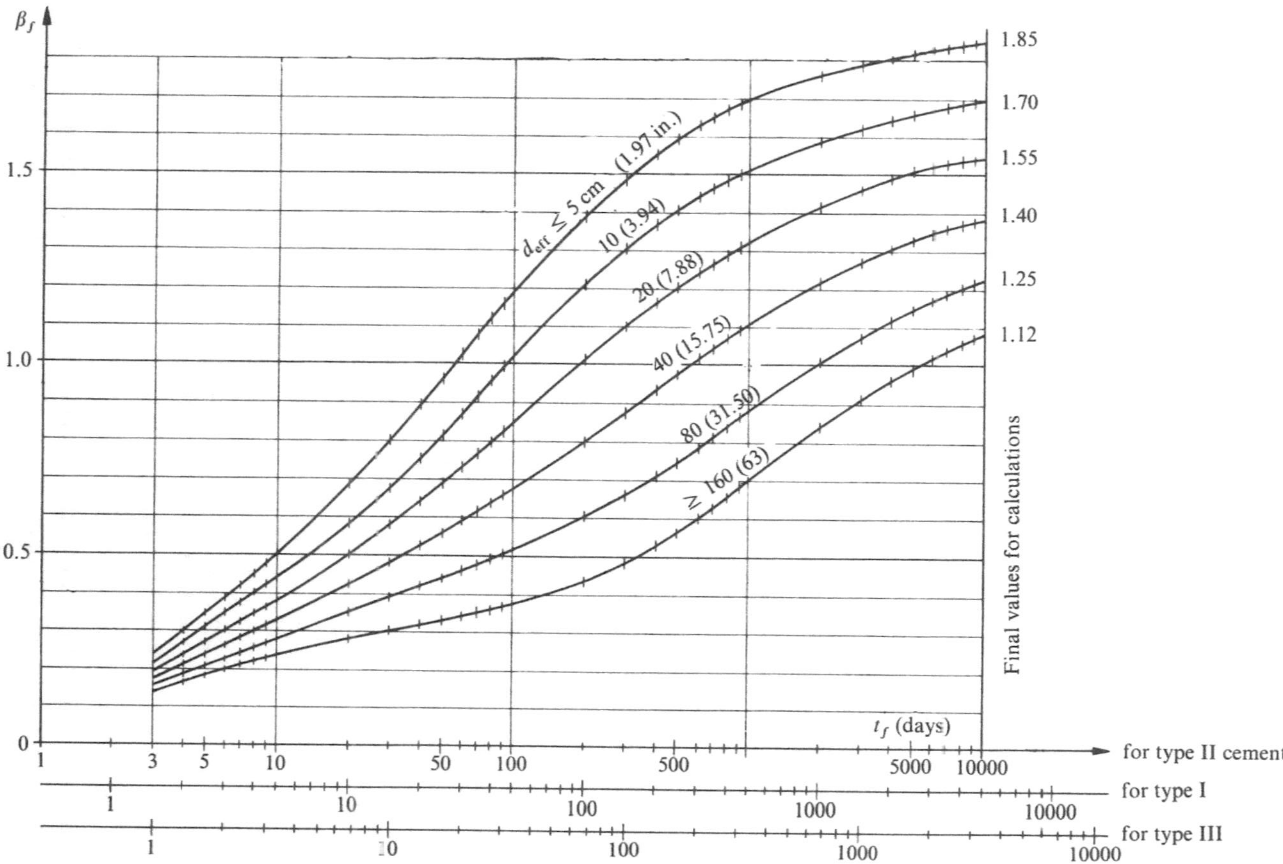

Figure 1.4. Coefficient β_f which depends on the effective age of concrete t_f. It describes the development of flow with time for various types of cement (see also Table 1.3).

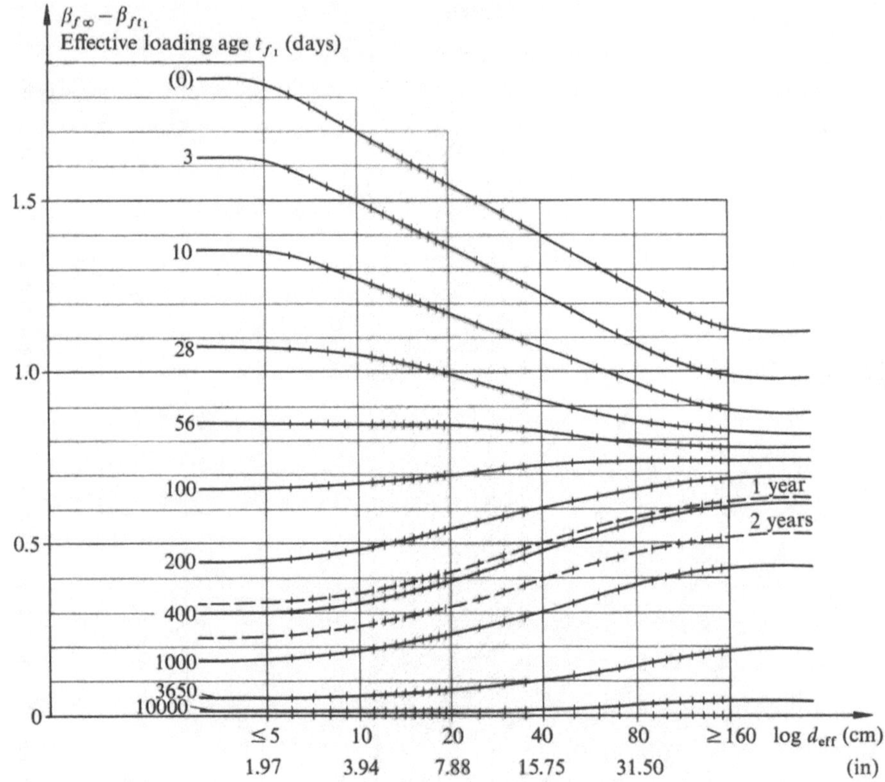

Figure 1.5. Diagram to determine the difference $\beta_{ft_2} - \beta_{ft}$ as a function of effective concrete age t_f. The values of the ordinates correspond to the final values $\beta_{f\infty} - \beta_{ft}$. Intermediate values $\beta_{ft_2} - \beta_{ft_1}$ may be obtained from the differences of the ordinates of the corresponding ages.

β_{ce} is a coefficient that takes into account the rate of hydration of the cement. We have

$$\begin{aligned} \beta_{ce} &= 1 \text{ for Type II cements,} \\ &= 2 \text{ for Type I cements,} \\ &= 3 \text{ for Type III cements.} \end{aligned}$$

$$\beta_T = (1/t) \sum [(T°\text{C} + 10\,°\text{C})/30°\,\text{C}]\Delta t.$$

β_T is a coefficient to take into account the influence of temperature on the strength development of concrete. T is the temperature of concrete in °C. For $T = 20\,°\text{C}$ (68 °F) = const, we have $\beta_T = 1$.
The coefficient β_T plays a role only when the concrete is subjected to temperatures which differ significantly from room temperature, as is the case for concrete construction in winter. As a rule, $\beta_T = 1$ and, consequently, $t_f = \beta_{ce} t$.

Δt is the number of days at concrete temperature T. (See also Eq. (B2.2))

Table 1.3. Tabulation of β_d, β_f, and β_s values.

Time in days	Values of β_d	Time in days	Values of β_d
1	0.280	500	0.945
2	0.300	1000	0.985
5	0.350	2000	1.000
10	0.400	5000	1.000
20	0.465	10000	1.000
50	0.580	20000	1.000
100	0.700	∞	1.000
200	0.830		

	Value of β_f versus effective thickness cm (in.)					
Time in days	≤5 (1.97)	10 (3.94)	20 (7.88)	40 (15.75)	80 (31.50)	≥160 (63)
3	0.240	0.210	0.190	0.170	0.155	0.140
5	0.345	0.310	0.270	0.235	0.210	0.185
10	0.505	0.440	0.380	0.328	0.280	0.235
20	0.685	0.575	0.500	0.420	0.350	0.280
50	0.964	0.810	0.690	0.562	0.443	0.330
100	1.195	1.025	0.850	0.680	0.520	0.375
200	1.395	1.215	1.020	0.800	0.603	0.435
500	1.600	1.413	1.208	0.980	0.750	0.566
1000	1.698	1.514	1.320	1.107	0.884	0.703
2000	1.762	1.589	1.416	1.217	1.010	0.842
5000	1.820	1.660	1.510	1.330	1.148	1.000
10000	1.846	1.695	1.545	1.383	1.225	1.085
20000	1.850	1.700	1.550	1.400	1.250	1.120
∞	1.850	1.700	1.550	1.400	1.250	1.120

	Value of β_s versus effective thickness cm					
Time in days	≤5 (1.97)	10 (3.94)	20 (7.88)	40 (15.75)	80 (31.50)	≥160 (63)
1	0.110	0.040	0.010	0.0	0.0	0.0
2	0.170	0.080	0.020	0.0	0.0	0.0
5	0.290	0.160	0.055	0.005	0.005	0.0
10	0.420	0.240	0.100	0.030	0.020	0.0
20	0.560	0.340	0.160	0.060	0.030	0.0
50	0.760	0.510	0.270	0.120	0.055	0.010
100	0.900	0.650	0.375	0.185	0.085	0.020
200	1.020	0.780	0.490	0.260	0.120	0.045
500	1.110	0.910	0.660	0.410	0.210	0.090
1000	1.160	0.980	0.770	0.550	0.340	0.175
2000	1.190	1.040	0.840	0.660	0.500	0.310
5000	1.200	1.050	0.885	0.750	0.660	0.510
10000	1.200	1.050	0.895	0.790	0.725	0.640
20000	1.200	1.050	0.900	0.800	0.750	0.700
∞	1.200	1.050	0.900	0.800	0.750	0.700

d_{eff} is the effective member thickness

$$d_{eff} = \beta_{eff} 2A/u. \tag{1.5}$$

β_{eff} is a coefficient to take into account the influence of relative humidity on the effective thickness d_{eff} (Table 1.2).

A is the area of the concrete cross section.

u is the portion of the cross section that is subjected to drying.

The data given in Table 1.2 require a critical comment. 40% as an average relative humidity for indoors is unrealistic. The following more realistic levels are recommended:

In water: 100%
In very humid air: 90%
Outdoors, generally: 70–80%
In normal rooms: 60%
In very dry rooms: 50%
In heavily heated rooms: 40%

1.3. Shrinkage (For Details, cf., Part B, Sections 3.4 and 4)

$$\varepsilon_{st} = \varepsilon_{s0}(\beta_{st} - \beta_{st_1}), \tag{1.6}$$

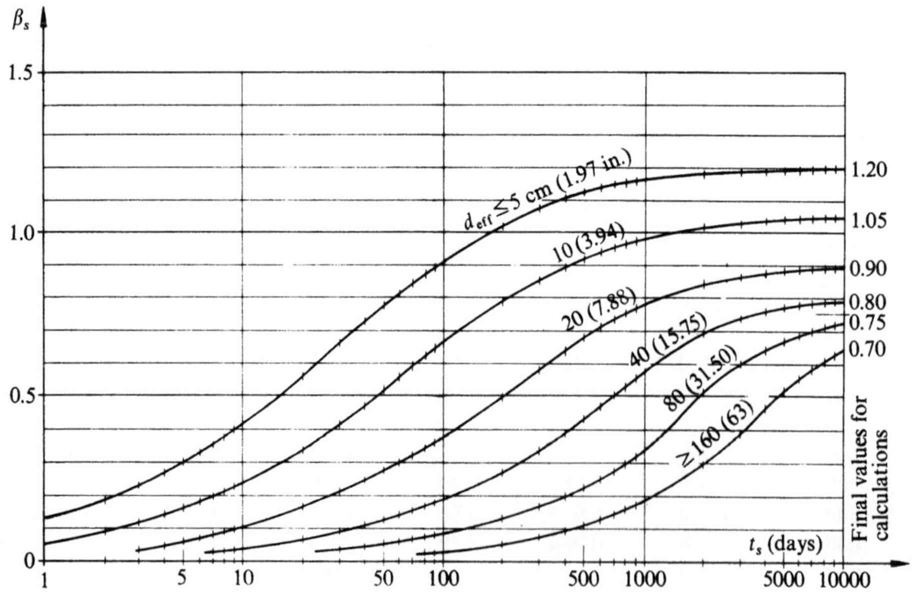

Figure 1.6. Coefficient β_s as a function of effective concrete age t_s to describe the development of shrinkage (see also Table 1.3).

where

ε_{s0} is the basic shrinkage strain in accordance with Table 1.2.
t_s is the effective age of concrete for shrinkage,

$$t_s = \beta_T t. \tag{1.7}$$

The influence of rate of hydration of cement on shrinkage is negligible. The influence of the temperature of concrete, however, should be taken into account. For this, the coefficient β_T can be used as described in Section 1.2.

β_s is the time development of shrinkage in accordance with Fig. 1.6. The determination of the difference $\beta_{st} - \beta_{st_1}$ is facilitated by Fig. 1.7.

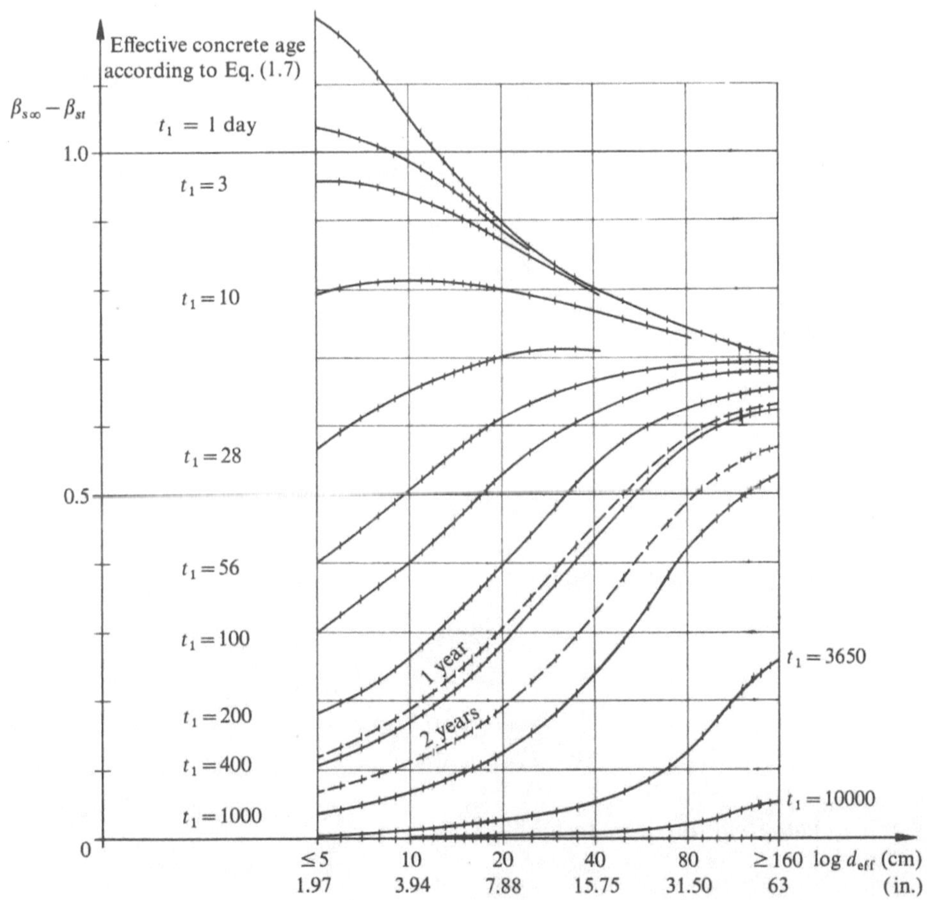

Figure 1.7. Diagram to determine the difference $\beta_{st_2} - \beta_{st}$. The values of the ordinates correspond to the final values $\beta_{s\infty} - \beta_{st}$. Intermediate values $\beta_{st_2} - \beta_{st}$ may be obtained from the differences of the ordinates of the corresponding effective ages.

1.4. Examples of Determination of the Effective Member Thickness in Accordance with Eq. (1.5)

The summary given in the following, demonstrates the dependence of the effective thickness d_{eff} on the volume/surface ratio expressed by the area/circumference ratio $2A/u$.

1.4.1. Beam

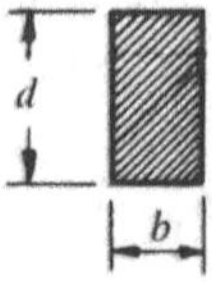

$$\frac{2A}{u} = \frac{d_{\text{eff}}}{\beta_{\text{eff}}} = \frac{2bd}{2(b+d)} = \frac{bd}{b+d} \text{ for } b = d \rightarrow \frac{d}{2}.$$

1.4.2. Wide Slab

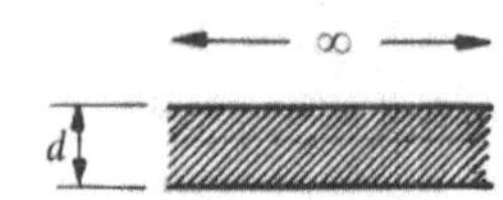

$$\frac{2A}{u} = \frac{d_{\text{eff}}}{\beta_{\text{eff}}} = \frac{2d\infty}{2\infty} = d.$$

1.4.3. Circular Section

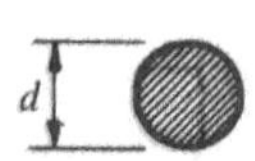

$$\frac{2A}{u} = \frac{d_{\text{eff}}}{\beta_{\text{eff}}} = \frac{2d^2\pi}{4d\pi} = \frac{d}{2}.$$

1.4.4. Circular Ring, Interior Sealed

$$\frac{2A}{u} = \frac{d_{\text{eff}}}{\beta_{\text{eff}}} \approx \frac{2d\pi t}{d\pi} = 2t.$$

1.4.5. Circular Ring, Interior Aerated

$$\frac{2A}{u} = \frac{d_{\text{eff}}}{\beta_{\text{eff}}} \sim \frac{2d\pi t}{2d\pi} = t.$$

1.4.6. T-Beam

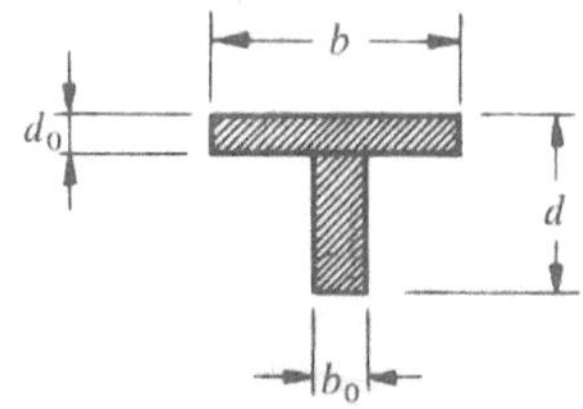

$$\frac{2A}{u} = \frac{d_{\text{eff}}}{\beta_{\text{eff}}} = \frac{2[bd_0 + (d - d_0)b_0]}{2b + 2d}$$

$$= \frac{bd_0 + db_0 - d_0 b_0}{b + d}$$

in slender T-beams, $d_0 b_0 \sim 0$.

1.4.7. Box Girder

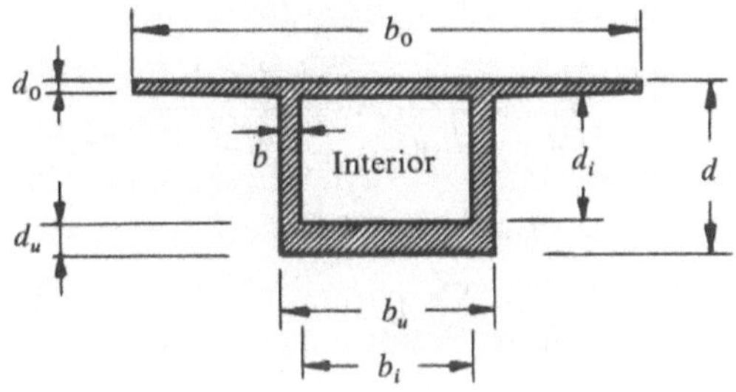

Interior fully aerated:

$$\frac{2A}{u} = \frac{b_0 d_0 + b_u d_u + 2bd_i}{b_0 + d + b_i + d_i}.$$

Interior poorly aerated (50%):

$$\frac{2A}{u} = \frac{b_0 d_0 + b_u d_u + 2bd_i}{b_0 + d + 0.5(b_i + d_i)}.$$

Interior sealed:

$$\frac{2A}{u} = \frac{b_0 d_0 + b_u d_u + 2bd_i}{b_0 + d}.$$

Numerical Example

For $b_0 = 10.00$ m (393.7 in.), $b_u = 5.00$ (1968), $b = 0.40$ (15.7), $d = 3.00$ (118.1), $d_0 = d_u = 0.25$ m (9.84 in.), we obtain the following values for $2A/u$: 0.29 m (11.4 in.) (Fully aerated); 0.35 m (13.8 in.) (poorly aerated); 0.44 m (17.3 in.) (sealed). Most box girders made must be rated in the poorly aerated category.

1.4.8. Structure with Varying Thickness

The object is to determine the mean thickness $d_{\text{eff}\,m}$, taking into account the creep inducing stresses.

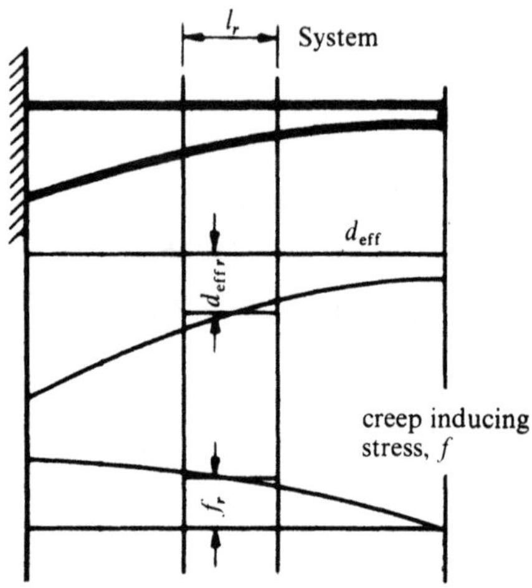

$$\frac{2A}{u} = \frac{d_{\text{eff}\,m}}{\beta_{\text{eff}}} = \frac{\sum_1^n d_{\text{eff}\,r} l_r f_r}{l f_m},$$

$$l = \sum_1^n l_r,$$

$$f_m = \frac{\sum_1^n f_r l_r}{l}.$$

1.5. Numerical Examples for the Determination of φ and ε_s

EXAMPLE 1 (Dry room). A beam (rectangular section 30 × 60 cm 11.8 × 23.6 in.) is to be loaded at the age of 7 days in a very dry room. The normal-slump concrete, hardened at normal temperature, has been made with a Type II cement. What will be the final creep and what final shrinkage should be expected?

From Table 1.2, $\varphi_{f0} = 3.0$, $\varepsilon_{s0} = 40 \times 10^{-5}$, and $\beta_{\text{eff}} = 1.0$.

From Eq. (1.5) $d_{\text{eff}} = \beta_{\text{eff}}\, 2A/u = 1.0(2)bd/2(b + d) = 20$ cm (7.87 in.).

From Eq. (1.4) $t_{f1} = t_{s1} = 7$ days, since $\beta_{ce} = 1$ and—under normal temperature conditions—$(T\,°\text{C} + 10\,°\text{C})/30\,°\text{C} = 1$; $t = \infty$.

From Eq. (1.3) and Figs. 1.3 and 1.4, $\varphi_\infty = 0.4\,(1.0) + 3.0\,(1.55 - 0.32) = 4.1$.

From Eq. (1.6) and Fig. 1.6, $\varepsilon_{s\infty} = 40 \times 10^{-5}\,(0.90 - 0) = 36 \times 10^{-5}$.

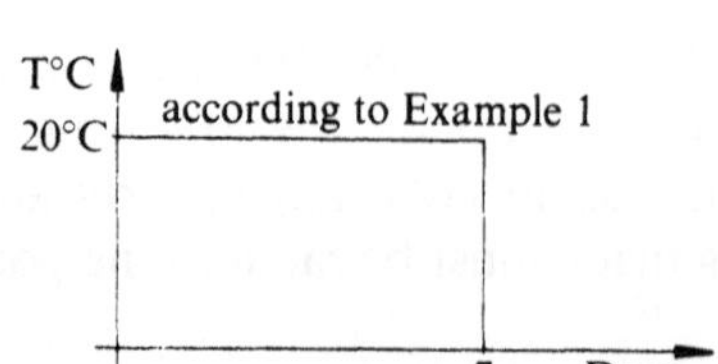

EXAMPLE 2 (The Influence of concrete temperature). Assuming, in Example 1, that a period of low temperatures retards hardening of the concrete after it is cast, the problem is to determine how long the application of load must be delayed in order not to deviate from the assumptions made in the structural analysis. It is only in such cases that the expression

$$\beta_T = (1/t) \sum \Delta t(T\,^\circ\mathrm{C} + 10\,^\circ\mathrm{C})/30\,^\circ\mathrm{C},$$

—which was disregarded in Example 1—is needed. Over the first 8 days, the following concrete temperatures were recorded: 18 °C, 14 °C, 5 × 2 °C. During the thaw that followed, the temperature rose to 10 °C and then to 20 °C. This gives

$$\sum \Delta t(T^\circ\mathrm{C} + 10\,^\circ\mathrm{C}) = [(1)28 + (1)24 + (1)18 + (5)12 + (1)20 + X(20 + 10) = 7(20 + 10).$$

$X = (210 - 150) : 30 = 2$. Thus, it takes $9 + 2 = 11$ days to reach the same level of maturity as in Example 1.

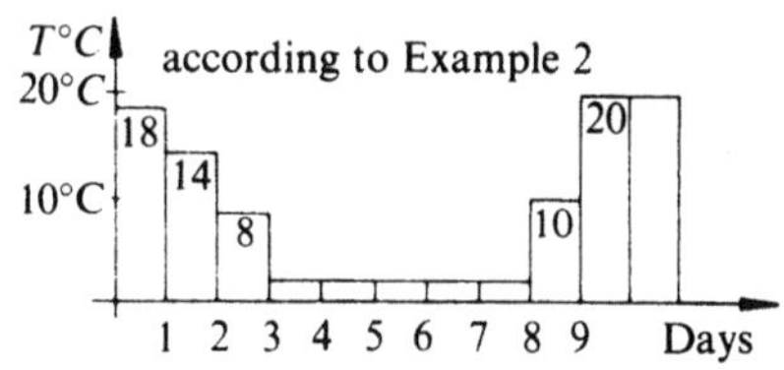

EXAMPLE 3 (Influence of the type of cement). What values are obtained in Example 1 if the concrete is made with a Type III rather than a Type II cement?

For Type III we have $\beta_{ce} = 3$. As a consequence $t_{f1} = 3(7) = 21$ days.

$$\varphi_\infty = 0.4(1.0) + 3.0(1.55 - 0.53) = 3.46 < 4.1.$$

ε_s does not change from Example 1, since $\beta_{ce} = 1$ for shrinkage regardless of the type of cement used.

EXAMPLE 4 (Load variation). Maintaining the assumptions of Example 1, let the question be asked, what strain characteristics should be used after 200 days for a load which acted only from the 7th to the 90th day?

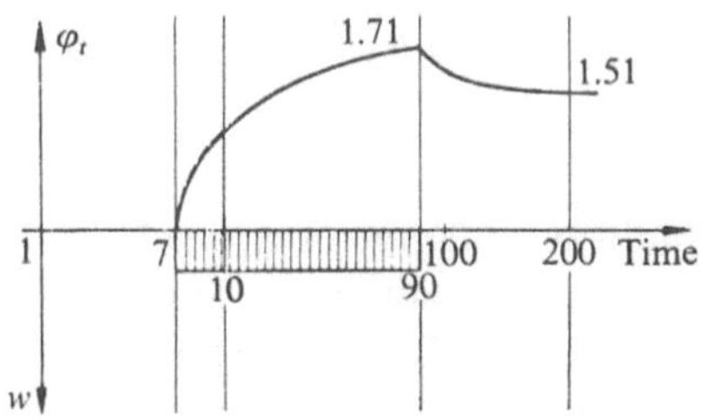

$$t_{f1} = 7,\ t_{f2} = 90,\ t_{f3} = 200,\ t_{f3} - t_{f1} = 83 \text{ days},\ t_{f3} - t_{f2} = 110 \text{ days},$$

$$\varphi_{200} = 0.4(0.68) + 3.0(0.83 - 0.32) - 0.4(0.68)0.74 = 0.27 + 1.44 - 0.20 = 1.51,$$

$$\varepsilon_{s200} = 40 \times 10^{-5}(0.50) = 20 \times 10^{-5}.$$

EXAMPLE 5 (Influence of relative humidity). What values would characterize Example 1 if the member were stored under very humid instead of dry conditions?

The reduced drying is taken into account by means of φ_{f0} and ε_{s0}, and the fictitious increase in d_{eff} by means of the coefficient β_{eff}.

$$\beta_{\text{eff}} = 5.0;\ d_{\text{eff}} = 5(20) = 100 \text{ cm } (39.37 \text{ in.});\ \varphi_{f0} = 1.3;\ \varepsilon_{s0} = 10 \times 10^{-5};$$

$$t_{f1} = t_{s1} = 7,$$

$$\varphi_{\infty} = 0.4(1.0) + 1.3(1.20 - 0.23) = 1.66,$$

$$\varepsilon_{s\infty} = 10 \times 10^{-5}(0.73 - 0) = 7.3 \times 10^{-5}.$$

EXAMPLE 6 (Influence of a change in relative humidity). Let the last question be what effect a change in the environmental conditions would have as, e.g., in the case of precast members which are prestressed for 1 day and have had 14 days of humid storage before reaching the storage area. It is advisable to account for the most unfavorable case where the installation in the structure is delayed for up to $\frac{1}{2}$ year. How does one determine the loss of prestress by the time of installation, and what values should one use for the subsequent stress conditions? Assumptions: $2\,A/u = 4$ cm (1.57 in.); Type III cement; average humidity is 90% between t_1 and t_2 and 70% thereafter.

$$d_{\text{eff}(t_2 - t_1)} = 4(5) = 20\,(7.87),\ d_{\text{eff}(t - t_2)} = 4(1.5) = 6 \text{ cm } (2.36 \text{ in.}),$$

with $\beta_{ce} = 3$ for creep.

$$t_{f1} = 3(1) = 3,\ t_{f2} = 3(1 + 14) = 45,\ t_{f3} = 3(360/2) = 540 \text{ days},$$

with $\beta_{ce} = 1$ for shrinkage.

$$t_{s1} = 1,\ t_{s2} = 15,\ t_{s3} = 180,$$

$$\begin{aligned}\varphi_{180}\ & 0.4(0.95) + 1.3(0.66 - 0.19) + 2.0(1.60 - 0.91) \\ &= 0.38 + 0.61 + 1.38 = 2.37,\end{aligned}$$

$$\begin{aligned}\varepsilon_{s180} &= 10 \times 10^{-5}(0.13 - 0) + 25 \times 10^{-5}(0.98 - 0.48) \\ &= 1.3 \times 10^{-5} + 12.5 \times 10^{-5} = 13.8 \times 10^{-5}.\end{aligned}$$

After installation, average humidity is assumed to be 40%. Then we have $d_{\text{eff}} = 4$ cm (1.57 in.) and

$$\varphi_{(180 \div \infty)} = 0.4(1.00) + 3.0(1.85 - 1.65) = 0.40 + 0.60 = 1.00,$$

$$\varepsilon_{s(180 \div \infty)} = 40 \times 10^{-5}(1.20 - 1.00) = 8.0 \times 10^{-5}.$$

If, e.g., the two load conditions, the member dead weight d_1 before installation and the total load $d_1 + d_2$ after installation, are treated separately, we have $\varphi_{180} = 2.37$ for d_1 and $\varphi_{(180+\infty)} = 1.0$ for d_1 and d_2, with the delayed elastic strain taken into account with d_1 from the first loading phase. One can also proceed by allowing d_1 to act up to $t = \infty$ and applying d_2 only during the period after installation. Then, $\varphi_{\infty} = 2.37 + 0.60 = 2.97$ for d_1 and $\varphi_{(180+\infty)} = 1.00$ for d_2.

In the first period up to 180 days, shrinkage is 13.8×10^{-5}. In the second period from $t = 180$ days to ∞, shrinkage increases by 8×10^{-5}. Thus, the total shrinkage is $(13.8 + 8) \times 10^{-5} = 21.8 \times 10^{-5}$. If the finished member were exposed from the first day to a humidity of only 40%, the total shrinkage would be $40 \times 10^{-5}(1.2 - 0.16) = 41.6 \times 10^{-5}$. The improvement in the first case is a result of the curing of the young concrete. In extreme cases, however, the procedure shown can lead to unreliable shrinkage values. For instance, 1000-day water storage followed by storage under 40% relative humidity would yield a shrinkage of $[10(0.17) - 40(1.2 - 1.17)] \times 10^{-5} \sim 0$. In such extreme cases, the procedure breaks down because the underlying values originated from tests under constant environment. One assumes that the completely water-saturated concrete undergoes greater shrinkage during the drying phase than indicated in the curve established for dry storage, and $\varepsilon_{s\infty}$ may be estimated as follows:

$$\left[10(0.17) - \left(40(1.20) - \frac{40(1.17) - 10(0.17)}{2}\right)\right] \times 10^{-5} = -23.7 \times 10^{-5}.$$

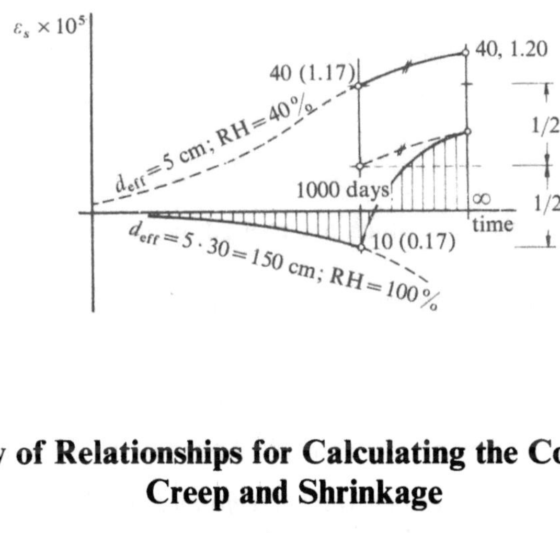

2. Summary of Relationships for Calculating the Consequences of Creep and Shrinkage

2.1. The Expanded Dischinger Equation (cf., Appendix II, Section 2.2)

2.1.1. A Structural Member which Creeps and an Elastic Member Are Coupled Locally (Example: Unbonded Prestressing)

In accordance with Appendix II, Section 2.2, the general equations are

$$S_{et} = S_{ewt} + \gamma_{sus} S_{esus0} + \gamma_r S_{er0} + \gamma_s S_{est}^{(e)}, \tag{2.1}$$

$$S_{ecst} = (\gamma_{sus} - 1) S_{esus0} + (\gamma_r - 1) S_{er0} + \gamma_s S_{est}^{(e)}. \tag{2.2}$$

Under prestressing, the subscript e is replaced by p:

$$S_{pt} = S_{pwt} + \gamma_{sus} S_{psus0} + \gamma_r S_{pr0} + \gamma_s S_{pst}^{(e)}, \tag{2.3}$$

where

S_{et}, S_{pt} is the total value of action effects, taking into account the effect of creep and shrinkage at time t in the elastic structural member (unstressed reinforcement s or prestressed tendon p) or the coupling force occurring at time t.

S_{ecst}, S_{pcst} is the total value of action effects induced in the elastic structural member by creep and shrinkage.

S_{ewt}, S_{pwt} are the action effects acting in the elastic member at time t as a result of loads applied only temporarily.

S_{esus0}, S_{psus0} are the action effects induced in the elastic member by sustained loads at time $t = 0$, such as

$$S_{esus0} = \frac{\delta_{c1sus}}{\delta_{c11} + \delta_{e11}}. \tag{2.4}$$

S_{er0}, S_{pr0} as above, but as a result of a very rapidly generated constraint, such as the pretensioning, p. e.g.,

$$S_{er0} = \frac{\delta_{c1r}}{\delta_{c11} + \delta_{e11}}. \tag{2.5}$$

$S_{est}^{(e)}, S_{pst}^{(e)}$ is an auxiliary action effect assumed to be purely elastic to take into account shrinkage, e.g.,

$$S_{est}^{(e)} = \frac{\delta_{c1st}}{\delta_{c11} + \delta_{e11}}. \tag{2.6}$$

$\gamma_{sus}, \gamma_r, \gamma_s$ are coefficients which are plotted against φ and α in Figs. 2.1, 2.2, and 2.3.

α is the stiffness ratio according to Dischinger:

$$\alpha = \frac{\delta_{c11}}{\delta_{c11} + \delta_{e11}}. \tag{2.7}$$

The terms δ are displacements as defined in Appendix II, Section 2.2.

2.1.2. *A Structural Member which Creeps and an Elastic Member Are Bonded Continuously*

The general equation has the same structure as Eq. (2.1). Only the action effects are replaced by stresses (cf., Appendix II, Section 2.1.3)

$$f_{et} = f_{ewt} + \gamma_{sus} f_{esus0} + \gamma_r f_{er0} - \gamma_s |\varepsilon_{st}| E_e(1 - \alpha) \tag{2.8a}$$

or, for prestressing,

$$f_{pt} = f_{pwt} + \gamma_{sus} f_{psus0} + \gamma_r f_{pp0} - \gamma_s |\varepsilon_{st}| E_p(1 - \alpha). \tag{2.8b}$$

If only the stresses induced by creep and shrinkage are to be determined, the following equation is used.

$$f_{ecst} = (\gamma_{sus} - 1) f_{esus0} + (\gamma_r - 1) \cdot f_{er0} - \gamma_s |\varepsilon_{st}| E_e(1 - \alpha) \tag{2.9}$$

$S_{e\,sust} = S_{e\,sus\,0} \cdot \gamma_{sus}$ or $f_{e\,sust} = f_{e\,sus\,0} \cdot \gamma_{sus}$

$$\gamma_{sus} = \frac{1}{\alpha} - \frac{1-\alpha}{\alpha(1 + 0.4 \cdot \beta_d \cdot \alpha)} \exp\left(-\alpha \frac{\varphi_t - 0.4 \cdot \beta_d}{1 + 0.4 \cdot \beta_d \cdot \alpha}\right) = \frac{1}{\alpha} - \frac{1-\alpha}{\alpha(1 + \varphi_d \cdot \alpha)} \exp\left(-\frac{\alpha \cdot \varphi_f}{1 + \alpha \cdot \varphi_d}\right)$$

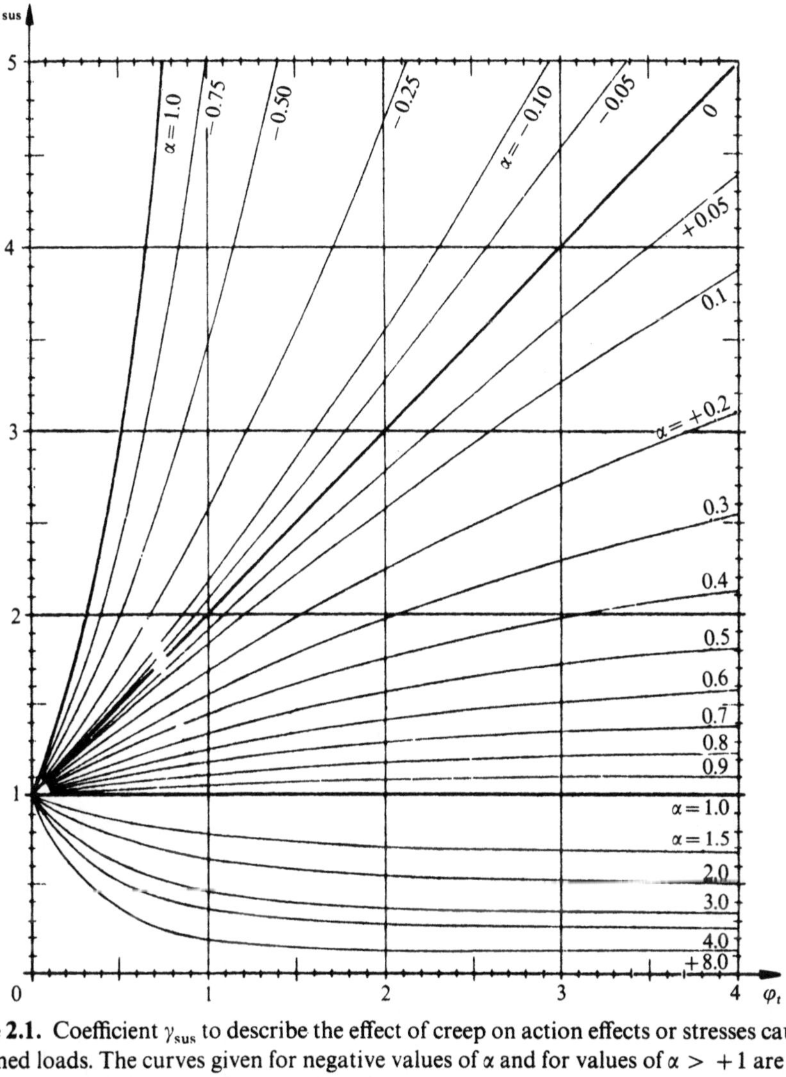

Figure 2.1. Coefficient γ_{sus} to describe the effect of creep on action effects or stresses caused by sustained loads. The curves given for negative values of α and for values of $\alpha > +1$ are needed only in the Dischinger/Kupfer procedure for members with different creep and shrinkage properties.

or, for prestressing,

$$f_{pcst} = (\gamma_{sus} - 1) f_{psus0} + (\gamma_r - 1) \cdot f_{pp}^{(0)}(1 - \alpha) - \gamma_s |\varepsilon_{st}| E_p(1 - \alpha), \quad (2.10)$$

where $f_{pp}^{(0)}$ is the steel stress prior to releasing the steel. In the case of post-tensioning, $f_{pp}^{(0)}$ is determined from

$$f_{pp}^{(0)} = f_{pp0} - n f_{cp0} = \frac{f_{pp0}}{1 - \alpha}. \quad (2.11)$$

$$S_{ert} = S_{er0} \cdot \gamma_r \quad \text{or} \quad f_{ert} = f_{er0} \cdot \gamma_r$$

$$\gamma_r = \frac{1}{1 + 0.4 \cdot \beta_d \cdot \alpha} \exp\left(-\alpha \frac{\varphi_t - 0.4 \cdot \beta_d}{1 + 0.4 \cdot \beta_d \cdot \alpha}\right) = \frac{1}{1 + \varphi_d \cdot \alpha} \exp\left(-\frac{\alpha \cdot \varphi_f}{1 + \alpha \cdot \varphi_d}\right)$$

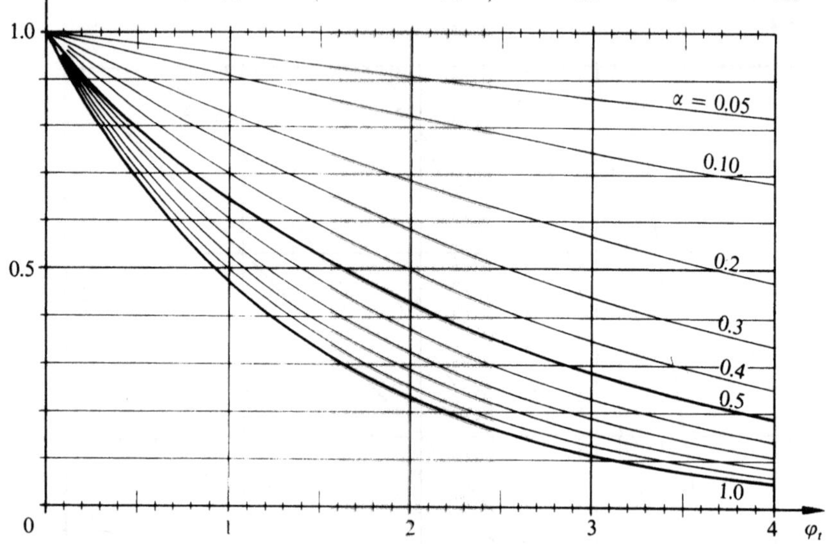

Figure 2.2. Parameter γ_r to describe the effect of creep on action effects or stresses caused by rapid restraint.

$$S_{est} = S_{est}^{(e)} \cdot \gamma_s = S_{es00}^{(e)} \cdot \frac{\varphi_t}{\varphi_\alpha} \gamma_s \quad \text{or} \quad f_{est} = \gamma_s |\varepsilon_{st}| \cdot E_e \cdot (1 - \alpha)$$

$$\gamma_s = \frac{1 - \exp\left(-\alpha \frac{\varphi_t - 0.4 \cdot \beta_d}{1 + 0.4 \cdot \beta_d \cdot \alpha}\right)}{\alpha(\varphi_t - 0.4 \cdot \beta_d)} = \frac{1 - \exp\left(-\frac{\alpha \cdot \varphi_f}{1 + \alpha \cdot \varphi_d}\right)}{\alpha \cdot \varphi_f}$$

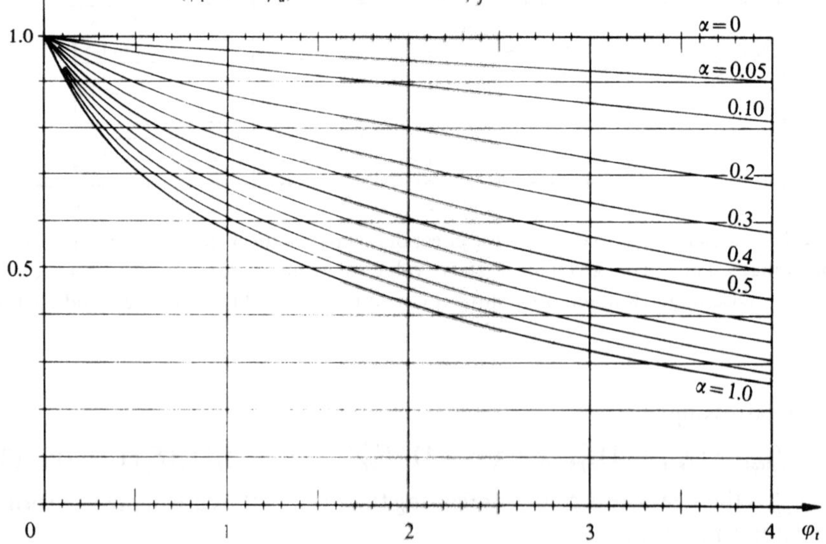

Figure 2.3. Coefficient γ_s to describe the effect of creep on restraint which develops similar to creep, e.g., shrinkage.

If one wants to proceed from the stresses in the neighboring concrete fiber, the following applies:

$$f_{pcst} = n\left[(\gamma_{\text{sus}} - 1)f_{c\text{sus}0} - (\gamma_r - 1)\frac{1-\alpha}{\alpha} f_{cp0} - \gamma_s(1-\alpha)|\varepsilon_s|E_c\right]. \quad (2.12a)$$

If $a \to 1$, it is appropriate to relate the creep and shrinkage problem entirely to the stresses in the concrete. We then have $\gamma_{\text{sus}} = 1$ and, consequently,

$$f_{ccst} = (\gamma_r - 1)f_{cp0} + \gamma_s|\varepsilon_{st}|E_c\alpha. \quad (2.12b)$$

The coefficients γ_{sus}, γ_r, and γ_s are obtained from Figs. 2.1–2.3. The formula for the stiffness ratio is

$$\alpha = n\frac{A_e}{A_{\text{id}}}\left(1 + \frac{A_{\text{id}}}{I_{\text{id}}} y_{\text{id}e}^2\right), \quad (2.13)$$

$$A_{\text{id}} = A_c + (n-1)A_e, \quad (2.14)$$

$$I_{\text{id}} = I_c + (n-1)A_e y_{ce} y_{\text{id}e}, \quad (2.15)$$

$$y_{c\,\text{id}} = \frac{1}{A_{\text{id}}}(n-1)A_e y_{ce}, \quad (2.16)$$

$$y_{\text{id}e} = y_{ce} - y_{c\,\text{id}}, \quad (2.17)$$

$$n = \frac{E_e}{E_c}, \quad (2.18)$$

where (Fig. 2.4)

- e as a subscript encompasses all purely elastic members placed in the same position, i.e., both the unstressed reinforcement (s) and the tendon (p). If there are only tendons, e is replaced by p.
- α is the stiffness ratio according to Dischinger [Eq. (2.13)].
- A_{id} is the idealized cross-sectional area [Eq. (2.14)].
- J_{id} is the idealized moment of inertia [Eq. (2.15)].
- y_{ce} is the distance between the tendon and the centroid of the concrete section.
- $y_{\text{id}e}$ is the distance between the tendon and the centroid of the idealized cross section.

In general, the notations have the same meaning as in Section 2.1.1. The general rule for the subscripts is as follows:

- t time under study ($t = t$),
- 0 start of the time scale (time $t = 0$),
- e acting in the elastic component,
- p as before, when prestressing tendon is used,
- w temporary effect,
- sus sustained effect,
- r rapidly generated constraint,
- p from prestressing,
- s shrinkage.

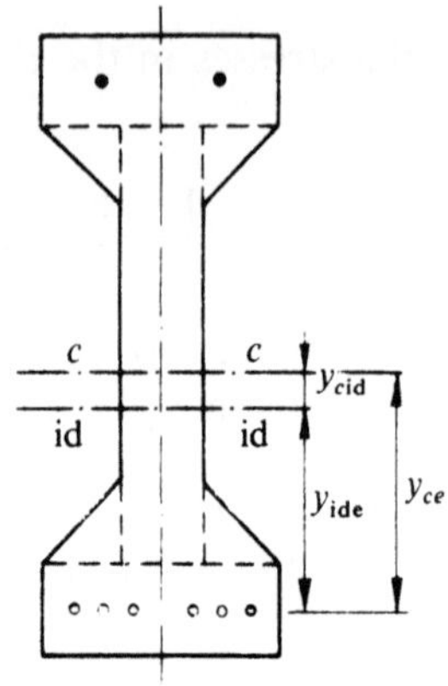

Figure 2.4.

2.1.3. Two Structural Members of Different Creep and Shrinkage Characteristics Are Coupled Locally (cf., Appendix II, *Sections 2.1.4 and 2.2.4)*

If the members are designated U and W, respectively, the coupling force X, in accordance with the Dischinger/Kupfer method (cf., Part C, Sections 2.1.4 and 2.2.4) is

$$X_t = X_{Wt} + \gamma_{\text{sus}\,U} X_{U0} \frac{1 + 0.4\alpha_U}{1.4} + \gamma_{\text{sus}\,W} X_{W0} \frac{1 + 0.4\alpha_W}{1.4} + \gamma_r X_{r0} + \gamma_s X_{st}^{(e)} \tag{2.19}$$

or, for creep and shrinkage only,

$$X_{cst} = \left(\gamma_{\text{sus}\,U} \frac{1 + 0.4\alpha_U}{1.4} - 1\right) X_{U0} + \left(\gamma_{\text{sus}\,W} \frac{1 + 0.4\alpha_W}{1.4} - 1\right) X_{W0} + \gamma_r X_{r0} + \gamma_s X_{st}^{(e)}. \tag{2.20}$$

Table 2.1

Coefficient	Guiding values α	φ
$\gamma_{\text{sus}\,U}$	$\alpha = \dfrac{\varphi_{UW} - 0.4}{1.4\varphi_U - 0.4\varphi_{UW} - 0.4}$	$\varphi = \varphi_U$
$\gamma_{\text{sus}\,W}$	$\alpha = \dfrac{\varphi_{UW} - 0.4}{1.4\varphi_W - 0.4\varphi_{UW} - 0.4}$	$\varphi = \varphi_W$
γ_r	$\alpha = 1$	$\varphi = \varphi_{UW}$
γ_s	$\alpha = 1$	$\varphi = \varphi_{UW}$

where $$\varphi_{UW} = \frac{\delta_{U1,1}\varphi_U + \delta_{W1,1}\varphi_W}{\delta_{U11} + \delta_{W11}}. \tag{2.21}$$

The coefficients γ may be determined from Figs. 2.1–2.3 with the help of guiding values found in Table 2.1.

2.2. The Trost (cf., Appendix II, Section 4) and Bažant (cf., Appendix II, Section 5) Methods

Using the TROST METHOD, it is appropriate, for better clarity, to start out with the equations from Sections 2.1.1 and 2.1.2, namely Eq. (2.1) for only localized coupling or Eq. (2.8) for continuous bonding, and to use γ coefficients. The equations now read

$$\gamma_{\text{sus}} = 1 + \frac{1-\alpha}{\alpha}\,\frac{\alpha\varphi}{1+\rho\alpha\varphi} = 1 + \frac{(1-\alpha)\varphi}{1+\alpha\rho\varphi}, \tag{2.22}$$

$$\gamma_r = 1 - \frac{\alpha\varphi}{1+\alpha\rho\varphi}, \tag{2.23}$$

$$\gamma_s = \frac{1}{1+\alpha\rho\varphi}, \tag{2.24}$$

where ρ is the so-called relaxation coefficient. Values of ρ given in the following Table 2.2 for $\alpha = 1$ in accordance with Appendix II, Section 4.2 differ from the data in [2] in that they take into account separation of the creep coefficient into components of delayed elastic strain and flow.

The stiffness ratio α is computed from Eq. (2.7) for localized coupling and from Eq. (2.13) for continuous bonding.

If, instead of setting β_d equal to unity, a parabolic increase of β_d for the range of $\varphi_{t=0}$ and $\varphi_{t=2}$ is assumed, then the values of ρ are found to be approximately 0.81 even for $\varphi < 2.0$. Thus, a constant value of $\rho = 0.8$ is a good overall approximation.

The BAŽANT METHOD involves the so-called aging coefficient χ which controls the creep problem by means of the effective modulus of elasticity

$$E_{\text{eff}} = \frac{E_c}{1+\chi\varphi}. \tag{2.25}$$

Some values for χ which depend on the thickness of member, loading age, and time under load are listed in Appendix II, Section 5. There, similarities between the Trost and Bažant methods are discussed.

Table 2.2 Relaxation coefficients ρ according to Trost.

φ_t	1.5	2.0	2.5	3.0	3.5	4.0	4.5	5.0
ρ	0.85	0.82	0.80	0.80	0.80	0.81	0.82	0.83

2.3. The Method of the Mean Creep-Inducing Stresses and Action Effects

This method is thoroughly analyzed in Appendix II, Section 3.2. The notations used in the equations grouped below are explained in Fig. 2.5 and in Appendix II, Section 3.2.

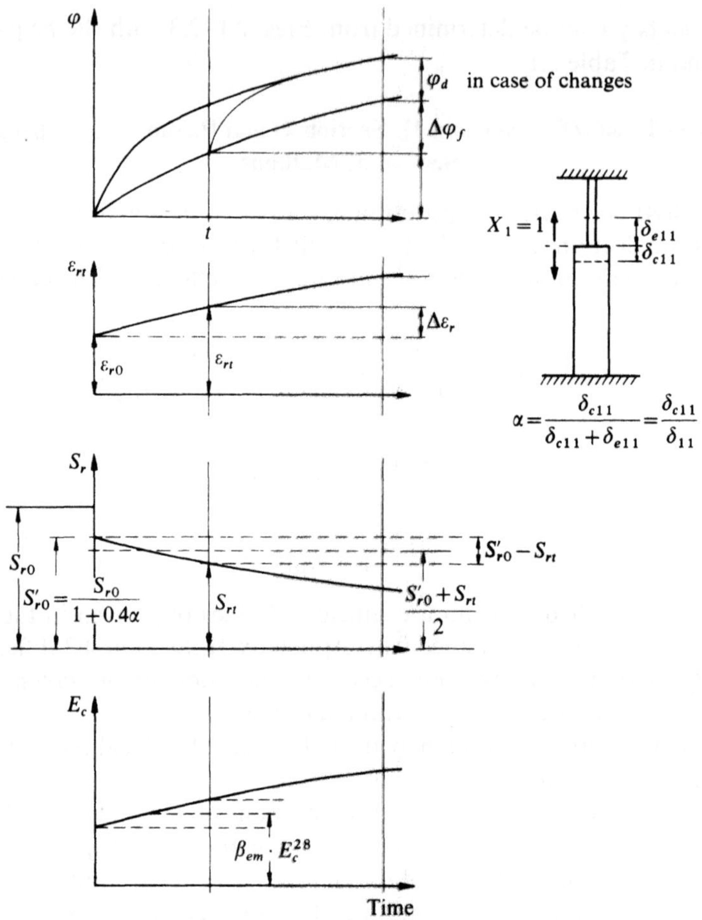

Figure 2.5.

(a) The compatibility condition reads

$$\frac{S'_{r0} + S_{rt}}{2} \Delta\varphi_f - (S'_{r0} - S_{rt})C_m - \frac{\Delta\varepsilon_r}{\delta_{c11}^{28}} = 0.$$

(b) The coefficients C_m and S'_{r0}, respectively, are defined as follows:

$$C_m = \frac{1}{\alpha} + 0.5\left(\frac{1}{\beta_{e0}} + \frac{1}{\beta_{et}}\right) - 0.6, \tag{2.26}$$

$$S'_{r0} = \frac{\varepsilon_{r0}}{C_0 \alpha \delta_{11}^{28}}. \tag{2.27}$$

(c) The general solution reads

$$S_{rt} = \frac{(C_m + \Delta\varphi_f/2)\varepsilon_{r0}/C_0 + \Delta\varepsilon_r}{(C_m + \Delta\varphi_f/2)\delta_{11}^{28}\alpha}. \tag{2.28}$$

Setting $\beta_e = 1$ and $\Delta\varepsilon_f = 0$, we obtain

$$S_{rt} = \frac{1 + 0.6\alpha - 0.5\alpha\varphi}{1.0 + 0.6\alpha + 0.08\alpha^2 + \alpha\varphi(0.5 + 0.2\alpha)} S_{r0} \tag{2.29a}$$

and, neglecting the delayed elastic strain,

$$S_{rt} = \frac{1 - 0.5\alpha\varphi}{1 + 0.5\alpha\varphi} S_{r0}. \tag{2.30}$$

For $\alpha = 1$, Eq. (2.29a) is simplified to

$$S_{rt} = \frac{1.6 - 0.5\varphi}{1.68 + 0.7\varphi}. \tag{2.29b}$$

Up to $\alpha\varphi < 1.5$, the use of Eqs. (2.29) results in an error of only about 10% with respect to the more accurate γ values found in Fig. 2.2. Equation (2.30) should be used only up to $S_{rt}/S_{r0} > 0.6$, i.e., $\alpha\varphi < 0.5$, since the error otherwise rapidly exceeds 10%.

(d) Adaptation of the general solution to the step-by-step calculation (cf., Comments, Appendix II, Section 3.2.4):

$$S_{ri} = \frac{(C_{mi} - 0.5\Delta\varphi_{fi})S_{r(i-1)} + \Delta\varepsilon_{ri}/(\delta_{11}^{28}\alpha)}{C_{mi} + 0.5\Delta\varphi_{fi}}. \tag{2.31}$$

2.4 The Busemann method

Appendix II, Section 6 clearly presents all the information required to apply this method.

3. Statically Indeterminate Coupling between a Flexible Steel Tension Member and a Concrete Structure

3.1. Example of Tied Arch

In this example, the concrete arch represents the structural member subject to shrinkage and creep. The steel tie which takes up the horizontal thrust, is the elastic coupling element. The force that is induced in it can be increased through prestressing. The prestraining required creates a constraint. The system is indeterminate to the first degree.

The horizontal thrust induced in such an arch by the sustained loads prior to prestressing is

$$H_{s\,\mathrm{sus}\,0} = \frac{\delta_{c11}}{\delta_{c11} + \delta_{s11}} H_{s0} = \alpha H_{s0}, \tag{3.1}$$

where H_{s0} represents the horizontal thrust in a two-hinge arch without tie rod and α is the stiffness ratio in accordance with Eq. (2.7) which is now developed further as follows:

$$\alpha = \frac{\delta_{c11}}{\delta_{c11} + \delta_{s11}} = \frac{\int y^2 \frac{ds}{E_c I_c} + \int \cos^2 \beta \frac{ds}{E_c A_c}}{\int y^2 \frac{ds}{E_c I_c} + \int \cos^2 \beta \frac{ds}{E_c A_c} + \frac{l}{E_s A_s}}. \tag{3.2}$$

As a rule, this value of α is of the order of 0.85–0.99 with a mean value of 0.90. For this mean value, and assuming that $\varphi_t = 2$, Figs 2.1–2.3 yield the following values to be inserted in Eq. (2.1):

$$H_{st} = H_{swt} + \gamma_{sus} H_{s\,sus\,0} + \gamma_r H_{sp0} + \gamma_s H_{sst}^{(e)} = H_{swt} + 1.09 H_{s\,sus\,0} + 0.25 H_{sp0} + 0.47 H_{sst}^{(e)}. \tag{3.3}$$

By prestressing, the tie rod force, H_{sp0}, in excess of $H_{s\,sus\,0}$ can be chosen such that the sum of the action effects due to dead weight d, prestress p, and shrinkage s is in large measure independent of time. However, this requires that the value of H_{sst} be known. In accordance with Section 2.1.1, Eq. (2.6), the latter was determined to be of a value that corresponds to $H_{ss\infty} \approx -0.15 H_{s\,sus\,0}$. The requirement stated above may be formulated as follows:

$$H_{s\,sus\,0} + H_{sp0} = 1.09 H_{s\,sus\,0} + 0.25 H_{sp0} - 0.47(0.15) H_{s\,sus\,0}.$$

We then obtain $H_{sp0} = 0.027 H_{s\,sus\,0}$.

One obtains a clearer view of the influence of larger φ-values by considering the limiting value of $\varphi = \infty$. Equation (2.23) and (2.24), Appendix II give us

$$\lim_{\varphi \to \infty} (H_{s\,sus} + H_{sp}) = \lim \left(\left(\frac{1}{\alpha} - \frac{1 - \alpha}{\alpha(1 + 0.4\alpha)} e^{-\alpha(\varphi_t - 0.4)/(1 + 0.4\alpha)} \right) H_{s\,sus\,0} + \frac{1}{1 + 0.4\alpha} e^{-\alpha(\varphi_t - 0.4)/(1 + 0.4\alpha)} H_{sp0} \right) = \frac{1}{\alpha} H_{s\,sus\,0}. \tag{3.4}$$

However, since $H_{s\,sus\,0} = \alpha H_{s0}$, the behavior of a tied arch, because of creep, is close to that of a two-hinge arch without a tie rod which rests against rigid abutments. This is understandable since a reduction in the stiffness of the arch must have the same effect as an increase in the stiffness of the tie rod.

3.2. Example of Prestressing without Bond

In the structural member shown in Fig. 3.1 as a conceptual model, the tie rod outside the concrete cross section is to be prestressed. The prestressing is created by prestraining the steel. The required tension force Z_p is also a constraining force. Therefore, Eq. (2.1) now reads:

$$Z_t = Z_{wt} + \gamma_{sus} Z_{sus\,0} + \gamma_r Z_{p0} + \gamma_s Z_{st}^{(e)}. \tag{3.5}$$

The coefficients γ for each particular instance are taken from Figs. 2.1–2.3, as a function of the stiffness ratio α in accordance with Eq. (2.7), and of φ_t.

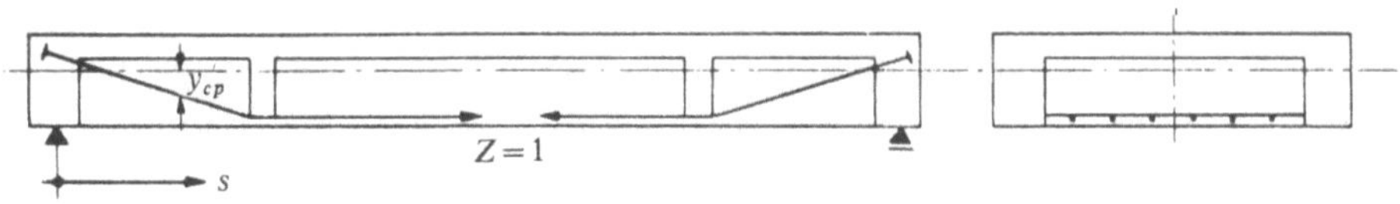

Figure 3.1.

The displacements required to calculate $\alpha = \delta_{c11}/(\delta_{c11} + \delta_{p11})$ are for an average slenderness of the member and disregarding the effect of the tie rod's inclination

$$\delta_{c11} = \int_0^l y_{cp}^2 \frac{dx}{E_c I_0} + \frac{l_c}{E_c A_c}, \tag{3.6}$$

$$\delta_{p11} \sim \frac{l_p}{E_p A_p}. \tag{3.7}$$

In normal prestressed concrete construction, the values obtained for α lie between 0.025 and 0.15.

The following is a numerical example for $\alpha = 0.05$ and $\varphi_t = 3.0$. Equation (2.1) and the γ values from Figs. 2.1–2.3 yield

$$Z_t = Z_{wt} + 3.6Z_{\text{sus}\,0} + 0.86Z_{p0} + 0.91Z_{st}^{(e)}.$$

4. Internally Statically Indeterminate Coupling between Bonded Reinforcement and Concrete

4.1. Validity Range

A system is internally indeterminate to the first degree only as long as all the reinforcement may be represented by one tendon. For the analysis of prestressing, tension members may be combined into one member only if they are in approximately the same location within the cross section and if they have the same prestress. On the other hand, when studying the effect of a sustained load or a constraint, including shrinkage, neighboring reinforcing bars may be combined into one tendon, regardless of the degree of prestressing.

For concentric loading, reinforcing bars distributed symmetrically through the cross section may, naturally, be combined into one tendon.

4.2. Stress Redistribution in a Longitudinally Reinforced Column

The given problem is to examine how the stresses in a concentrically loaded reinforced concrete column are distributed between concrete and steel when the load is applied and at time $t = \infty$.

Given

$$A_c = 25(25) = 625\ \text{cm}^2\ (96.83\ \text{in}^2),$$
$$A_s = 4 \text{ bars } 14\ \text{mm} = 6.15\ \text{cm}^2\ (0.953\ \text{in}^2.)\ f_y = 220\ \text{N/mm}^2$$
$$\rho = A_s/A_c = 6.15/625 = 0.0098,$$
$$f'_{c28} = 25\ \text{N/mm}^2,$$
$$n = E_s/E_c = 2\,10\,000F/30\,000F = 7,$$
concentric sustained load $F = 530$ kN (117.8×10^3 lb).

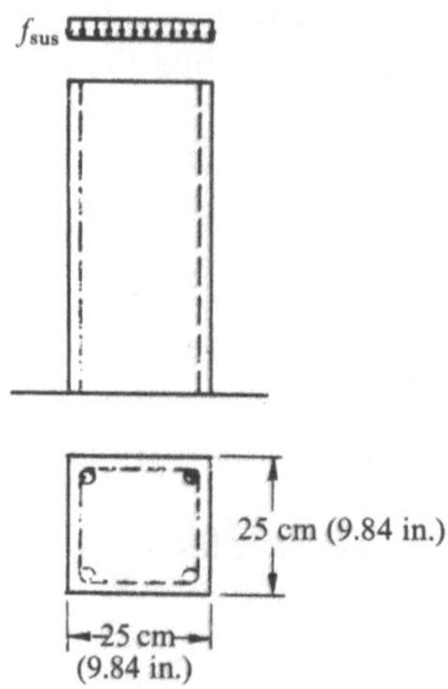

Figure 4.1.

Since under concentric load $y_{ce} = y_{\text{id}\,e}$, Eq. (2.13) for α reads

$$\alpha = n\frac{A_s}{A_{\text{id}}} = \frac{nA_s}{A_c + (n-1)A_s} \approx \frac{nA_s}{A_c + nA_s} = \frac{n\rho}{1 + n\rho} = \frac{0.0098(7)}{1 + 0.0098(7)} = 0.064. \tag{4.1}$$

Determination of the creep coefficient and of shrinkage in accordance with Section 1.2 and 1.3 yielded

$$\varphi = 3; \qquad \varepsilon_s = -20 \times 10^{-5},$$
$$|\varepsilon_s| E_s = 42\ \text{N/mm}^2.$$

The stress distribution at time $t = 0$ (see foregoing derivation of α) is

$$-f_{c0} = \frac{F}{A_c(1 + n\rho)} = \frac{F_c}{A_c} = \frac{(1-\alpha)F}{A_c} = 8.0\ \text{N/mm}^2\ (1147\ \text{psi}),$$

$$-f_{s0} = \frac{F_s}{A_s} = \frac{\alpha F}{A_s} = nf_c = 56.0\ \text{N/mm}^2\ (8030\ \text{psi}).$$

The stress distribution at time $t = \infty$, in accordance with Eq. (2.8a), is

$$f_{s\infty} = \gamma_{\text{sus}} f_{s0} - \gamma_s |\varepsilon_s| E_s (1 - \alpha) = -3.54(56.0) - 0.91(42.0)(1 - 0.064)$$
$$= -235.0\ \text{N/mm}^2\ (33.7 \times 10^3\ \text{psi}).$$

Thus, steel with a yield strength of 220 N/mm^2 would yield.

$-f_{c\infty} = (F - f_{c\infty}A_s)/A_c = [5300 - 235.0(6.15)/625 = 6.2$ N/mm^2 (88.9 psi).

The distribution of the stress and force in steel and concrete, in kN and %, is given in the following table:

	Time $t = 0$			Time $t = \infty$		
	f (N/mm^2)	F (kN)	%	f (N/mm^2)	F (kN)	%
Concrete	−8.0	496	93.6	−6.2	385	72.7
Steel	−56.0	34	6.4	−235.0	145	27.3
Total		530	100		530	100

4.3. Bonded Prestressing with a Single Tendon

4.3.1. Calculations Using the Expanded Dischinger Equation

Since prestressing corresponds to a constraint exerted upon the concrete cross section, Eq. (2.10) applies as follows:

$$f_{pcs} = (\gamma_{sus} - 1)f_{p\,sus\,0} + (\gamma_r - 1)f_{pp0} - \gamma_s|\varepsilon_{st}|E_p(1 - \alpha).$$

If one wants to start from the stresses in the neighboring concrete fibers, Eq. (2.12a) applies as follows:

$$f_{pcs} = n\left[(\gamma_{sus} - 1)f_{p\,sus\,0} - (\gamma_r - 1)\frac{1 - \alpha}{\alpha} f_{cp0} - \gamma_s(1 - \alpha)|\varepsilon_{st}|E_c\right].$$

Numerical Example (see Fig. 4.2)

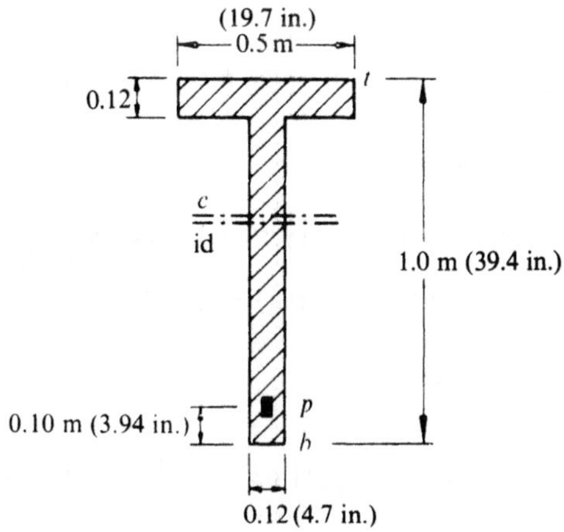

Figure 4.2.

Given:

$$f_{c28} = 45\ \text{N/mm}^2,\ E_c = 37\,000\ \text{N/mm}^2\ (5.3 \times 10^6\ \text{psi})\ n = \frac{210\,000}{37\,000} = 5.68;$$

$$\varphi = 2.0,\ \varepsilon_s = -14 \times 10^{-5};$$

$$A_c = 0.166\ \text{m}^2\ (257\ \text{in.}^2),\ I_c = 0.0165\ \text{m}^4\ (0.396 \times 10^5\ \text{in.}^4),$$

$$A_p = 0.0012\ \text{m}^2\ (1.86\ \text{in.}^2);$$

$$f_{pp}^{(0)} = 900\ \text{N/mm}^2\ (1.29 \times 10^5\ \text{psi}),\ y_{cp} = 0.522\ \text{m}\ (20.55\ \text{in.}).$$

From Eqs. (2.13)–(2.18)

$$A_{\text{id}} = A_c + (n-1)A_p = 0.1716\ \text{cm}^2\ (0.0266\ \text{in.}^2);$$

$$y_{c\,\text{id}} = \frac{1}{A_{\text{id}}}(n-1)A_p y_{cp} = 0.017\ \text{m}\ (0.669\ \text{in.});$$

$$y_{\text{id}\,p} = y_{cp} - y_{c\,\text{id}} = 0.503\ \text{m}\ (19.8\ \text{in.});$$

$$I_{\text{id}} = I_c + (n-1)A_p y_{cp} y_{ip} = 0.0180\,\text{m}^4 (0.043 \times 10^6\ \text{in.}^4);$$

$$n\frac{A_p}{A_{\text{id}}}\left(1 + \frac{A_{\text{id}}}{I_{\text{id}}} y_{\text{id}\,p}^2\right) = 0.136;$$

$$f_{cp0} = -\frac{\alpha}{n} f_{pp}^{(0)} = -21.6\ \text{N/mm}^2\ (3097\ \text{psi});$$

$$f_{pp0} = (1-\alpha) f_{pp}^{(0)} = 778\ \text{N/mm}^2 (11.16 \times 10^4\ \text{psi});$$

$$M_{\text{sus}} = 400\ \text{kNm}\ (3.5 \times 10^6\ \text{in.lb});$$

$$f_{c\,\text{sus}\,0} = \frac{M_{\text{sus}}\, y_{\text{id}\,p}}{I_{\text{id}}} = 11.2\ \text{N/mm}^2\ (1606\ \text{psi});$$

$$f_{p\,\text{sus}\,0} = n f_{c\,\text{sus}\,0} = 63.5\ \text{N/mm}^2\ (9106\ \text{psi}).$$

The solution from Eq. (2.10) is

$$f_{pcs} = (2.50 - 1)63.5 + (0.77 - 1)778.0 - 0.86|\varepsilon_{st}|E_p(1 - 0.136)$$
$$= 105.8\ \text{N/mm}^2\ (1.52 \times 10^4\ \text{psi}) = 13.6\%\ \text{of}\ f_{pp0}.$$

The solution from Eq. (2.12a) is

$$f_{pcs} = 5.68\left[(2.50 - 1)11.2 + (0.77 - 1)21.6\frac{1 - 0.136}{0.136} - 0.86|\varepsilon_{s,t}|E_c(1 - 0.136)\right]$$

$$= 105.8\ \text{N/mm}^2\ (1.52 \times 10^4\ \text{psi}).$$

4.3.2. *Calculation in Accordance with the Method of the Mean Creep-Inducing Stress*

Expanding the method of the mean creep-inducing stress described in Appendix II, Section 3.2, one can derive a simple approximate formula to estimate the prestress loss in prestressed concrete. The advantage of this method is the fact that α need not be determined.

The compatibility conditions for the tendon fiber is

$$\frac{f_{cp\,\mathrm{sus}\,0} + f_{ccs}/2}{E_c}\varphi + \frac{f_{ccs}}{E_c} - \frac{f_{pcs}}{E_p} + \varepsilon_s = 0, \tag{4.2}$$

where the following ratio applies:

$$\frac{f_{ccs}}{f_{cp0}} = \frac{f_{pcs}}{f_{pp0}}, \; f_{ccs} = \frac{f_{pcs}}{f_{pp0}} f_{cp0}$$

(see Fig. 4.3).

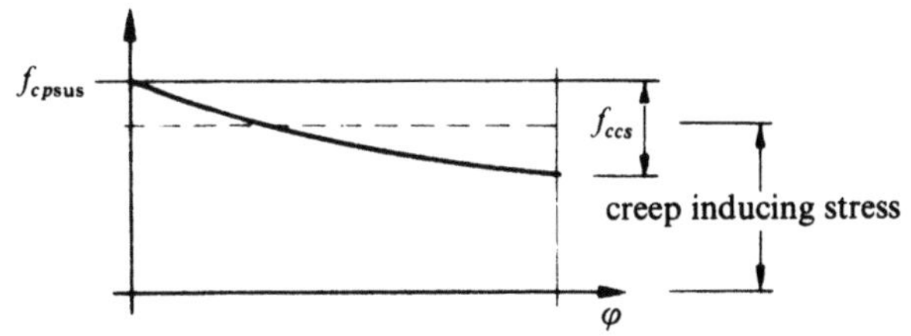

Figure 4.3.

Applying Eq. (4.3) and multiplying by E_p, we get

$$nf_{cp\,\mathrm{sus}\,0}\varphi + n\frac{f_{pcs}}{f_{pp0}} f_{cp0}(1 + \varphi/2) - f_{pcs} + \varepsilon_s E_p = 0,$$

$$f_{pcs} = \frac{-|\varepsilon_s|E_p + n\varphi f_{cp\,\mathrm{sus}\,0}}{1 - n(f_{cp0}/f_{pp0})(1 + \varphi/2)}. \tag{4.3}$$

This equation had already been recommended two decades ago [3].

In this method, the creep effect from sustained loading, from the prestress constraint, and from shrinkage is dealt with in a similar way. As was shown in the preceding sections, this can only be an approximation. However, the method is simple to handle, and it yields reasonably accurate results.

Numerical Example

The input data correspond to the example in Section 4.3.1:

$$f_{pcs} = \frac{-14 \times 10^{-5}(210\,000) + 5.68(2)(11 - 21)}{1 + 5.68(21.6/778.0)(1 + 2/2)} = 112 \text{ N/mm}^2 \; (1.6 \times 10^4 \text{ psi}),$$

$$\triangleq 14.4\,\% \text{ of } f_{pp}.$$

The agreement with the result obtained in Section 4.3.1 is satisfactory. The loss of prestress is slightly too large, as can be anticipated from Appendix II, Fig. 8.1.

4.3.3. *Influence of the Creep of Steel*

As long as the maximum admissible prestress—disregarding transient overstresses for the partial compensation of friction—is limited to 0.55 f_{tu} (tensile strength) or 0.75 f_y (yield strength) and as long as there are no elevated temperatures acting over an extended period, relaxation of the prestressing steel is of no significance as a rule. However, if the stress levels are raised to adapt to the regulations of other countries, or if higher temperatures are not just transient—a rise to 40 °C is sufficient—it is no longer possible to disregard the relaxation of the prestressing steel.

As stated in Section 5 of Part B, the deformation characteristics of prestressing steel cannot be described in terms of generally applicable rules, as is the case for concrete. Consequently, the necessary data must be determined experimentally for each type of steel.

The compatibility conditions for prestress loss in a concentrically prestressed concrete member are as follows, using the method of the mean creep-inducing stress (see Fig. 4.4):

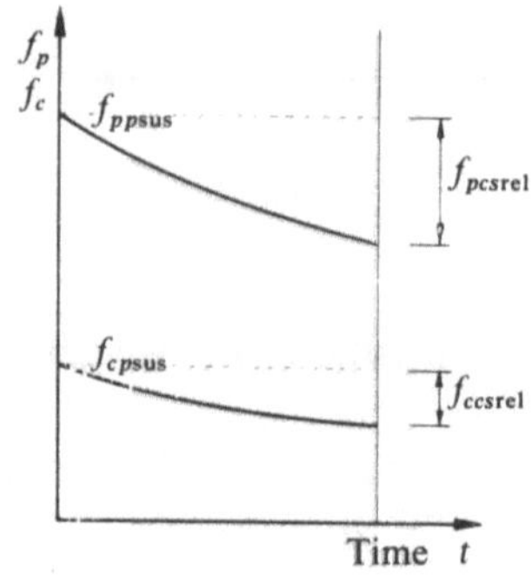

Figure 4.4.

$$\frac{f_{cp\,sus} + f_{ccs\,rel}/2}{E_c}\varphi + \frac{f_{ccs\,rel}}{E_c} - \frac{f_{pcs\,rel}}{E_p} - \varepsilon_{p\,rel} + \varepsilon_s = 0. \tag{4.4}$$

From

$$\frac{f_{ccs\,rel}}{f_{cp0}} = \frac{f_{pcs\,rel}}{f_{pp0}}$$

we get

$$f_{ccs\,rel} = \frac{f_{pcs\,rel}}{f_{pp0}} f_{cp0}.$$

Applying Eq. (4.4) and multiplying by E_p, we get

$$nf_{cp\,sus}\varphi + n\frac{f_{pcs\,rel}}{f_{pp0}} f_{cp0}\left(1 + \frac{\varphi}{2}\right) - f_{pcs\,rel} - E_p\varepsilon_{p\,rel} + E_p\varepsilon_s = 0,$$

$$f_{pcs\,rel} = \frac{-|\varepsilon_s|E_p - \varepsilon_{p\,rel}E_p + n\varphi f_{cp\,sus}}{1 - n(f_{cp0}/f_{pp0})(1 + \varphi/2)}. \tag{4.5}$$

$\varepsilon_{p\,rel}$ is the time-dependent inelastic strain of the prestressing steel. As a first approximation, one can use the relaxation value corresponding to the initial stress $f_{pp\,sus}$. It is better, however, to start with a value $f_{p\,rel}$ which takes into account the true development of the steel stress as it is affected by concrete creep and shrinkage. Furthermore, since the relaxation values increase faster than linearly as the steel stress rises, it is recommended, in determining $\varepsilon_{p\,rel}$, that one assumes a steel stress somewhat above the mean value:

$$f_{p\,rel} = f_{ppd} - 0.3f_{pcs}.$$

The estimated value f_{pcs} is subsequently checked with Eq. (4.5). A similar procedure is described in [18].

For the example given in Section 4.3.1, we will now take into account, in addition, the influence of creep of the prestressing steel. The stress in an annealed steel, $f_y = 135$, $f_{tu} = 150\ \text{N/mm}^2$ at time $t = 0$, in accordance with Section 4.3.1 is $778.0 + 63.5 = 841.5\ \text{N/mm}^2 = 0.57f_{tu}$. In experiments, a relaxation loss after 1000 and 500 000 hr (6 weeks and 60 years, respectively) of 0.6% and 1.6%, respectively, of the initial stress has been observed.

In accordance with Eq. (4.5), the overall stress loss—if the relaxation loss is set as $\varepsilon_{p\,rel}E_p$ under the initial stress $0.57f_{tu}$—is

$$f_{pcs\,rel} = \frac{-|\varepsilon_s|E_p - \varepsilon_{p\,rel}E_p + n\varphi f_{cp\,sus}}{1 - nf_{cp0}/f_{pp0}(1 + \varphi/2)}$$

$$= \frac{-14 \times 10^{-5}210\,000 \quad 0.6 \text{ resp. } 1.6(841.5) + 5.68(2)(11.2 - 21.6)}{1 + 5.68(21.6/778.0)(1 + 2/2)}$$

$$= 117.0 \text{ resp. } 125.5\ \text{N/mm}^2 \text{ or } 15\% \text{ resp. } 16.1\% \text{ of } f_{p,p0}$$

$$(1.68 \times 10^4 \text{ and } 1.8 \times 10^4 \text{ psi}).$$

It can clearly be seen that when the level of prestressing is low, the losses due to prestressing steel relaxation can be disregarded. This is no longer true if the prestressing level is high.

4.4. Partial Prestressing

Frequently, a distinction is made only between structures with full prestressing, with limited prestressing, i.e., an upper limit of allowable tensile stress, and without prestressing (reinforced concrete). More recently, attention has also been paid to the range of partial prestressing which is situated between limited and no prestressing. Particularly in cases of high live-load/dead-load ratios, partial prestressing presents considerable advantages. When applying limited prestressing, one would have to refrain from making full use of the height of a particular beam, i.e., prestress would have to be applied with reduced eccentricity. Otherwise, the working load would create unacceptably high tensile stresses. With partial prestressing, one gives up an upper limit for the tensile stresses, but, owing to the considerably higher susceptibility to corrosion of the prestressing steels, one must make sure that the cracks induced by the live load remain closed under the sustained load. Under certain conditions, it may also be necessary to verify that the allowable range of stress is not exceeded for the prestressing system chosen in order to avoid fatigue problems.

By introducing partial prestressing, one can gain the following advantages:

With respect to fully prestressed concrete:
- Utilization even when the proportion of life load is high.
- Easier placement of the reduced number of anchorings.
- More favorable behavior under high constraint owing to reduced stiffness *EI*.

With respect to reinforced concrete:
- Utilization of the advantages of high-strength steel.
- Reduction of deflections.
- Prevention, in large measure, of cracks under working conditions.

In the area of partial prestressing, no new approaches are needed in order to make allowance for creep and shrinkage. Since these influences are observed only under sustained loads, the equations developed for prestressed structural members can be applied without modification, as long as one maintains the requirement that the cracks occurring under total loading remain closed under this load condition.

It is thus superfluous to present a separate example for the case of partial prestressing, since Section 4.3.1 will serve for this case. In exceptional cases where creep acts on cracked sections, one must proceed as explained in Section 14.3.1.3.

5. Two-fold Internally Indeterminate Coupling between a Rigid Reinforcement Bonded to Concrete: Solution by the Busemann Method

In the foregoing sections, it was assumed that the reinforcement is not rigid. However, in the analysis of stress redistributions in a composite beam, a considerable role is played not only by the area of the structural steel member, but also by its moment of inertia.

For the mathematical treatment of this problem, there are solutions, e.g., by Sattler [4, and 19], which demand a considerable computational effort. This is why we are presenting only the graphic method developed by Busemann [5 and 6], using an example which leads to a stunningly simple solution. The derivation of this method has been described in Appendix II, Section 6, where the equations to be used are grouped in a clear fashion.

The characteristics of the cross section corresponding to Fig. 5.1 are

$$\text{Concrete:}\quad A_c = 1.083\ \text{in}^2;\ I_c = 0.184\ \text{m}^4;\ r_c = \sqrt{\frac{0.184}{1.083}} = 0.412\ \text{m (1.35 ft)};$$

$$\text{Steel:}\quad A_s = 0.0925\ \text{m}^2,\ I_s = 0.070\ \text{m}^4;\ r_s = \sqrt{\frac{0.070}{0.0925}} = 0.870\ \text{m (2.85 ft)}.$$

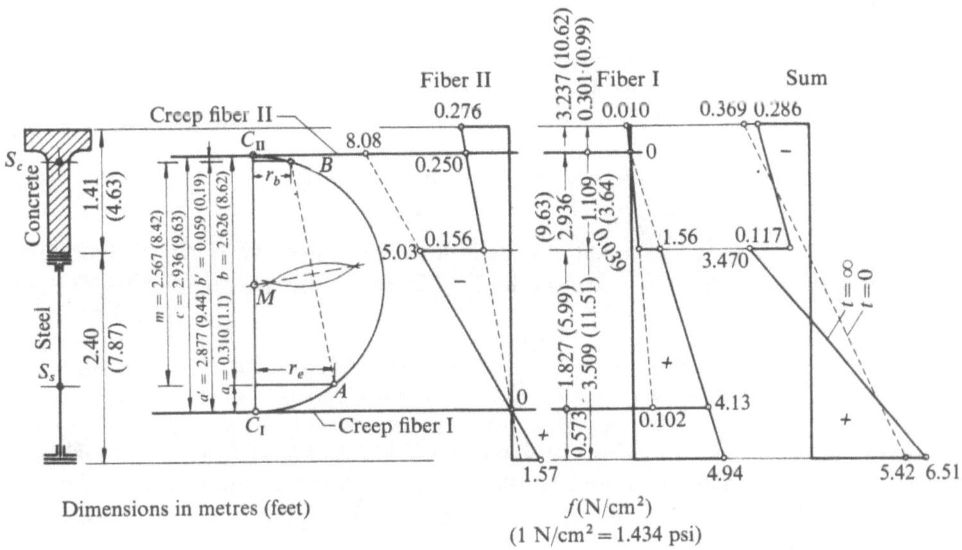

Figure 5.1.

The distance m between the centers of gravity is 2.567 m (8.42 ft).

$$\varkappa = r_s^2 + m^2 - r_c^2 = 0.757 + 6.589 - 0.170 = 7.176, \qquad \text{(AII, 6.3)}$$

$$b' = \frac{\sqrt{\varkappa^2 + 4m^2 r_c^2} - \varkappa}{2m} = 0.059(0.19\ \text{ft}). \qquad \text{(AII, 6.4)}$$

In this method the position and the characteristics of the cross section of two so-called creep fibers are calculated and the action effects are distributed among those fibers. Thus, each creep fiber can be delt with as being a simply internally indeterminate system.

$$a'b' = r_c^2 \rightarrow a' = r_c^2/b' = 2.877\ \text{m (9.44 ft)}, \qquad \text{(AII, 6.1)}$$

$$c = a' + b' = 2.936\ \text{m (9.63 ft)}, \qquad \text{(AII, 6.7)}$$

$$a = a' - m = 0.310\ \text{m (1.1 ft)}, \qquad \text{(AII, 6.6)}$$

$$b = c - a = 2.626\ \text{m (8.62 ft)}.$$

The distribution of the concrete and steel areas between the two creep fibers CI and CII results in:

Portion of CI:

$$A_{cI} = A_c \frac{b'}{c} = 0.0218 \text{ m}^2 \ (0.235 \text{ ft}^2), \qquad \text{(AII, 6.9)}$$

$$A_{sI} = A_s \frac{b}{c} = 0.0827 \text{ m}^2 \ (0.890 \text{ ft}^2)$$

Portion of CII:

$$A_{cII} = A_c \frac{a'}{c} = 1.0612 \text{ m}^2 \ (11.42 \text{ ft}^2),$$

$$A_{sII} = A_s \frac{a}{c} = 0.0098 \text{ m}^2 \ (0.105 \text{ ft}^2). \qquad \text{(AII, 6.10)}$$

In the following, we examine the stresses for the moment $M = 10$ kNm induced by a sustained load, which creates a couple of forces in the two creep fibers. We get

$$N_I = N_{II} = \pm \frac{10}{2.936} = \pm 3.44 \text{ kN (764.4 lb)}. \qquad \text{(AII, 6.11)}$$

If $n = E_s/E_c = 10$, we get, for the idealized areas of the creep fibers resisting these forces, their reinforcement ratio and the α values corresponding to Eq. (2.13):

For fiber I:

$$A_{cI} + nA_{sI} = 0.0218 + 10(0.0827) = 0.85 \text{ m}^2,$$

$$\rho_r = \frac{0.0827}{0.0218} = 3.79; \qquad \alpha_I = \frac{10(3.79)}{1 + 10(3.79)} = 0.974.$$

For fiber II:

$$A_{cII} + nA_{sII} = 1.0612 + 10(0.0098) = 1.16 \text{ m}^2,$$

$$\rho_{II} = \frac{0.0098}{1.0612} = 0.0092; \qquad \alpha_{II} = \frac{10(0.0092)}{1 + 10(0.0092)} = 0.084.$$

Thus, the creep fiber stresses are

At time $t = 0$:

$$\left.\begin{aligned} &\text{In CI concrete:} \quad f_{c0} = \frac{N_I}{A_I} = \frac{3.44 \times 10^3}{0.85 \times 10^4} = 0.405 \text{ N/cm}^2, \\ &\text{steel:} \quad f_{s0} = nf_{c0} = 10(0.405) = 4.05 \text{ N/cm}^2. \end{aligned}\right\} \text{tension}$$

$$\left.\begin{aligned} &\text{In CII concrete:} \quad f_{c0} = \frac{N_{II}}{A_{II}} = \frac{3.44 \times 10^3}{1.16 \times 10^4} = 0.297\ \text{N/cm}^2, \\ &\text{steel:} \quad f_{s0} = nf_{s0} = 10(0.297) = 2.97\ \text{N/cm}^2, \end{aligned}\right\}\ \text{compression}$$

At time $t = \infty$, with $\varphi = 2$, in accordance with Eq. (2.8):

In CI:
$$f_{st} = \gamma_{\text{sus}} f_{s\,\text{sus}\,0} = 1.02(4.05) = 4.13\ \text{N/cm}^2,$$

$$f_{ct} = f_{c0} + (f_{s0} - f_{st})\rho_I = 0.405 + (4.05 - 4.13)3.79 = 0.102\ \text{N/cm}^2.$$

In CI:
$$f_{st} = \gamma_{\text{sus}} f_{s\,\text{sus}\,0} = -2.72(2.97) = -8.08\ \text{N/cm}^2,$$

$$f_{ct} = 0.297 - (8.08 - 2.97)0.0092 = -0.250\ \text{N/cm}^2.$$

These stresses are shown in Fig. 5.1 for each respective creep fiber. Since the two creep fibers do not affect each other, the stress occuring simultaneously in the other creep fiber is always equal to 0.

Linear extrapolation yields the extreme fiber stresses in N/cm^2:

Fiber I		Fiber II	Total
$t = 0$:			
top	$f_{ct} = -\dfrac{0.405}{2.936}0.301 = -0.042$	$f_{ct} = -\dfrac{0.297}{2.936}3.237 = -0.327$	$\sum f_c = -0.369$
bottom	$f_{sb} = \dfrac{4.05}{2.936}3.509 = 4.84$	$f_{sb} = \dfrac{2.97}{2.936}0.573 = 0.58$	$\sum f_s = 5.42$
$t = \infty$			
top	$f_c = -\dfrac{0.102}{2.936}0.301 = -0.010$	$f_c = -\dfrac{0.250}{2.936}3.237 = -0.276$	$\sum f_{ct} = -0.286$
bottom	$f_{cb} = \dfrac{0.102}{2.936}1.109 = 0.039$	$f_{cb} = -\dfrac{0.250}{2.936}1.827 = -0.156$	$\sum f_{cb} = -0.117$
top	$f_{st} = \dfrac{4.13}{2.936}1.109 = 1.560$	$f_{st} = -\dfrac{8.08}{2.936}1.827 = -5.030$	$\sum f_{st} = -3.470$
bottom	$f_{sb} = \dfrac{4.13}{2.936}3.509 = 4.940$	$f_{sb} = \dfrac{8.08}{2.936}0.573 = 1.570$	$\sum f_{sb} = 6.510$

Stresses induced by other effects—such as fast-rising constraints or shrinkage—are determined in the same way. The action effects are distributed between the two creep fibers, and the influence of creep is determined with Eq. (2.8) using the coefficients γ_r or γ_s from Fig. 2.2 or 2.3.

6. Two-fold Statically Indeterminate Interaction between Two Unstressed or Prestressed Bonded Tendons and Concrete

6.1. General

In many cases, particularly when the live-load component is large, top as well as bottom reinforcement must be used to take up varying moments. Then, Eqs. (2.10) and (4.3) are no longer valid. However, they can be used as approximations for the prestress loss in the main tendon when the force in the adjoining tendon is small.

A rigorous solution to the problem leads to coupled linear differential equations and to somewhat involved formulas. This is why preference usually goes to graphic methods which are largely based on iteration or step-by-step integration.

These methods are based on the assumption that in the analysis of stresses due to prestressing, the tension bars may be combined into one bar and that their stresses are approximately equal. In the case of unequal prestressing, or when the influence of the unstressed reinforcement next to the prestressing tendons can no longer be disregarded, the tension force must be related to the entire steel cross section. The loss of prestress caused by creep and shrinkage then induces compression stresses in the nonprestressed reinforcement (cf., Section 4.2).

6.2. Iteration Procedure with the Stiffness Ratios and a Resulting Closed Approximation

If the prestressing tendons can be combined into tendons 1 and 2, the stiffness ratio α_{12} then characterizes the effect of tendon 2 on fiber 1. As usual, the first subscript of the stiffness ratio designates the location of the effect; the second subscript indicates the cause. By analogy with Section 2.1.2, the required stiffness ratios, which permit a significant simplification of the analysis, can be written as follows:

$$\alpha_{11} = n\frac{A_{p1}}{A_{\text{id}}}\left(1 + \frac{A_{\text{id}}}{I_{\text{id}}}y_{\text{id}\,1}^2\right), \tag{6.1}$$

$$\alpha_{12} = n\frac{A_{p2}}{A_{\text{id}}}\left(1 + \frac{A_{\text{id}}}{I_{\text{id}}}y_{\text{id}}y_{\text{id}\,2}\right), \tag{6.2}$$

$$\alpha_{22} = n\frac{A_{p2}}{A_{\text{id}}}\left(1 + \frac{A_{\text{id}}}{I_{\text{id}}}y_{\text{id}\,2}^2\right), \tag{6.3}$$

$$\alpha_{21} = n\frac{A_{p1}}{A_{\text{id}}}\left(1 + \frac{A_{\text{id}}}{I_{\text{id}}}y_{\text{id}\,2}y_{\text{id}\,1}\right) \tag{6.4}$$

(cf., Fig. 6.1).

This approximation method is based on the concept that concrete stress is constantly changing in each fiber, from the initial value $f_{cp0} + f_{c\,\text{sus}\,0}$ up to the final value $f_{cp0} + f_{c\,\text{sus}\,0} + f_{ccs}$. It is then possible to seek a mean creep-inducing stress f_{cm} between the initial and final values, for which the creep strain of concrete may be determined by using the equation valid for constant stress $\varepsilon_c = (f_{cm}/E_c)\varphi$. The inelastic strains, ε_c and ε_s, of the concrete must be shared by the bonded prestress tendons. This yields the following steel stresses expressed as a fraction of the prestress prior to releasing the initial prestress:

$$f_{pcs}^{(0)} = \left(\frac{f_{cm}}{E_c}\varphi + \varepsilon_s\right)E_p = n\varphi f_{cm} + \varepsilon_s E_p. \tag{6.5}$$

As has been done several times in this collection of problems, a very good approximation can be obtained by setting

$$f_{cm} = f_{cp0} + f_{c\,\text{sus}\,0} + \tfrac{1}{2}f_{ccs}. \tag{6.6}$$

In evaluating the foregoing equations, it must be remembered that compression stresses f and shrinkage ε_s must be considered negative.

(a) If the iteration approach is chosen, an estimated mean value between f_{c_1m} and f_{c_2m} is selected for the first step, such as

$$f_{cm} \approx 0.8(f_{cp0} + f_{c\,\text{sus}\,0}). \tag{6.7}$$

This is used to calculate the stresses $f_{p_1cs}^{(0)}$ and $f_{p_2cs}^{(0)}$ with Eq. (6.5), and with Eqs. (6.1)–(6.4) the concrete stresses caused by creep and shrinkage.

$$f_{c_1cs} = -\frac{1}{n}(f_{p_1cs}^{(0)}\alpha_{11} + f_{p_2cs}^{(0)}\alpha_{12}), \tag{6.8a}$$

$$f_{c_2cs} = -\frac{1}{n}(f_{p_1cs}^{(0)}\alpha_{21} + f_{p_2cs}^{(0)}\alpha_{22}). \tag{6.8b}$$

It is now possible to check with Eq. (6.6) the estimated values f_{cm} obtained with Eq. (6.7) and carry out a second calculation with corrected values. However, even the results of the first calculation are frequently satisfactory.

Example. This example involves a pretensioned girder; the problem is to determine the creep and shrinkage loss under dead weight d_1 and the prestressing p for the time up to installation which takes place 180 days after casting. For dimensions, see Fig. 6.1.

$$A_c = 148.5\ \text{cm}^2\ (23.02\ \text{in}^2),$$

$$A_{id} = 155\ \text{cm}^2\ (24.03\ \text{in}^2)\ \text{in accordance with (2.14)},$$

$$A_{p1} = 1.2\ \text{cm}^2\ (0.186\ \text{in}^2),$$

$$A_{p2} = 0.4\ \text{cm}^2\ (0.06\ \text{in}^2),$$

$$I_c = 8314\ \text{cm}^4\ (119.7\ \text{in}^4),$$

$$I_{id} = 8832\ \text{cm}^4\ (212.1\ \text{in}^4)\ \text{in accordance with Eq. (2.1.5)},$$

$$n = \frac{200\,000}{40\,000} = 5.$$

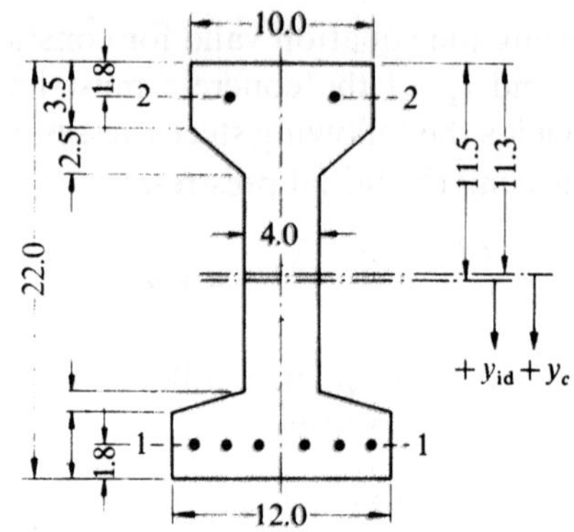

Figure 6.1.

From Example 6, Section 1.5, $\varphi_1 = 2.37$ and $\varepsilon_s = 13.8 \times 10^{-5}$, and Eqs. (6.1)–(6.4) yield

$$\alpha_{11} = +0.0902,$$

$$\alpha_{22} = +0.0342,$$

$$\alpha_{12} = -0.0062,$$

$$\alpha_{21} = -0.0186.$$

The stress prior to tension release is

$$f_{p1p}^{(0)} = 980.0 \text{ N/cm}^2 \text{ (140 532 psi)},$$

$$f_{p2p}^{(0)} = 300.0 \text{ N/cm}^2 \text{ (43 020 psi)}.$$

The concrete stresses due to prestress are

$$f_{c1p0} = -\frac{1}{n}(f_{s1p}^{(0)}\alpha_{11} + f_{s2p}^{(0)}\alpha_{12}) = -17.3 \text{ N/mm}^2 \text{ (2481 psi)},$$

$$f_{c2p0} = -\frac{1}{n} f_{s2p}^{(0)}\alpha_{21} + f_{s2p}^{(0)}\alpha_{22}) = 1.6 \text{ N/mm}^2 \text{ (229 psi)}.$$

The concrete stresses due to dead weight are

$$f_{c1d10} = 0.7 \text{ N/mm}^2 \text{ (100 psi)},$$

$$f_{c2d10} = -0.8 \text{ N/mm}^2 \text{ (115 psi)}.$$

The initial stress for creep is

$$f_{c1p+\text{sus}} = 17.3 + 0.7 = -16.6 \text{ N/mm}^2 \text{ (2380 psi)},$$

$$f_{c2p+\text{sus}} = 1.6 - 0.8 = 0.8 \text{ N/mm}^2 \text{ (115 psi)}.$$

The estimated mean value of the creep-inducing stress from Eq. (6.7) is

$$f_{c1m} = 0.8\,(-17.3 + 0.7) = -133\ \text{N/mm}^2\ (1907\ \text{psi}),$$

$$f_{c2m} = 0.8(1.6 - 0.8) = 6\ \text{N/mm}^2\ (86\ \text{psi}).$$

We will show only the second step in the iteration, for which Eq. (6.6) yields the following starting values

$$f_{c1m} = -14.8\ \text{N/mm}^2\ (2122\ \text{psi}),$$
$$f_{c2m} = 0.5\ \text{N/mm}^2\ (72\ \text{psi}).$$

Thus, with Eq. (6.5), we have

$$\frac{1}{n} f_{p1cs}^{(0)} = -14.8(2.37) - 13.8 \times 10^{-5}(40\,000) = -40.596\ \text{N/mm}^2\ (5821\ \text{psi}),$$

$$\frac{1}{n} f_{p2cs}^{(0)} = 0.5(2.37) - 5.25 = 4.335\ \text{N/mm}^2\ (623\ \text{psi});$$

and with Eqs. (6.8),

$$f_{c1cs} = 40.596(0.0902) - 4.335(0.0062) = +3.635\ \text{N/mm}^2\ (521\ \text{psi}),$$

$$f_{c2cs} = -40.596(0.0186) + 4.335(0.0342) = -0.607\ \text{N/mm}^2\ (87\ \text{psi}).$$

Checking the mean values of the creep-inducing stress,

$$f_{c1m} = -17.3 + 0.7 + \tfrac{1}{2}(3.635) = -14.8\ \text{N/mm}^2\ (2122\ \text{psi}),$$

$$f_{c2m} = 1.6 - 0.8 - \tfrac{1}{2}(0.607) = +0.5\ \text{N/mm}^2\ (72\ \text{psi}).$$

Stresses in the tension member caused by creep and shrinkage under $p + d_1$ are

$$f_{p1cs} = -n\left(\frac{1}{2} f_{p1cs}^{(0)} - f_{c1cs}\right) = -5(40.596 - 3.635) = -184.8\ \text{N/mm}^2\ (26\,500\ \text{psi}),$$

$$f_{p2cs} = -5(4.335 + 0.607) = -24.72\ \text{N/mm}^2\ (3545\ \text{psi}).$$

Assuming a linear stress distribution, the load condition $p +$ sus yields the following extreme fiber stresses in concrete:

$$f_{cbcs} = +4.05\ \text{N/mm}^2\ (581\ \text{psi}),$$

$$f_{ctcs} = -1.02\ \text{N/mm}^2\ (146\ \text{psi}).$$

(b) The coupled differential equations referred to in Section 6.1 can be transformed—as shown in Section 8 for a two-fold externally statically indeterminate system—into a system of linear equations by the choice of a mean creep-inducing stress.

$$f_{c1m} = \frac{f_{c1\,sus} - \varepsilon_s E_c(\alpha_{11} + \alpha_{12})/2 - f_{c2m}\alpha_{12}\varphi/2}{1 + \alpha_{11}\varphi/2},$$

$$f_{c2m} = \frac{f_{c2p\,sus} - \varepsilon_s E_c(\alpha_{21} + \alpha_{22})/2 - f_{c1m}\alpha_{21}\varphi/2}{1 + \alpha_{22}\varphi/2}.$$

Inserting the values from the preceding example,

$$f_{c1m} = \frac{-16.6 + 13.8 \times 10^{-5}(4 \times 10^4)(0.0902 - 0.0062)/2 - f_{c2m}0.0062(2.37/2)}{1 + 0.0902(2.37/2)}$$

$$= -14.80 + 0.00665 f_{c2m}$$

and

$$f_{c2m} = \frac{0.8 + 13.8 \times 10^{-5}(4 \times 10^4)(-0.0186 + 0.0342)/2 - f_{c1m}0.0186(2.37/2)}{1 + 0.0342(2.37/2)}$$

$$= 0.843 + 0.021 f_{c1m}$$

The first equation directly yields $f_{c1m} = -14.8$ N/mm², since the second term is very small. We then have $f_{c2m} = 0.843 - 0.021(14.8) = 0.533$ N/mm². The agreement with the result obtained by the iteration process is very good.

6.3. The Method of Step-By-Step Integration

6.3.1. No Consideration of the Influence of Delayed Elastic Strain

It may be preferable to utilize a different approximation method—particularly in complex cases—which achieves the desired purpose with relative speed and adequate precision, even when various causes of creep must be taken into consideration. The creep and shrinkage processes are divided into several steps, under the assumption that the creep-inducing stresses remain constant within each step.

If, e.g., the creep coefficient $\varphi = 4$ and $\varepsilon_s = 30 \times 10^{-5}$, one can select four steps with $\Delta\varphi = 1$ and $\Delta\varepsilon_s = 7.5 \times 10^{-5}$. Using Eq. (6.5), we get the change of stress

$$\Delta f_{pcs}^{(0)} = \Delta\varphi n f_{cm} + \Delta\varepsilon_s E_p$$

in the prestressing tendon, and we use

$$\Delta f_{ccs} = f_{cp} \frac{\Delta f_{pcs}^{(0)}}{f_{pp}^{(0)}} = f_{cp} \frac{\Delta f_{pcs}}{f_{pp}} \tag{6.10}$$

to obtain the change in the concrete stress Δf_{ccs}. The change in stress caused by the first step is now added to the initial stress, and the new stress values are used for the second step.

The accuracy of this method can be enhanced by estimating the value of Δf_{ccs} and setting $f_{cm} = f_c + \frac{1}{2}\Delta f_{ccs}$ as creep-inducing stress, as was done in Section 6.2. However, we forego this process here for the sake of simplicity.

Example (For step-by-step integration). The example treated in Section 6.2 is repeated. Integration proceeds in three steps.

$$\Delta\varphi = \frac{\varphi}{3} = \frac{2.37}{3} = 0.79; \qquad \Delta\varepsilon_s = \frac{\varepsilon_s}{3} = \frac{13.8}{3} \times 10^{-5} = 4.6 \times 10^{-5}.$$

The tension member stress is

$$f_{p1p}^{(0)} = 980 \text{ N/mm}^2 \text{ (140 532 psi)},$$

$$f_{p2p}^{(0)} = 300 \text{ N/mm}^2 \text{ (43 020 psi)}.$$

The concrete stress in a composite cross section is (cf., the Example in Section 6.2)

$$f_{c1p0} = -17.3\ \text{N/mm}^2\ (2481\ \text{psi}),$$
$$f_{c2p0} = +1.6\ \text{N/mm}^2\ (229\ \text{psi}),$$
$$f_{c1p+\text{sus}} = -16.6\ \text{N/mm}^2\ (2380\ \text{psi}),$$
$$f_{c2p+\text{sus}} = +0.8\ \text{N/mm}^2\ (115\ \text{psi}).$$

Step 1.

$$\Delta f^{(0)}_{p1cs} = -0.79(5)16.6 - 4.6 \times 10^{-5}(200\,000) \text{ using Eq. (6.5)}$$
$$= -65.6 - 9.2 = -74.8\ \text{N/mm}^2\ (10\,726\ \text{psi}),$$

$$\Delta f^{(0)}_{p2cs} = 0.79(5)0.8 - 9.20 = -6.0\ \text{N/mm}^2 (860\ \text{psi}),$$

$$\Delta f_{c1cs} = f_{c1p}\frac{f^{(0)}_{p1cs}}{f^{(0)}_{p1p}} = -17.3\frac{(-74.8)}{980.0} = +1.32\ \text{N/mm}^2\ (189.3\ \text{psi}) \text{ using Eq. (6.10)}$$

$$\rightarrow f_{c1p+\text{sus}} = -16.6 + 1.32 = -15.28\ \text{N/mm}^2\ (2191\ \text{psi}),$$

$$\Delta f_{c2cs} = f_{c2p}\frac{f^{(0)}_{p2cs}}{f^{(0)}_{p2p}} = 1.6\frac{-6.0}{300.0} = -0.03\ \text{N/mm}^2\ (43\ \text{psi})$$
$$\rightarrow f_{c2p+\text{sus}} = 0.80 - 0.03 = 0.77\ \text{N/mm}^2\ (110\ \text{psi}).$$

Step 2.

$$\Delta f^{(0)}_{p1cs} = -0.79(5)15.28 - 4.6 \times 10^{-5}(200\,000) = -60.4 - 9.2 = -69.6\ \text{N/mm}^2\ (9981\ \text{psi}),$$

$$\Delta f^{(0)}_{p2cs} = 0.79(5)0.77 - 9.2 = -6.2\ \text{N/mm}^2\ (889\ \text{psi}),$$

$$\Delta f_{c1cs} = -17.3\frac{-69.6}{980.0} = 1.23\ \text{N/mm}^2\ (176\ \text{psi})$$

$$\rightarrow f_{c1p+\text{sus}} = -15.28 + 1.23 = -14.05\ \text{N/mm}^2\ (2015\ \text{psi}),$$

$$\Delta f_{c2cs} = +1.6\frac{-6.2}{300.0} = -0.03\ \text{N/mm}^2\ (43\ \text{psi})$$

$$\rightarrow f_{c2p+\text{sus}} = 0.77 - 0.03 = 0.74\ \text{N/mm}^2\ (106\ \text{psi}).$$

Step 3.

$$\Delta f^{(0)}_{p1cs} = -0.79(5)14.05 - 9.2 = -63.7\ \text{N/mm}^2\ (9135\ \text{psi}),$$

$$\Delta f^{(0)}_{p2cs} = 0.79(5)0.74 - 9.2 = -6.3\ \text{N/mm}^2\ (903\ \text{psi}),$$

$$\Delta f_{c1cs} = -17.3\frac{-63.7}{980.0} = +1.12\ \text{N/mm}^2\ (161\ \text{psi})$$

$$\rightarrow f_{c1p+\text{sus}} = -14.05 + 1.12 = -12.93\ \text{N/mm}^2\ (1854\ \text{psi}),$$

$$\Delta f_{c2cs} = 1.6\frac{-6.3}{300.0} = -0.03\ \text{N/mm}^2\ (4\ \text{psi})$$

$$\rightarrow f_{c2p+\text{sus}} = 0.74 - 0.03 = 0.71\ \text{N/mm}^2\ (102\ \text{psi}).$$

The creep and shrinkage reduction of stress in the concrete fibers is

$$f_{c1cs} = 1.32 + 1.23 + 1.12 = +3.67\ \text{N/mm}^2\ (526\ \text{psi}),$$

$$f_{cucs} = 4.05\ \text{N/mm}^2\ (581\ \text{psi}),$$

$$f_{c2cs} = -0.03 - 0.03 - 0.03 = -0.09\ \text{N/mm}^2\ (13\ \text{psi}),$$

$$f_{c0cs} = -0.47\ \text{N/mm}^2\ (67\ \text{psi}).$$

The creep and shrinkage reduction of stress in the tension member under $p + d_1$ is

$$f_{p1cs}^{(0)} = -74.8 - 69.6 - 63.7 = -2081\ \text{N/mm}^2\ (29\,842\ \text{psi}),$$

$$f_{p1cs} = -208.1 + 3.67(5) = -189.7\ \text{N/mm}^2\ (27\,203\ \text{psi}),$$

$$f_{p2cs}^{(0)} = -6.0 - 6.2 - 6.3 = -18.5\ \text{N/mm}^2\ (2653\ \text{psi}),$$

$$f_{p2cs} = -18.5 - 0.09(5) = -19.0\ \text{N/mm}^2\ (2725\ \text{psi}).$$

These values are only slightly improved when four steps are used and each iteration step starts with the mean creep-induced stress. In this way, we get $f_{cucs} = 4.02\ \text{N/mm}^2$ (576 psi) and $f_{c0cs} = -0.46\ \text{N/mm}^2$ (66 psi).

6.3.2. *Allowing for the Influence of the Delayed Elastic Strain*

In this iteration method, the effect of reversible component φ_d of the creep coefficient $\varphi = \varphi_d + \varphi_f$ may be taken into account. Thus, the accuracy may be enhanced. However, this can be of interest only in the case of constraint coming from near-rigid supports, which is particularly reduced by creep (cf., Section 9). The principle of the method has been analyzed in Appendix II, Section 3.1. This method will not be discussed here.

7. Redistribution Moments in Prestressed Structures which Are Supported Statically Indeterminate. Examples of Internally and Externally Statically Indeterminate Systems

7.1. Introduction

Section 3.2 dealt with the case of an unbonded prestressed beam which is supported externally statically determinate, but which is internally statically indeterminate. Now we will examine the case of a post-tensioned beam under externally statically indeterminate support conditions.

In the system treated in Section 3.2—under externally statically determinate support conditions—prestressing created no action effects and only an internal stress state which is the result of the constraint between concrete and steel created by prestressing. In systems which are supported externally statically indeterminate, the strain forced upon the concrete cross section by prestressing usually can not take place unimpeded. As a rule, it creates so-called redistribution action effects. Part A, Section 1, dealt at length with the significance of the difference between internal stresses, imposed action effects, and redistribution moments.

For any given prestressed concrete structure which is supported statically indeterminate, it is of course possible to find a pattern for the tendons in which prestressing will not generate redistribution action effects. There exists then—as under statically determinate support conditions—a pure imposed stress condition. In such cases, the preferred expression is "prestressing without redistribution." In the usual tendon patterns, however, there occur, in most cases, small redistribution moments. However, there are also cases in which a marked redistribution of the action effects is created intentionally by prestressing in order to achieve more favorable stress conditions. This is very frequently the case in prestressed shells, for instance.

Creep and shrinkage reduce the prestress force and consequently the accompanying internal stress condition. The redistribution action effects computed for time $t = 0$ change over the course of time in approximate proportion to the mean loss of prestress.

7.2. Redistribution Moments in Prestressing without Bond

The following example illustrates the computation of the redistribution action effects induced by prestressing. In the case of the two-span beam represented in Fig. 7.1, a distinction is made between the open system and the statically determinate basic system. The "open system" is created when the tendon is cut at

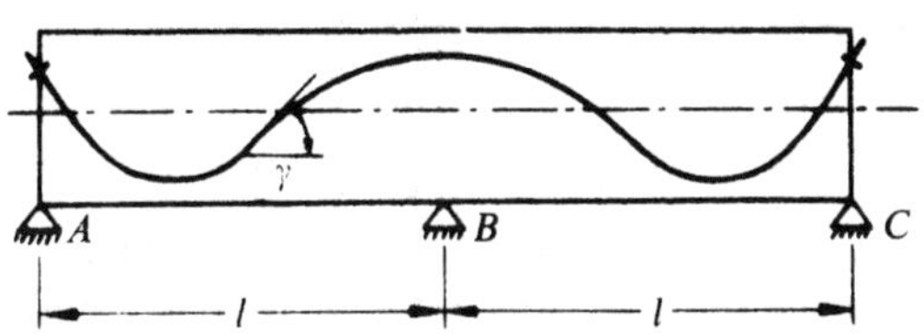

Figure 7.1. Example of load condition prestressing: The prestressing tendon is still unbonded.

one point (system (0)). It is still externally statically indeterminate. In order to transform it into a "statically determinate basic system," it is given an additional degree of freedom, e.g., by inserting a hinge over the middle support (system (00)).

This two-fold statically indeterminate system is now analyzed in two steps. In the first step, the hinge is eliminated over the center support for $F_{p1} = 1$, thus solving a fundamental problem that occurs under all load conditions. This problem is simplified if the hinge above the center support is located at the level of the tendon axis, as was done in Fig. 7.2. The second step is then the transformation of the open system (0) into a closed system. This yields the value of the tendon force F_p.

Under the effect of $F_{p1} = 1$, the concrete beam in the statically indeterminate basic system (00) is subjected only to imposed action effects, while the statically indeterminate action effects which convert system (00) into system (0) correspond to the redistribution moment triggered by $F_{p1} = 1$. We would have a case of "prestressing without redistribution" if the action effects became equal to 0.

(a)

(b)

(c)

(d)

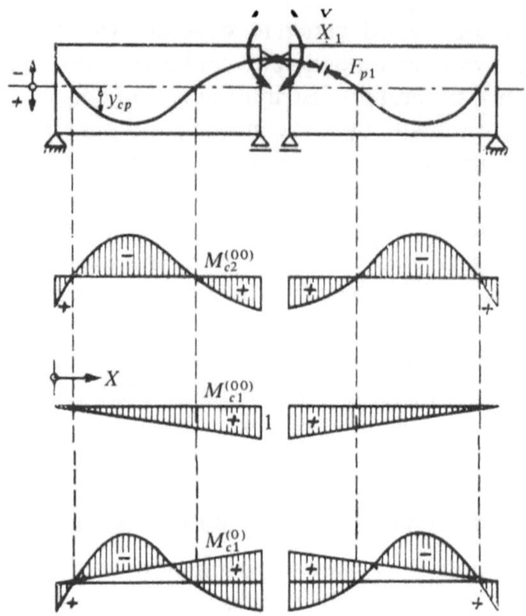

Figure 7.2.

For the two-span continuous beam, both externally and internally statically indeterminate to the first degree, the statically indeterminate variables are as follows (see Fig. 7.1):

$$X_2 = F_{p1} \quad \text{and} \quad X_1 = M_{cB},$$

which is the moment at center support for the concrete beam.

Contrary to procedures followed up to now (cf., [3] and [9]), X_1 and X_2 have been switched in the following. It seems more logical to assign X_1 to the first step and X_2 to the second step. All the more so, as X_2 can be freely chosen and only X_1 needs to be determined.

First Step

The following is a calculation of the action effects for $X_2 = F_{p1} = 1$ by resolving the externally static indeterminacy for the internally still-open system.

The "open" basic system is made statically determinate with the hinge over the center support at the level of the tendon axis (tendon not bonded), see Fig. 7.2a.

The condition $X_2 = 1$:

F_{p1} varies, owing to friction, and the following is valid only at a selected point (see Fig. 7.2b):

$$\begin{aligned} F_{p1} &= X_2 = 1, \\ M_{c2}^{(00)} &= y_{c2}(F_{p1}), \\ N_{c2}^{(00)} &= -\cos\gamma(F_{p1}), \\ V_{c2}^{(00)} &= -\sin\gamma(F_{p1}), \end{aligned} \tag{7.1}$$

$$\delta_{c12} = \frac{\int y_{cp} F_{p1} M_{c1}^{(00)}\, ds}{E_c I_c}. \tag{7.2}$$

The condition $X_1 = 1$ (see Fig. 7.2c):

$$M_{c1}^{(00)} = \frac{x}{l};\ V_{c1}^{(00)} = -\frac{1}{l};\ N_{c1}^{(00)} = 0, \qquad (7.3)$$

$$\delta_{c11} = \frac{\int (M_{c1}^{(00)})^2\, ds}{E_c I_c}.$$

Open system (see Fig. 7.2d):

$$X_1 = -\frac{\delta_{c12}}{\delta_{c11}}, \qquad (7.4)$$

$$\begin{aligned} M_{c1}^{(0)} &= M_{c2}^{(00)} + X_1 M_{c1}^{(00)} \\ &= M_{c2}^{(00)} + M_1^{(0)} \end{aligned} \qquad (7.5)$$

$$N_{c1}^{(0)} = N_{c2}^{(00)} + X_1 N_{c1}^{(00)} = N_{c2}^{(00)} + N_1^{(0)} = -\cos\gamma F_{p1} \sim 1, \qquad (7.6a)$$

$$V_{c1}^{(0)} = V_{c2}^{(00)} + X_1 V_{c1}^{(00)} = V_{c2}^{(00)} + V_1^{(0)} = -\sin\gamma\, F_{p1} - \frac{X_1}{l}. \qquad (7.6b)$$

Second Step

The following is the solution of the internal static indeterminancy by closure of the tendon for the sustained load condition sus.

For the externally statically indeterminate but internally still "open" system (see Fig. 7.3a)

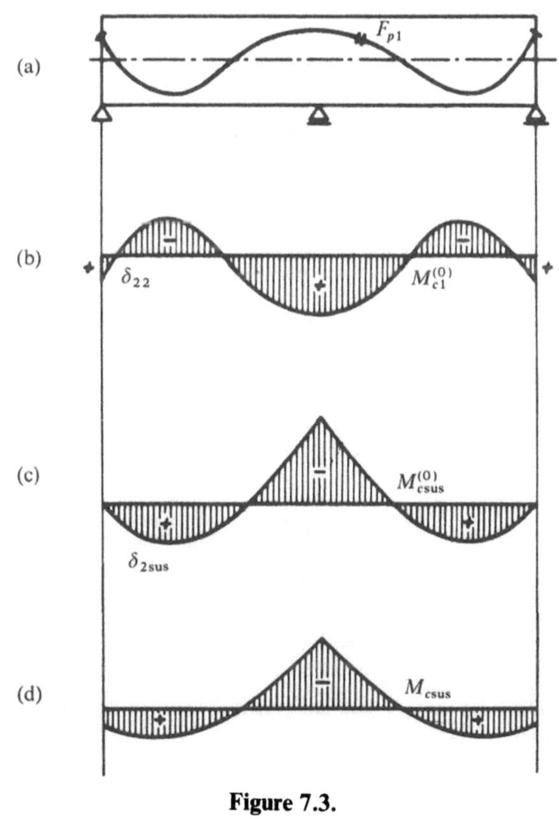

Figure 7.3.

The condition $X_2 = F_{p1} = 1$ (see Fig. 7.3b):

$$M_{c1}^{(0)} = M_{c2}^{(00)} + X_1 M_{c1}^{(00)}, \tag{7.5}$$

$$N_{c1}^{(0)} = -F_{p1} \cos \gamma = 1, \tag{7.6a}$$

$$V_{c1}^{(0)} = -F_{p1} \sin \gamma - \frac{X_1}{l}, \tag{7.6b}$$

$$\delta_{c22} = \frac{\int (M_{c1}^{(0)})^2 \, ds}{E_c I_c} + \frac{\int (N_{c1}^{(0)})^2 \, ds}{E_c A_c} + \overbrace{\frac{\int (V_{c1}^{(0)})^2 \, ds}{G_c A_c}}^{\sim 0}, \tag{7.7}$$

$$\delta_{p22} \frac{\int F_{p1}^2 \, ds}{E_p A_p} \cong F_{pm}^2 \frac{1}{E_p A_p}, \tag{7.8}$$

$$\delta_{22} = \delta_{c22} + \delta_{p22}, \alpha = \frac{\delta_{c22}}{\delta_{22}}. \tag{7.9}$$

The sustained load condition, sus (see Fig. 7.3c):

$$M_{c\,\mathrm{sus}}^{(0)}, N_{c\,\mathrm{sus}}^{(0)}, V_{c\,\mathrm{sus}}^{(0)};$$

$$\delta_{2\,\mathrm{sus}} = -\int M_{c\,\mathrm{sus}}^{(0)} M_{c1}^{(0)} \frac{ds}{E_c I_c} - \overbrace{\int N_{c\,\mathrm{sus}}^{(0)} N_{c1}^{(0)} \frac{ds}{E_c A_c}}^{=0} - \overbrace{\int V_{c\,\mathrm{sus}}^{(0)} V_{c1}^{(0)} \frac{ds}{G_c A_c}}^{\sim 0}. \tag{7.10}$$

Closed system (see Fig. 7.3d):

$$X_2 = F_{\mathrm{sus}} = -\frac{\delta_{2\,\mathrm{sus}}}{\delta_{22}}, \tag{7.11}$$

$$M_{c\,\mathrm{sus}} = M_{c\,\mathrm{sus}}^{(0)} + F_{\mathrm{sus}} M_{c1}^{(0)}, \tag{7.12}$$

$$N_{c\,\mathrm{sus}} = N_{c\,\mathrm{sus}}^{(0)} - F_{\mathrm{sus}} N_{c1}^{(0)}, \tag{7.13a}$$

$$V_{c\,\mathrm{sus}} = V_{c\,\mathrm{sus}}^{(0)} - F_{\mathrm{sus}} V_{c1}^{(0)}. \tag{7.13b}$$

Thus, the following applies to the individual load conditions.
Load condition prestressing:

The tendon force F_p is chosen in accordance with the particular requirements. Then,

$$\begin{aligned} M_{cp} &= F_p(M_{c2}^{(00)} + X_1 M_{c1}^{(00)}) \\ N_{cp} &\cong -F_p, \\ V_{cp} &= -F_p\left(\sin\gamma + \frac{X_1}{l}\right). \end{aligned} \tag{7.14}$$

In these equations, the terms F_p, X_1, and $M_{c1}^{(00)}$ yield the redistribution moments; $M_1^{(0)}$ and $(F_p X_1)/l$ provide the corresponding shear forces. The remaining terms $F_p M_{c2}^{(00)}$ correspond to an imposed stress condition. This breakdown is significant for the ultimate load at the level of the tendon. It is valid only for the particular location of the hinge which thus creates the primary system.

Appropriate placement of the tendon can result in $M_1^{(0)} = X_1 M_{c1}^{(00)} = 0$. Such is the case when, applying Eq. (7.2), $\delta_{c12} = 0$ (prestressing without redistribution).

However, large redistribution moments can also be generated intentionally. They can become so large that they completely balance the moment at the center support, which is created at the ultimate external load, so that the two spans then behave like freely supported beams (e.g., $1.75M_{\text{sus}} + 1.3M_1^{(0)} = 0$). As a rule, however, only a limited deviation from prestressing without redistribution is feasible. The reason is that during the transition from live load to failure load, a major deviation leads to a pronounced change in the direction and distribution of action effects and can, thus, render dimensioning more difficult.

Load Conditions: Dead Weight, Live Load, and Sustained Load

The action effects $M_{c\,\text{sus}}^{(0)}$ are to be computed for the statically indeterminate open system. For the dead-weight and live-load load conditions, only the subscripts need be switched in Eqs. (7.10)–(7.13).

Shrinkage-Induced Action Effects

The action effects $M_{cs}^{(0)}$ and $N_{cs}^{(0)}$ induced by shrinkage must also be computed for the statically indeterminate open system when shrinkage is restrained by the supports. (However, this is not the case for the example shown in Fig. 7.1). The displacement reads:

$$\delta_{10} = \delta_{1s} = -\varepsilon_s l - \int M_{cs}^{(0)} M_{c1}^{(0)} \frac{ds}{E_c I_c} - \int N_{cs}^{(0)} N_{c1}^{(0)} \frac{ds}{E_c A_c}. \tag{7.15}$$

Superposition of Action Effects

Tables 7.1 and 7.2 present a clear listing of the equations for the action effects.

For purposes of stress analysis one can combine the action effects induced by prestressing and the redistribution action effects for any given cross section (cf., Table 7.1); for the determination of ultimate loads, only the redistribution action effects acting in the overall cross section are significant (cf., Table 7.2.)

In regard to the formulas for load conditions, prestressing, creep and shrinkage, it should be noted that the only imposed load included by them is the case of

Table 7.1. Summary of the equations to be used in prestressing without bond for the calculation of action effects in concrete cross sections under live load. In applying them, it must be kept in mind that $F_s^{(e)}$ is usually negative. The γ coefficients are taken from Figs. 2.1–2.3 after determining α and φ. For sustained and live loads one need only substitute subscripts in the equations listed for sustained loads, sus.

Loading	Action effects	Equation
	$N_{cp} = N_{c1}^{(0)} F_p$	(7.16a)
p	$M_{cp} = M_{c1}^{(0)} F_p$	(7.16b)
	$V_{cp} = V_{c1}^{(0)} F_p$	(7.16c)
	$N_{ccs} = (\gamma_{sus} - 1) N_{c\,sus} + (\gamma_r - 1) N_{cp} + \gamma_s F_s^{(e)} N_{c1}^{(0)}$	(7.16d)
$c + s$	$M_{ccs} = (\gamma_{sus} - 1) M_{c\,sus} + (\gamma_r - 1) M_{cp} + \gamma_s F_s^{(e)} M_{c1}^{(0)}$	(7.16e)
	$V_{ccs} = (\gamma_{sus} - 1) V_{c\,sus} + (\gamma_r - 1) V_{cp} + \gamma_s F_s^{(e)} V_{c1}^{(0)}$	(7.16f)
	$N_{cpcs} = (\gamma_{sus} - 1) N_{c\,sus} + \gamma_r N_{cp} + \gamma_s F_s^{(e)} N_{c1}^{(0)}$	(7.16g)
$p + c + s$	$M_{cpcs} = (\gamma_{sus} - 1) M_{c\,sus} + \gamma_r M_{cp} + \gamma_s F_s^{(e)} M_{c1}^{(0)}$	(7.16h)
	$V_{cpcs} = (\gamma_{sus} - 1) V_{c\,sus} + \gamma_r V_{cp} + \gamma_s F_s^{(e)} V_{c1}^{(0)}$	(7.16i)
	$N_{c\,sus} = N_{c\,sus}^{(0)} + N_{c1}^{(0)} F_{sus}$	(7.16k)
sus	$M_{c\,sus} = M_{c\,sus}^{(0)} + M_{c1}^{(0)} F_{sus}$	(7.16l)
	$V_{c\,sus} = V_{c\,sus}^{(0)} + V_{c1}^{(0)} F_{sus}$	(7.16 m)

prestressing with steel tendons. Especially where an additional constraint is applied, such as may be created in a statically indeterminate structure by the choice of a particular construction method, this must be taken into consideration.

7.3. Redistribution Moments in Post-Tensioned Systems

In general, prestressing is applied to the system prior to bonding. Prestressing activates the dead weight. It is therefore appropriate to combine prestress force F_p and the tendon force F_{sus} induced by dead weight (load condition sus + p). This simplifies the computation process described in Section 7.2. The prestressing force F_{sus} is defined as a target for the load condition sus + p. Static indeterminacy is thus reduced by one degree. Frequently, a unit load condition, such as $F_p + F_{sus} = 1000$ MN, is analysed (see Section 7.4) and the final value of F_p is determined only after the results for the unit load condition are available.

The equations listed in Section 7.2 are then simplified as follows:

F_{p+d_1} is no longer introduced as a statically indeterminate variable (d_1 is the weight of structure, a component of sus), but rather either selected from the outset or introduced as a provisional operator, such as $F_0 = \max F_{p+d_1}$.

Table 7.2. Summary of the equations to be used in prestressing without bond and statically indeterminate systems for the calculation of action effects for total cross sections (a section which comprises concrete and tendon) and needed to calculate ultimate loads. In most cases $F_s^{(r)}$ is negative. For sustained and for live loads the subscript sus must be substituted.

Loading	Action effects	Equation
	$N_p = N_1^{(0)} F_p$	(7.17a)
p	$M_p = M_1^{(0)} F_p$	(7.17b)
	$V_p = V_1^{(0)} F_p$	(7.17c)
	$N_{cs} = (\gamma_{sus} - 1)N_{sus} + (\gamma_r - 1)N_p + \gamma_s F_s^{(e)} N_1^{(0)}$	(7.17d)
$c + s$	$M_{cs} = (\gamma_{sus} - 1)M_{sus} + (\gamma_r - 1)M_p + \gamma_s F_s^{(e)} M_1^{(0)}$	(7.17e)
	$V_{cs} = (\gamma_{sus} - 1)V_{sus} + (\gamma_r - 1)V_p + \gamma_s F_s^{(e)} V_1^{(0)}$	(7.17f)
	$N_{pcs} = (\gamma_{sus} - 1)N_{sus} + \gamma_r N_p + \gamma_s F_s^{(e)} N_1^{(0)}$	(7.17g)
$p + c + s$	$M_{pcs} = (\gamma_{sus} - 1)M_{sus} + \gamma_r M_p + \gamma_s F_s^{(e)} M_1^{(0)}$	(7.17h)
	$V_{pcs} = (\gamma_{sus} - 1)M_{sus} + \gamma_1 V_p + \gamma_s F_s^{(e)} V_1^{(0)}$	(7.17i)
	$N_{sus} = N_{c\,sus}^{(0)} + N_1^{(0)} F_{sus}$	(7.17k)
sus	$M_{sus} = M_{c\,sus}^{(0)} + M_1^{(0)} F_{sus}$	(7.17l)
	$V_{sus} = V_{c\,sus}^{(0)} + V_1^{(0)} F_{sus}$	(7.17m)

X_1 replaces the externally statically indeterminate variable and, with an appropriate choice of the basic system (cf., Section 7.2), it already constitutes the statically indeterminate part of the redistribution moment

$$X_1 = \frac{-\int F_0 y_{cp} M_{c1}^{(00)}(ds/E_c I_c)}{\int (M_{c1}^{(00)})^2 (ds/E_c I_c)}. \tag{7.18}$$

M_{po} are the final moments under load condition F_0, the first term representing the constraint moments and the second term the redistribution moments induced by them—subject to an appropriate choice of the basic system.

$$M_{po} = -F_0 y_{cp} + X_1 M_{c1}^{(00)}. \tag{7.19}$$

Following subsequent bonding, this leaves only the influence of live load, as well as of creep and shrinkage, to be considered.

Most prefabricated members are pretensioned, to be sure. However, initially they are usually supported under externally statically determinate conditions. Consequently, the statements above still apply.

As a rule, then, creep and shrinkage effects can be investigated entirely in the new system formed after bonding. However, since not only the action effects induced by sustained load and the moment induced by prestressing are of variable magnitude in the individual cross sections, but also the prestressing forces—owing to the friction loss occurring during pretensioning—different creep and shrinkage losses F_{cs} are generated in each investigated cross section. This also affects the redistribution moments. In general, however, one can make do with the simplified assumption that the redistribution moments decrease within each step proportionately to the newly computed prestressing force by a factor of $F_{p\,sus\,cs}/F_{p\,sus}$. A verification of the agreement between the results obtained by this process and the compatibility conditions is needed, at most, only when an exceptionally large redistribution of forces was sought.

7.4. Numerical Example of Redistribution Moments

A continuous bridge structure over two 50-m spans (162 ft) is to be conceived as a double-webbed T-beam with a height $h = 2.5$ m, (8.2 ft) (see Figs. 7.4–7.6).

The following is not a complete analysis; only the most significant problems are treated here.

(a) *System*

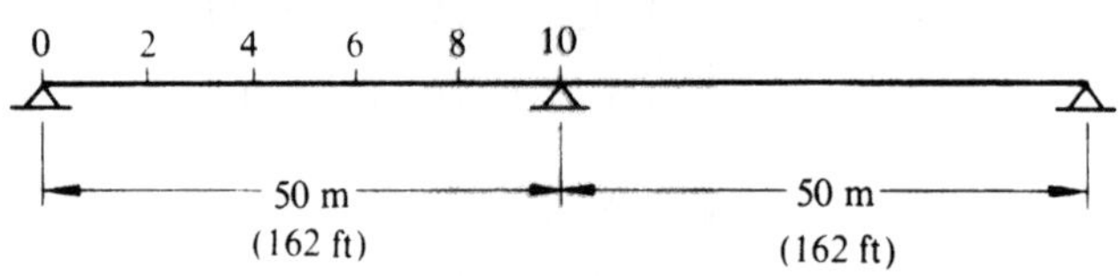

Figure 7.4.

(b) *Cross-Section* (*See Fig. 7.5*) *and Concrete Characteristics*
Concrete: $f'_c = 45$ N/mm^2 (Type I cement) DYWIDAG tendons: $d = 32$ mm, St 850/1050 (yield strength 850 N/mm^2, tensile strength 1050 N/mm^2); deformed reinforcing steel ($f_y = 420$ N/mm^2; $f_{tu} = 500$ N/mm^2); $2A/u = 35$ cm (13.78 in.); 80% rel. H. $\beta_{eff} = 2.4$; $\beta_{ce} = 2$; $d_{eff} = 84$ cm (33.07 in.). Age at loading is 7 days; $t_f = 14$; $\varphi_{f0} = 1.7$; $\beta_f = 0.94$; $\varphi = 0.4 + 0.94(1.7) = 2.0$. The shrinkage in accordance with Fig. 1.7: $\varepsilon_{s\infty} = 0.70(20) \times 10^{-5} = 14 \times 10^{-5}$.

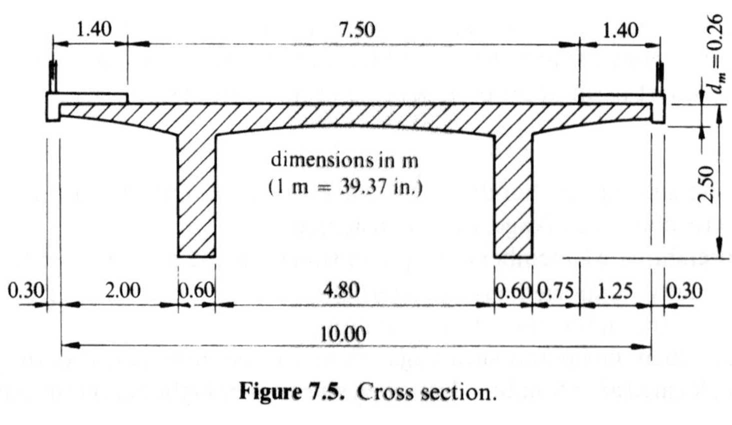

Figure 7.5. Cross section.

(c) *Moments of Inertia and Section Modulus of the Concrete Cross Section*

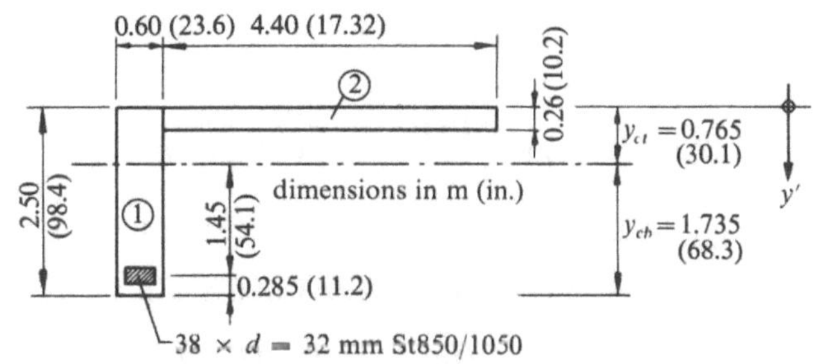

Figure 7.6. Equivalent cross section.

Cross section characteristics.

	A	y'	Ay'	$\Delta y = y_{ct} \to y'$	$A\Delta y^2$	$I' = \frac{bd^3}{12}$
	m^2	m	m^3	m	m^4	m^4
①	1.5	1.25	1.875	−0.485	0.353	0.781
②	1.144	0.13	0.149	0.635	0.461	0.006
Σ	2.644		2.024		0.814	0.787

$$y_{ct} = \frac{\sum Ay'}{\sum A} = \frac{2.024}{2.644} = 0.765 \text{ m (30.1 in.)},$$

$$I_c = \sum I' + \sum A\Delta y^2 = 0.787 + 0.814 = 1.601\ m^4 \text{ (185.5 ft}^4\text{)},$$

$$y_{cu} = d - y_{cO} = 2.5 - 0.765 = 1.735 \text{ m (68.3 in)},$$

$$W_t = \frac{I_c}{y_{ct}} = 2.093 \text{ m}^3 \text{ (73.9 ft}^3\text{)},$$

$$W_b = \frac{I_c}{y_{cb}} = 0.923 \text{ m}^3 \text{ (32.6 ft}^3\text{)},$$

$$y_p = y_{cb} - 0.285 = 1.450 \text{ m (57.1 in)},$$

$$W_p = \frac{I_c}{y_p} = 1.104 \text{ m}^3 \text{ (39.0 ft}^3\text{)}.$$

Idealised cross section characteristics (effect of concrete age and steel disregarded): $n = 210\,000/371\,000 = 5.68$;

Eq. (2.14):

$$A_{id} = 2.644 + (5.68 - 1)0.000804(38) = 2.787 \text{ m}^2 \ (30.0 \text{ ft}^2),$$

Eq. (2.16):

$$\Delta y_{c\,id} = 1.450 \frac{0.143}{2.787} = 0.074 \text{ m (2.9 in.)},$$

$$y_{id\,t} = 0.765 + 0.074 = 0.839 \text{ m (33 in.)},$$

$$y_{id\,b} = 1.735 - 0.074 = 1.661 \text{ m (65.4 in.)};$$

$$y_{id\,p} = 1.450 - 0.074 = 1.376 \text{ m (54.2 in.)},$$

Eq. (2.15):

$$I_{id} = 1.601 + 0.143(1.45)1.376 = 1.89 \text{ m}^4 \ (219 \text{ ft}^4),$$

$$W_{id\,t} = \frac{1.89}{0.839} = 2.25 \text{ m}^3 \ (79.5 \text{ ft}^3);$$

$$W_{id\,p} = \frac{1.89}{1.376} = 1.37 \text{ m}^3 \ (48.4 \text{ ft}^3),$$

$$W_{id\,b} = \frac{1.89}{1.661} = 1.14 \text{ m}^3 \ (40.3 \text{ ft}^3).$$

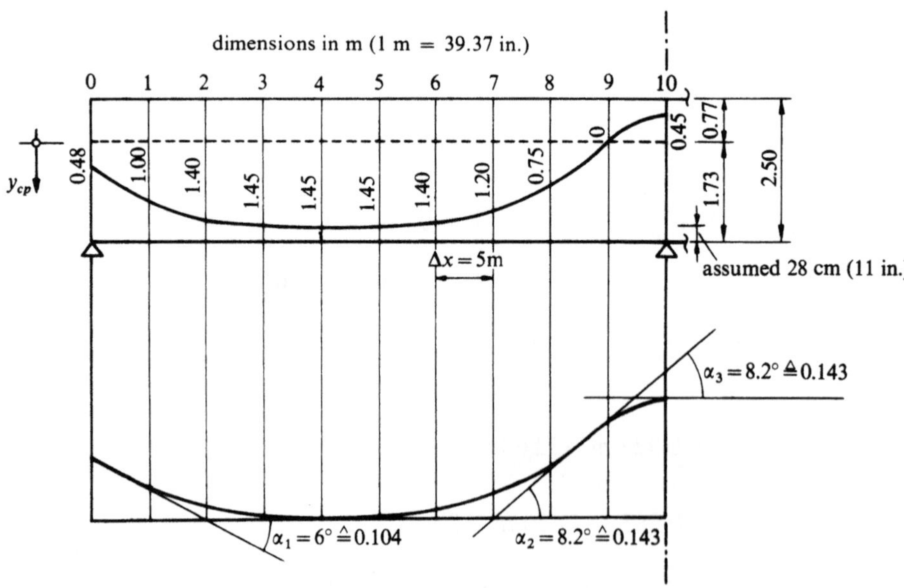

Figure 7.7.

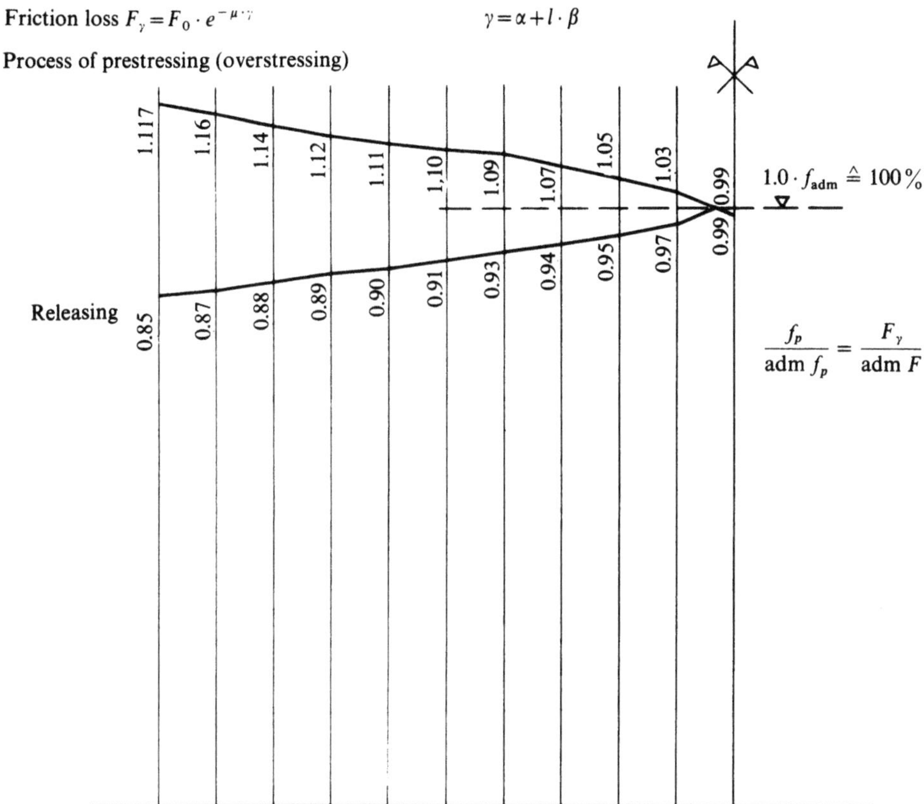

	0	1	2	3	4	5	6	7	8	9	10 Section
$\gamma_{prestressing}$ °	3.5	7.0	10.5	12.0	14.6	18.1	21.6	25.1	27.7	37.4	
$\gamma_{releasing}$ °	35.4	31.9	28.4	24.9	23.4	20.8	17.3	13.8	10.3	7.7	

Figure 7.8.

(d) *Tendon Pattern; Friction Losses*

A fairly pronounced redistribution is aimed at so as to shift the bending moments from the supports (rectangular cross sections), that are less well suited for flexural moments, towards mid-span (T-beams) (see Fig. 7.7). Thus the web width can be kept constant along the bridge.

For the prestressing system selected, the unintended angular change of the tendon is $\beta = 0.30°/\text{m}$, whereas the coefficient of friction is $\mu = 0.26$.

The tensile force distribution shown in Fig. 7.8 is generated with an overtensioning to max $f_p = 0.8 f_s = 0.8(8.5) = 68.00\ \text{kN/cm}^2$ (9.75×10^4 psi) and subsequent release to the admissible steel stress of adm $f_p = 0.55(10.5) = 57.75\ \text{kN/cm}^2$ (8.28×10^4 psi).

$$\frac{\max f_p}{\text{adm } f_p} = \frac{68.00}{57.75} = 1.177$$

Table 7.3. Numerical integration according to Simpson

Sect.	F_0	y_{cp}	M_1	$M_p = -F_0 y_{cp}$	λ	$M_1^2\lambda$	$M_p M_1 \lambda$	$M_1 X_1$	M_{p0}
0	0.85	0.48	0	−0.408	1	0	0	0	−0.408
1	0.87	1.00	0.1	−0.870	4	0.04	−0.348	0.115	−0.755
2	0.88	1.40	0.2	−1.232	2	0.08	−0.493	0.230	−1.002
3	0.89	1.45	0.3	−1.290	4	0.36	−1.548	0.345	−0.945
4	0.90	1.45	0.4	−1.305	2	0.32	−1.044	0.460	−0.845
5	0.91	1.45	0.5	−1.320	4	1.00	−2.640	0.575	−0.745
6	0.93	1.40	0.6	−1.302	2	0.72	−1.562	0.689	−0.613
7	0.94	1.20	0.7	−1.128	4	1.96	−3.158	0.804	−0.324
8	0.95	0.75	0.8	−0.713	2	1.28	−1.141	0.919	+0.206
9	0.97	0.0	0.9	0	4	3.24	0	1.034	+1.034
10	~1.00	−0.45	1.00	+0.450	1	1.00	+0.450	1.148	+0.703
						10.000	−11.484		

(e) *Computation of Prestressing-Induced Moments with Eq.* (*7.18*), *Taking into Account the Variation of the Tensile Force Along the Tendon*

$$M_{po} = -F_0 y_{cp} + \frac{\int F_0 y_{cp} M_{c_1}^{(00)}\, ds}{\int (M_{c_1}^{(00)})^2\, ds} \frac{x}{l} = -M_p + \frac{\int M_p M_1\, ds}{\int M_1^2\, ds} M_1$$

Table 7.3 gives

$$X_1 = \frac{\sum M_p M_1 \lambda}{\sum M_1^2 \lambda} = \frac{11.484}{10.0} = 1.148 \text{ kNm for } F_{0,10} = 1 \text{ kN},$$

$$M_p = -F_0 y_{cp} \triangleq M_{c_2}^{(00)} \text{ [Eq. (7.1)]},$$

$$X_1 M_1 \triangleq X_1 M_{c_1}^{(00)} = M_1^{(0)},$$

$$M_{po} = -F_0 y_{cp} + X_1 M$$

$$\triangleq M_{c_1}^{(0)} = M_{c_2}^{(00)} + M_1^{(0)} \text{ [Eq. (7.5)]}.$$

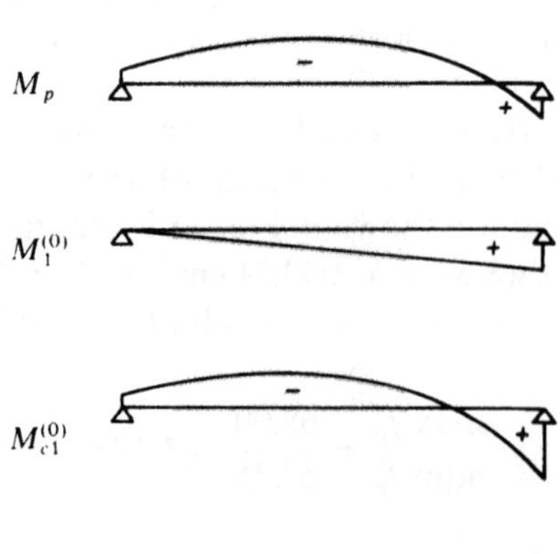

Table 7.4. Summary of the stresses in cross section 4 (N/mm^2). 1 N/mm^2 = 145 psi

Load conditions	f_{cp} [b]	f_o	f_u	f_p
d_1: M_{d_1} = +11 570 kNm	+8.3	−5.5[a]	+12.5[a]	—
d_2: M_{d_2} = +1980 kNm	+1.5	−0.9	+1.7	+8.5
w: max Mw = +8430 kNm	+6.1	−3.7	+7.3	+34.6
min Mw = −1990 kNm	−1.4	+0.9	−1.7	−8.0
N_p = −15 890 kN	−5.7	−6.1[a]	−6.1[a]	
M_p = −14 900 kNm	−10.9	+7.1[a]	−16.1[a]	
p:	−16.6	+1.0	−22.2	0.9(0.55)1050.0 = +520.0
$cs \triangleq 15\%$ of p	+2.5	−0.2	+3.3	−78.9
$d_1 + p$	−8.3	−4.5	−9.7	+520.0
$d_1 + d_2 + p$	−6.8	−5.4	−8.0	+528.5
$d_1 + d_2 + p + cs$		−5.3	−4.7	+450.3
$d_1 + d_2 + p + cs$ + max $w/2$		−7.1	−1.1	+467.6
$d_1 + d_2 + p + cs$ + min $w/2$		−9.0	+2.6	+484.9
$d_1 + d_2 + p$ + min $w/2$		−5.0	−8.8	+524.5
$d_1 + d_2 + p$ + min w		−4.5	−9.7	+520.5
adm. f in accordance with DIN 4227		−16.0	−18.0 +3.2	+577.5

[a] Computed with A_c and W_{ct} or W_{cb}.

[b] Concrete stress at level of tendon, computed with W_{idp}.

(f) *Stresses Induced by Creep and Shrinkage*

Often it is customary and a justified simplification for the analysis of load conditions $p + d_1$ in an unbonded system to proceed from a plain concrete cross section, and to use the idealized cross section characteristics only for load conditions, such as d, w and $c + s$, which occur after bonding. However, the tensile force along the tendon varies, owing to the friction occuring during prestressing. The influence of the change of slope of the tendon can usually be disregarded ($\cos \alpha \sim 1$). For the load conditions that are applied to the total cross section, the tensile forces F_{w+p} and F_{c+s} are different in each cross section. In determining the accompanying redistribution moments, Section 7.3 should be adhered to.

An evaluation has shown that by setting F_{p+d_1} = 17 650 kN, one can expect the stresses to remain within admissible limits. For cross section 4 of the span, a stress analysis is carried out for working load conditions, and creep and shrinkage losses are determined.

Cross Section 4
Load condition $p + d_1$:

$$N_{p+d_1} = 17\,650(0.90) = 15\,890 \text{ kN } (3531 \text{ kip}),$$
$$M_p = -17\,650(0.845) = 14\,900 \text{ kNm } (13.0 \times 10^7 \text{ in. lb}).$$

Load condition $c + s$:

$$f_{pcs} = \frac{\varepsilon_s E_p + n\varphi f_{cp\,\text{sus}}}{1 - n(f_{cp}/f_{pp})(1 + \varphi/2)}. \tag{4.3}$$

Load condition d_2: $M_{d_2} = 1980$ kNm (1.7×10^7 in.lb),
Load condition w (live load): $M_w = 8430$ kNm (7.3×10^7 in. lb).
According to Table 7.4 the tensile stress at the level of the tendon fiber ($d_1 + d_2 = \text{sus}$): $f_{cpd} = f_{p+d_1+d_2} = 6.8$ N/mm^2

$$f_{pcs} = \frac{-14 \times 10^{-5}(2.1 \times 10^5) - 5.68(2.0)6.8}{1 + 5.68(16.6/520.0)(1 + 2.0/2)} = 78.2 \text{ N/mm}^2 \; (11.2 \times 10^3 \text{ psi}).$$

The creep loss is

$$\frac{f_{pcs}}{f_{pp}} = \frac{78.2(100)}{520} = 15\%:$$

8. Systems with Multiple Degrees of Indeterminacy

8.1. Overview

In this section, we will show how one can avoid having to solve the coupled differential equations which normally result for systems with multiple degrees of indeterminacy. The approaches particularly suited for this purpose have been described in Appendix II, Section 7.1. The two examples presented below of externally statically indeterminate cases are intended to illustrate in detail how to proceed.

The first example (see Fig. 8.1) involves a four-span continuous concrete beam. It rests on steel columns whose flexibility is described by the spring constants c. In this example, only the method for solving the problem under the load condition resulting from the settlement of support 2 is presented. Under sustained loading, the distribution of the action effects is modified in a similar manner as in the example treated in Section 3.1; with the support conditions selected here, the load condition shrinkage has no effect.

The initial solution uses the method of mean creep-inducing stress. Then the differences which arise when the Trost method is applied to the computation are shown.

A complete numerical analysis is given only for the second, simpler example, which differs from the first one only by being confined to a symmetrical beam over three spans (see Fig. 8.2).

8.2. Four-Span Continuous Beam on Elastic Supports

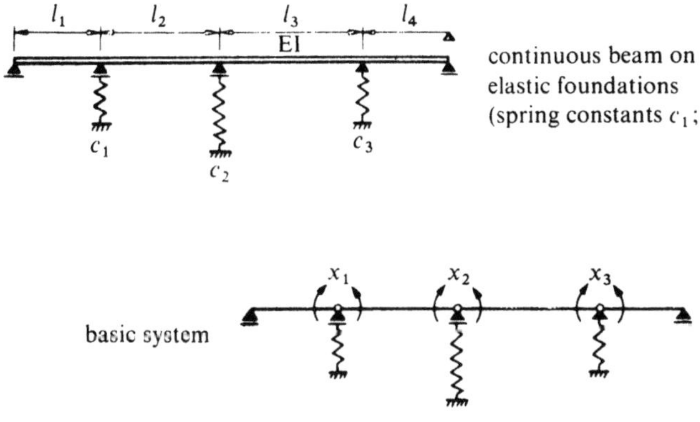

Figure 8.1a.

The displacements for the load conditions sustained load and support settlement are as follows:

For the load condition $X_1 = 1$:

$$\delta_{11} = \overbrace{\frac{l_1 + l_2}{3EI}}^{\delta_{c11}} + \overbrace{\left(\frac{1}{l_1} + \frac{1}{l_2}\right)c_1\left(\frac{1}{l_1} + \frac{1}{l_2}\right) + \frac{1}{l_2^2}c_2}^{\delta_{e11}},$$

$$\delta_{12} = \delta_{21} = \overbrace{\frac{l_2}{6EI}}^{\delta_{c12}} \overbrace{- \left(\frac{1}{l_1} + \frac{1}{l_2}\right)\frac{c_1}{l^2} - \frac{1}{l_2}\frac{c_2}{l_2} - \frac{1}{l_2}\frac{c_2}{l_3}}^{\delta_{e12}}$$

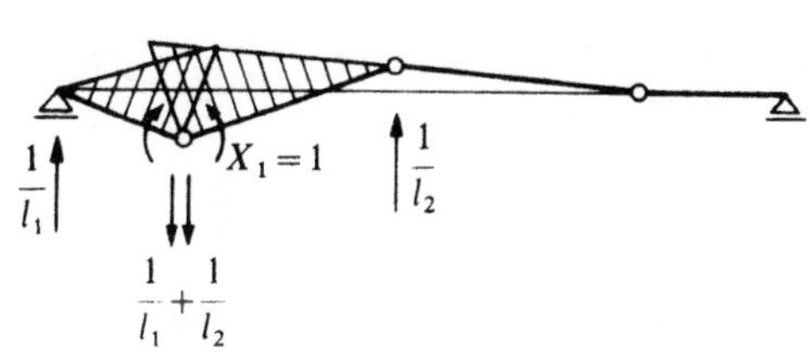

Figure 8.1b.

For the load condition $X_2 = 1$:

$$\delta_{22} = \overbrace{\frac{l_2 + l_3}{3EI}}^{\delta_{c22}} + \overbrace{\frac{1}{l_2^2} c_1 + \left(\frac{1}{l_2} + \frac{1}{l_3}\right) c_2 \left(\frac{1}{l_2} + \frac{1}{l_3}\right) + \frac{1}{l_3^2} c_3}^{\delta_{e22}},$$

$$\delta_{23} = \delta_{32} = \overbrace{\frac{l_3}{6EI}}^{\delta_{c23}} + \overbrace{\frac{1}{l_3}\frac{c_2}{l_3} - \frac{1c_3}{l_3 l_3} - \left(\frac{1}{l_3} + \frac{1}{l_4}\right)\frac{c_3}{l_4}}^{\delta_{e23}}.$$

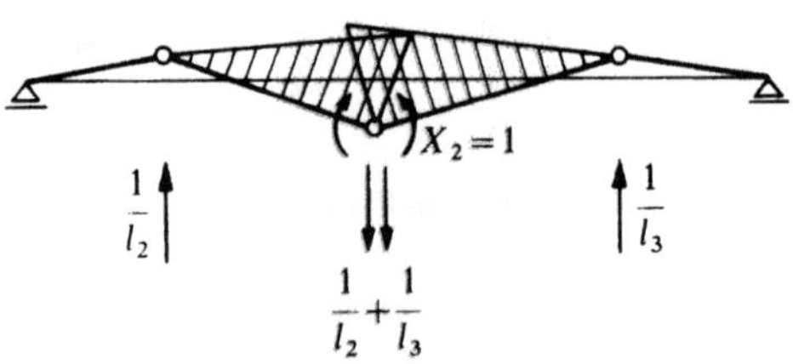

Figure 8.1c.

For the load condition $X_3 = 1$:

$$\delta_{33} = \overbrace{\frac{l_3 + l_4}{3EI}}^{\delta_{c33}} + \overbrace{\frac{1}{l_3^2} c_2 + \left(\frac{1}{l_3} + \frac{1}{l_4}\right) c_3 \left(\frac{1}{l_3} + \frac{1}{l_4}\right)}^{\delta_{e33}}$$

$$\delta_{13} = \delta_{31} = \overbrace{0}^{\delta_{c13}} + \overbrace{\frac{1}{l_3} c_2 \left(\frac{1}{l_2} + \frac{1}{l_3}\right)}^{\delta_{e13}}.$$

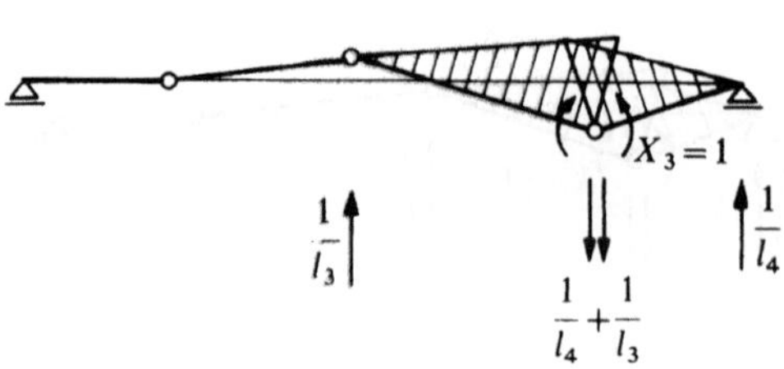

Figure 8.1d.

$$\alpha_{11} = \frac{\delta_{c11}}{\delta_{c11} + \delta_{e11}}, \qquad \alpha_{12} = \frac{\delta_{c12}}{\delta_{c12} + \delta_{c12}},$$

$$\alpha_{22} = \frac{\delta_{c22}}{\delta_{c22} + \delta_{e22}}, \qquad \alpha_{23} = \frac{\delta_{c23}}{\delta_{c23} + \delta_{e23}}, \qquad \alpha_{33} = \frac{\delta_{c33}}{\delta_{c33} + \delta_{e33}}.$$

For the load condition support settlement Δ_2:

$$\delta_{10} = -\frac{\Delta_2}{l_2},$$

$$\delta_{20} = \frac{\Delta_2}{l_2} + \frac{\Delta_2}{l_3},$$

$$\delta_{30} = -\frac{\Delta_2}{l_3}.$$

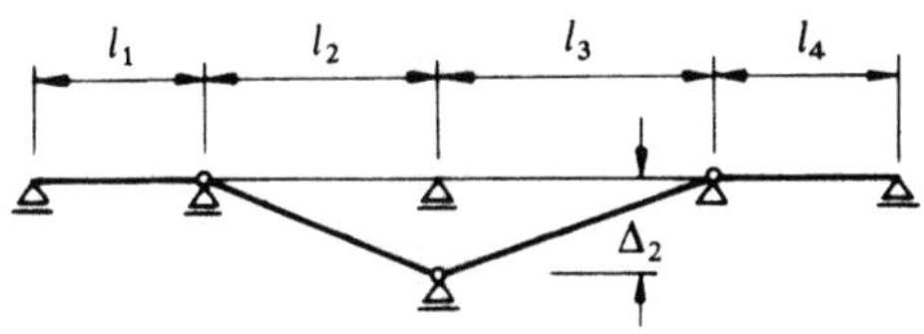

Figure 8.1e.

8.2.1. *Analysis by the Method of Mean Creep-Inducing Action Effects*

The action effects for the initial condition are obtained in the usual manner with the following deformation matrix:

$$\begin{aligned} X_{10}\delta_{11} + X_{20}\delta_{12} + X_{30}\delta_{13} &= \delta_{10}, \\ X_{10}\delta_{12} + X_{20}\delta_{22} + X_{30}\delta_{23} &= \delta_{20}, \\ X_{10}\delta_{13} + X_{20}\delta_{23} + X_{30}\delta_{33} &= \delta_{30}. \end{aligned} \tag{8.1}$$

The action effects for the final condition are determined on the basis of the same matrix [Eq. (8.1)]. Each individual term must be supplemented by only two additional components which take into account the elastic deformation $(X_t - X_0)\delta$ accompanying the change $X_t - X_0$, as well as the creep deformation $\frac{1}{2}(X_t + X_0)\delta_c\varphi$. In this simple formulation, however, the influence of delayed elasticity is still neglected (cf. Section 2.3). This results in the simple rule that each term $X_0\delta$ appearing in Eq. (8.1) must be replaced by the following expression:

$$X_0\delta + (X_t - X_0)\delta + \frac{X_t + X_0}{2}\delta_c\varphi \tag{8.2}$$

Thus, the following deformation matrix results for the final condition:

$$X_{10}\delta_{11} + (X_{1t} - X_{10})\delta_{11} + \frac{X_{1t} + X_{10}}{2}\delta_{c11}\varphi + X_{20}\delta_{12} + (X_{2t} - X_{20})\delta_{12}$$

$$+ \frac{X_{2t} + X_{20}}{2}\delta_{c12}\varphi + X_{30}\delta_{13} + (X_{3t} - X_{30})\delta_{13} + \frac{X_{3t} + X_{30}}{2}\delta_{c13}\varphi$$

$$= \delta_{10}(1 + \varphi), \tag{8.3a}$$

$$X_{10}\delta_{12} + (X_{1t} - X_{10})\delta_{12} + \frac{X_{1t} + X_{10}}{2}\delta_{c12}\varphi + X_{20}\delta_{22} + (X_{2t} - X_{20})\delta_{22}$$

$$+ \frac{X_{2t} + X_{20}}{2}\delta_{c22}\varphi + X_{30}\delta_{23} + (X_{3t} - X_{30})\delta_{23} + \frac{X_{3t} + X_{30}}{2}\delta_{c23}\varphi$$

$$= \delta_{20}(1 + \varphi), \tag{8.3b}$$

$$X_{10}\delta_{13} + (X_{1t} - X_{10})\delta_{13} + \frac{X_{1t} + X_{10}}{2}\delta_{c13}\varphi + X_{20}\delta_{23} + (X_{2t} - X_{20})\delta_{23}$$

$$+ \frac{X_{2t} + X_{20}}{2}\delta_{c23}\varphi + X_{30}\delta_{33} + (X_{3t} - X_{30})\delta_{33} + \frac{X_{3t} + X_{30}}{2}\delta_{c33}\varphi$$

$$= \delta_{30}(1 + \varphi). \tag{8.3c}$$

This system of linear equations is also to be solved in the usual manner for X_{1t}, X_{2t}, and X_{3t} after substituting for X_{10}, X_{20}, and X_{30} the values determined for the initial condition.

8.2.2. *Computation by the Methods of Trost or Bažant*

For this method, too, the matrix for the final condition can be used directly. One merely has to take into account the formulation for the effect of creep which is different from the method described in Section 8.2.1. The fundamental equation (4.1) given in Appendix II, Section 4.1, reads:

$$f_0(1 + \varphi_t) + (f_t - f_0)(1 + \rho\varphi_t) = E_c\varepsilon_t$$

In terms of action effects, one gets by analogy:

$$X_0(\delta_1 + \delta_{c1}\varphi) + (X_t - X_0)\delta_1 + (X_t - X_0)\delta_c\rho\varphi_t = \delta_0(1 + \varphi). \tag{8.4}$$

In order to obtain the new matrix for the final condition, one need merely supplement each term $X_0\delta$ in Eq. (8.1) in accordance with Eq. (8.4).

8.3. Elastically Supported Symmetrical Three-Span Continuous Beam

The symmetry of the structure permits considerable simplification of the computation by breaking down the loads into a symmetrical case (1) and an asymmetrical case (2).

8.3.1. Solution with Mean Creep-Inducing Action Effects

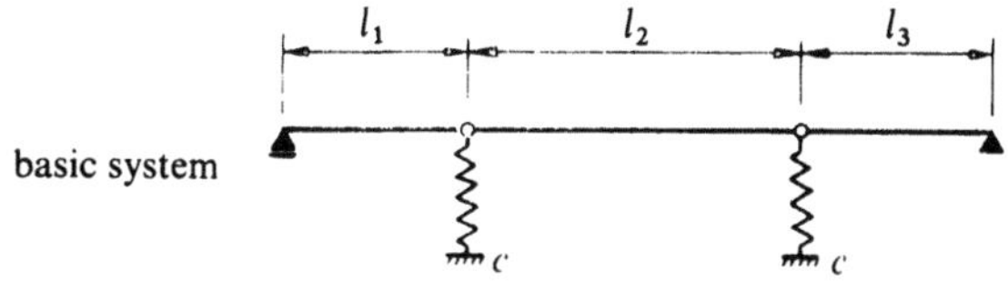

Figure 8.2a.

For load condition $X_1 = 1$:

$$\delta_{11} = \delta_{c11} + \delta_{e11},$$

$$\delta_{12} = \delta_{21} = 0,$$

$$\alpha_{11} = \delta_{c11}/\delta_{11}.$$

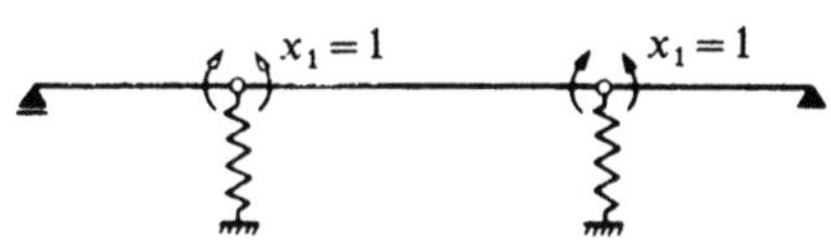

Figure 8.2b.

For load condition $X_2 = \pm 1$:

$$\delta_{22} = \delta_{c22} + \delta_{e22}$$

$$\alpha_{22} = \frac{\delta_{c22}}{\delta_{22}}$$

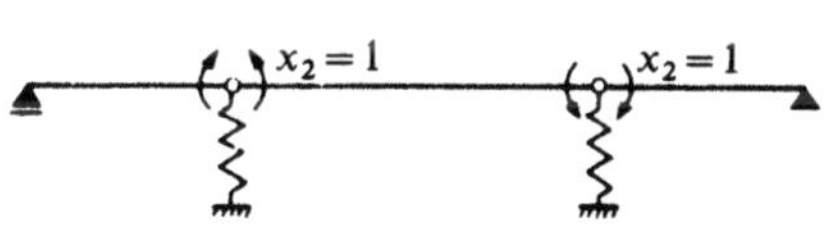

Figure 8.2c.

For the load condition of differential settlement of supports:

$$\delta_{10} = \frac{-(\Delta_1 + \Delta_2)}{l_1},$$

$$\delta_{20} = -(\Delta_1 - \Delta_2)\left(\frac{1}{l_1} + \frac{2}{l_2}\right).$$

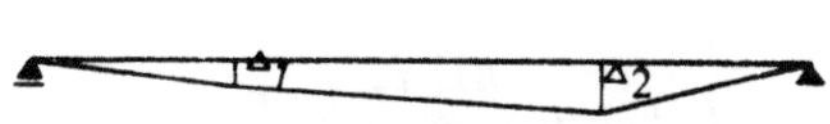

Figure 8.2d.

Since this is a sudden support settlement, the right-hand side of Eq. (8.3) is merely δ_{10}, and we obtain from Eqs. (8.2) and (8.3)

$$X_{10}\delta_{11} + (X_{1t} - X_{10})\delta_{11} + \frac{X_{1t} + X_{10}}{2}\delta_{c11}\varphi = \delta_{10},$$

$$X_{1t} + \frac{X_{1t} + X_{10}}{2}\alpha_{11}\varphi = \frac{\Delta_1 + \Delta_2}{l_1\delta_{11}} = X_{10},$$

$$X_{1t} = X_{10}\frac{1 - \alpha_{11}(\varphi/2)}{1 + \alpha_{11}(\varphi/2)}. \tag{8.5}$$

By the same approach, we get, for X_{2t},

$$X_{2t} = X_{20}\frac{1 - \alpha_{22}(\varphi/2)}{1 + \alpha_{22}(\varphi/2)}. \tag{8.6}$$

It was to be expected that Eqs. (8.5) and (8.6) would have exactly the same form as Eq. (2.30) given in Section 2.3. Inasmuch as $\alpha\varphi > 0.5$, they are replaced by the more accurate Eq. (2.29a):

$$X_{1t} = X_{10}\frac{1.+ 0.6\alpha_{11} - 0.5\alpha_{11}\varphi}{1.0 + 0.6\alpha_{11} + 0.08\alpha_{11}^2 + \alpha\varphi(0.5 + 0.2\alpha_{11})}, \tag{8.7}$$

$$X_{2t} = X_{20}\frac{1 + 0.6\alpha_{22} - 0.5\alpha_{22}\varphi}{1.0 + 0.6\alpha_{22} + 0.08\alpha_{22}^2 + \alpha\varphi(0.5 + 0.2\alpha_{22})}. \tag{8.8}$$

Numerical Example

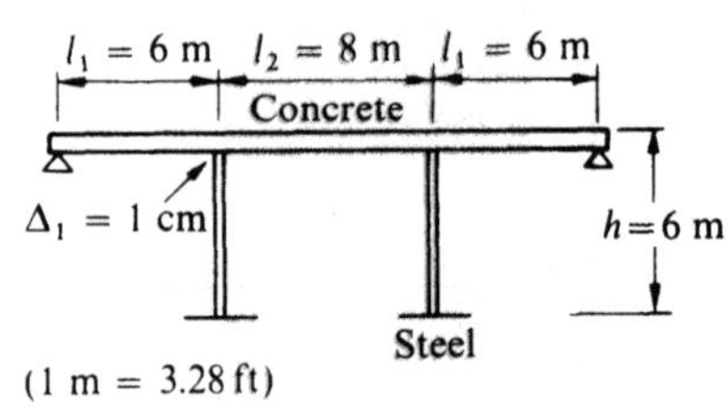

Figure 8.3

$$I_c = 0.015\ \mathrm{m}^4,\ \varphi = 1.5$$

$$E_c = 37\,000\ \mathrm{N/mm}^2,$$

$$A_{\mathrm{steel}} = 0.003\ \mathrm{m}^2,$$

$$E_s = 210\,000\ \mathrm{N/mm}^2,$$

$$E_c I_c c = \frac{hE_cI_c}{A_{\mathrm{steel}}E_s} = 5.27$$

$$10E_cI_c\delta_{11} = \frac{2l_1}{3} + l_2 + 2\frac{c}{l_1^2}E_cI_c$$

$$= 12.00 + 0.29 = 12.29$$

$$\alpha_{11} = \frac{\delta_{1c}}{\delta_{11}} = \frac{12.00}{12.29} = 0.98,$$

$$10E_cI_c\delta_{22} = \frac{2l_1}{3} + \frac{l_2}{6} + 2\left(\frac{1}{l_1} + \frac{2}{l_2}\right)\left(\frac{c}{l_1} + \frac{2c}{l_2}\right)E_cI_c$$

$$= 5.33 + 1.83 = 7.16,$$

$$\alpha_{22} = \frac{\delta_{2c}}{\delta_{22}} = \frac{5.33}{7.16} = 0.75,$$

$$E_cI_c\delta_{10} = E_cI_c\frac{\Delta_1}{l_1} = 92, \qquad X_{10} = \frac{\delta_{10}}{\delta_{11}} = 75 \text{ kNm},$$

$$E_cI_c\delta_{20} = E_cI_c\Delta_1\left(\frac{1}{l_1} + \frac{2}{l_2}\right) = 232, \qquad X_{20} = 324.0 \text{ kNm}.$$

Eq. (8.5) is no longer valid when $\alpha_{11}\varphi = 0.98(1.5) = 1.47 > 0.5$ (cf., Section 2.3). It would yield too small a value for X_{1t}. Using Eq. (3.17) from Appendix II,

$$X_{1t} = 75\left(\frac{1 - 1.47/4}{1 + 1.47/4}\right)^2 = 16.1 \text{ kNm},$$

a value that is still too small because it disregards the delayed elasticity.

It is only Eq. (8.7), derived from Eq. (2.29), that yields the more accurate value:

$$X_{1t} = 75\,\frac{1 + 0.6(0.98) - 0.5(0.98)1.5}{1.0 + 0.6(0.98) + 0.08(0.98)^2 + 0.98(1.5)[0.5 + 0.2(0.98)]} = 24.0 \text{ kNm}.$$

Using Eq. (8.8):

$$X_{2t} = 324.0\,\frac{1 + 0.6(0.75) - 0.5(0.75)1.5}{1.0 + 0.6(0.75) + 0.08(0.75)^2 + 0.75(1.5)[0.5 + 0.2(0.75)]}$$

$$= 130.0 \text{ kNm}.$$

8.3.2. *Solution According to Trost*

Using the displacements and stiffness ratios α derived in Section 8.2, one gets, on the basis of Eq. (8.4):

$$X_{10}\delta_{11} + X_{10}\delta_{c11}\varphi + (X_{1t} - X_{10})\delta_{11} + (X_{1t} - X_{10})\delta_{c11}\rho\varphi = \delta_{10},$$

$$X_{20}\delta_{22} + X_{20}\delta_{c22}\varphi + (X_{2t} - X_{20})\delta_{22} + (X_{2t} - X_{20})\delta_{c22}\rho\varphi = \delta_{20},$$

and from this

$$X_{1t} = X_{10}\left(1 - \frac{\alpha_{11}\varphi}{1 + \alpha_{11}\rho\varphi}\right) = X_{10}\frac{1 - \alpha_{11}(1 - \rho)\varphi}{1 + \alpha_{11}\rho\varphi}, \tag{8.9}$$

$$X_{2t} = X_{20}\left(1 - \frac{\alpha_{22}\varphi}{1 + \alpha_{22}\rho\varphi}\right) = X_{20}\frac{1 - \alpha_{22}(1 - \rho)\varphi}{1 + \alpha_{22}\rho\varphi}. \tag{8.10}$$

Numerical Example; Assumptions as in Section 8.3.1

In accordance with Eq. (8.9), we have for $\rho = 0.80$ (cf., Section 2.2)

$$X_{1t} = 75\,\frac{1 - 0.98(0.20)1.5}{1 + 0.98(0.80)1.5} = 24.2 \text{ kNm}.$$

From Eq. (8.10):

$$X_{2t} = 324.0\,\frac{1 - 0.75(0.20)1.5}{1 + 0.75(0.80)1.5} = 133.0 \text{ kNm}.$$

This example shows very clearly the advantage of the very simple Trost formula, which yields entirely satisfactory results for constraints up to about $\alpha\varphi = 3$.

8.3.3. Comparison of Results Obtained by Various Methods

Table 8.1. Moments in kNm and in % of $M_{t=0}$.

	$t = 0$	From Eqs. (8.7) and (8.8)	From Trost
X_1	75	24.0 kNm = 32%	24.2 kNm = 32%
X_2	324	130.0 kNm = 40%	133.0 kNm = 41%

9. Reduction of Constraints which Are Imposed through Rigid or Near-Rigid Supports

9.1. Introduction

9.1.1. The Constraint which Develops Suddenly

Most constraints of this type are unintentional, such as the result of non-uniform foundation settlements. They are very markedly reduced by creep. This is desirable in the case of settlement, but it is undesirable in those cases where the constraint was created on purpose in order to improve the distribution of action effects. For this reason, an artificially created constraint is used only in exceptional cases when the supports are near rigid.

Today, for instance, we have abandoned the technique of prestressing the crown of fixed end arches in order to reduce bending moments. There are cases, however, where the development of such a constraint is unavoidable, as shown in the example in Section 9.4 with prefabricated members.

Constraint which acts from rigid supports is reduced rapidly. This is illustrated by the following examples: In reinforced concrete a tension member acts similar to a rigid support if its cross-section is very large. In this case, $\delta_{e11} \sim 0$ and $\alpha = 1$. If this value is entered in Fig. 2.2, it becomes apparent that even for $\varphi = 2$, about $\frac{3}{4}$ of the original constraint is lost.

However, conditions are entirely different when the artificially created constraint acts against elastic supports, as is the case for prestressing with high-strength tendons. Then, the order of magnitude of α is about 0.05, and for $\varphi = 2$, the loss is only on the order of 10 %.

9.1.2. The Gradually Developing Constraint

When the constraint develops only gradually, as is the case for example, for foundation settlement in cohesive soils, it is not reduced as much. If it is assumed, for example, that the time development of constraint is similar to that of creep and shrinkage, Fig. 2.3 indicates that for $\alpha = 1$ and $\varphi = 2$, the original value is reduced only by 57 %.

The same is true, of course, when the constraint is induced by shrinkage. Because of their great importance, shrinkage problems are dealt with separately in Section 10.

9.2. The Consequences of a Fast Developing Constraint as Illustrated by the Example of Foundation Settlement in Non-cohesive Soils

A constraint is generated only by the non-uniform components of settlement or displacement of foundations. The associated action effects have the value X_{r0} at the time $t = 0$, and they are reduced in accordance with the equation $X_{rt} = \gamma_r X_{r0}$. The extent to which this takes place depends on the creep coefficient φ and the stiffness ratio α. One can almost always assume the soil to be near rigid (the effect of soil characteristics can, however, just as readily be taken into account in X_{r0}, see [17]) and therefore $\alpha = 1$ can be set in accordance with the information given in Section 9.1. The initially high constraint is then reduced rapidly.

Figure 9.1 shows an example which is valid for $2A/u = 20$ cm and a relative humidity of 40 % or 90 %, with the provision that the suddenly occurring constraint

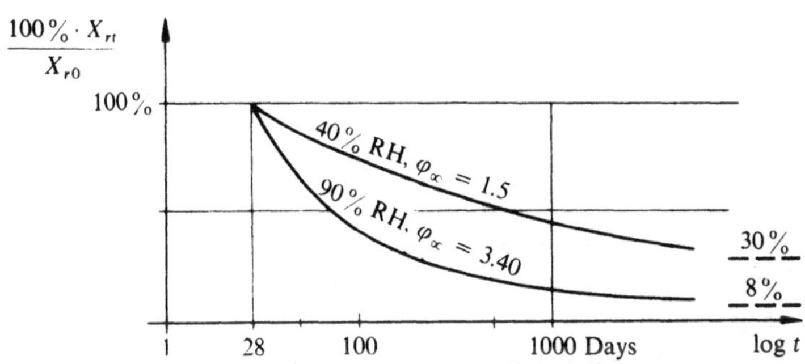

Figure 9.1.

(e.g., a foundation settlement in non-cohesive soil) starts to act on concrete that is 28 days old.

9.3. The Consequences of a Gradually Developing Constraint as Illustrated by the Example of Foundation Settlement in Cohesive Soil

9.3.1. Overview

Appendix II, Section 7.3 explains how to proceed in such cases. It is essential, above all, that the maximum value of the constraint no longer occurs at time $t = 0$ as would be the case in non-cohesive soil. When the support settlement develops more slowly than creep, the maximum value occurs at $t = \infty$. If it develops more rapidly, intermediate values are high, but they decrease again markedly for large values of t.

One disadvantage of the mothods of Dimel, Trost, and Mainz, described in Appendix II, is the fact that one is forced to mathematically couple the time-dependent development of the support settlement—which is a function of foundation dimensions and of the compositions of the subsoil—with the course of creep—which is affected by entirely different factors, such as the type of cement, climatic conditions, and dimensions of the structure. Appendix II, Section 7.3 calls attention to the problems arising from this contradiction. Such a conflict can be avoided by using the method of the mean creep-inducing action effects described in Section 2.3. This approach is applied below to derive coefficients which, by corresponding more closely to reality, lead to the solution as rapidly as the previous methods.

These techniques, which are described in Section 9.3.3–9.3.6, are illustrated by examples in which the settlement diminishes gradually in accordance with the expression $1 - e^{0.0069t}$ (t in days). The relative humidity is about 70%, $2A/u$ measures 10, 20, and 80 cm, and the loading age is 28 days. The corresponding values of φ are 2.44, 2.34, and 2.10. The results are subsequently compared.

9.3.2. Solution Obtained on the Assumption that the Development of Constraint and Creep in Time Are Similar

In many cases, one can assume as a simplification that settlement and flow develop similarly over time. Then

$$X_{rt} = \gamma_s X_{rt}^{(e)}, \tag{9.1}$$

where

$$X_{rt}^{(e)} = -\frac{\delta_{1rt}}{\delta_{11}}. \tag{9.2}$$

The auxiliary term $X_{rt}^{(e)}$ corresponds to the action effects computed by disregarding the effect of creep. The ratio γ_s can be taken from Fig. 2.3.

If, as in Fig. 9.1, one assumes that the structural members affected by the settlement are 28 days old when the process starts, and that the same exposure conditions are valid, one then obtains the curve shown in Fig. 9.2 for the constraint induced by the foundation settlement. Comparing it with Fig. 9.1, it becomes clear

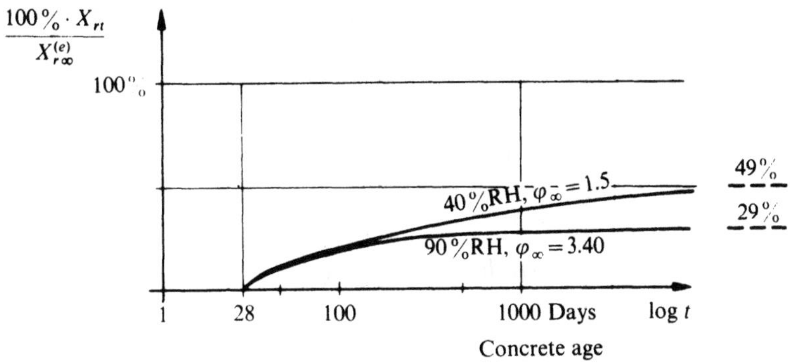

Figure 9.2.

that the consequences of a given settlement are considerably greater when the settlement develops gradually rather than suddenly.

9.3.3. Solution by the Dimel Method

It can be seen from Fig. 7.2 of Appendix II that for the conditions assumed in the examples of Section 9.3.1 [i.e., $\varphi = 2.44$ (or 2.34 or 2.10) and $\lambda \sim 6$), the maximum value for the constraint occurs after 4 (or 6 or 8, respectively) months and measures 0.68 (or 0.70 or 0.72, respectively) $\times X_{r\infty}^{(e)}$. Subsequently, the constraint decreases again and at time $t = \infty$ it reaches 0.20 (or 0.23 or 0.25, respectively) $\times X_{r\infty}^{(e)}$. Here, $X_{r\infty}^{(e)}$ is the constraint computed by disregarding creep and is therefore identical to the value that would result from sudden settlement at time $t = 0$.

It should be kept in mind that the coefficient c underlying Fig. 7.2 in Appendix II gives a result which, in accordance with Dimel, is approximately valid only for very thick structural elements (cf., Appendix II, Section 7.3).

9.3.4. Solution by the Trost and Mainz Method

For this method described in Appendix II, Section 7.3.2, auxiliary values were developed which make it possible to take into account concrete age at the onset of settlement on the basis of the CEB–FIP recommendations of 1970. For $\beta_d = 1$ (loading age equal to 28 days and $\varphi_N = \varphi_\infty/\beta_d = \varphi_\infty$) and the coefficient $\bar{\lambda} \sim 1.5$ ([7], Fig. 3) characterizing the assumed settlement curve, one obtains from the table (not reproduced here) of ρ values ([7], Table I) for $\varphi_N = 2.44$ (or 2.34 or 2.10, respectively) the values $\rho = 0.79$ (or 0.78 or 0.77, respectively) and thus:

$$X_{rt} = X_{rt}^{(e)} \frac{1}{1 + \rho\varphi_t} = f(t)^{\bar{\lambda}} \frac{X_{r\infty}^{(e)}}{1 + \rho\beta_d\varphi_N f(t)}, \tag{9.3}$$

$$X_{r\infty} = X_{r\infty}^{(e)} \frac{1}{1 + 0.79(1.0)2.44} = 0.34X_{r\infty}^{(e)} \text{(or } 0.36\,X_{r\infty}^{(e)} \text{ or } 0.38\,X_{r\infty}^{(e)}\text{, respectively).}$$

The development in time can be determined from the equation:

$$\frac{X_{rt}}{X_{r\infty}^{(e)}} = f(t)^{\bar{\lambda}} \frac{1}{1 + \rho\beta_d\varphi_N f(t)} \tag{9.4}$$

It is plotted in [7] for $\varphi_N = 2.0$ and shows that the maximum value for $\bar{\lambda} > 1$ occurs at time $t = \infty$.

The difference from the result obtained by the Dimel method is due only partially to the fact that the tables developed by Trost and Mainz and their graphic representations are still based on the old CEB–FIP provisions. It is much more significant that the computational assumptions of Trost and Mainz are for thin structural members, while those of Dimel are for thick members.

9.3.5. Application of the Method of the Mean Creep-Inducing Action Effects

Application of this method obviates the need to couple settlement and creep rates by a preset function. The settlement curve can be freely chosen and the creep curve can be taken into account to the extent that it is controlled by the effective thickness of the structure and its effective age at the time the constraint begins. The following example for such a numerical computation is valid for average conditions: $2A/u = 20$ cm; relative humidity 70%; Type II cement, onset of constraint at age $t_1 = 28$ days; the settlement curve is characterized by a half-life $T \sim 100$ days (for definition, see Section 9.3.6). This corresponds to the second example referred to in Section 9.3.1. Figure 9.3 shows the time development of a and φ_f. With regard to φ_d, it is assumed that the delayed elastic strain in each step occurs immediately. The influence of the increase of the modulus of elasticity E_c with age is taken into account, although it is not very large when the constraint starts only at a concrete age of 28 days. In order to obtain relative values, Eq. (2.31) is used as follows:

$$\frac{S_{ri}}{S_{r\infty}^{(e)}} = \frac{C_{mi} - 0.5\Delta\varphi_{fi}}{C_{mi} + 0.5\Delta\varphi_{fi}} \frac{S_{r(i-1)}}{S_{r\infty}^{(e)}} + \frac{\Delta a_{ri}/(a_{r\infty}\alpha)}{C_{mi} + 0.5\Delta\varphi_{fi}}, \tag{9.5}$$

$$S_{r\infty}^{(e)} = \frac{a_{r\infty}}{\delta_{11}^{28}}; \quad \text{since } \alpha = 1, \text{ we have } C_m = 0.5\left(\frac{1}{\beta_{e(i-1)}} + \frac{1}{\beta_{ei}}\right) + 0.4 \approx \frac{1}{\beta_{em}} + 0.4.$$

In Table 9.1 the computation is carried out in 6 steps. The values used for $\Delta\varphi_f$ and Δa were taken from Fig. 9.3.

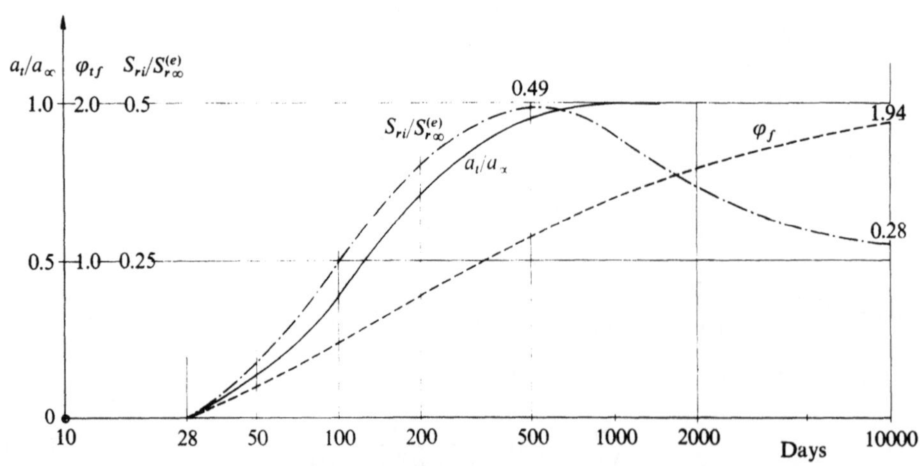

Figure 9.3. Development of creep, settlement $f \triangleq a$, and action effects for $d_{\text{eff}} = 300$ mm, (11.8 in.) $RH = 70\%$.

Table 9.1 $2A/u = 20$ cm; $d_{\text{eff}} = 30$ cm; relative humidity = 70%.

Days	φ_f	$\Delta\varphi_f$	a_t/a_∞	$\Delta a_t/a_\infty$	β_e	β_{em}	$\frac{1}{\beta_{em}}$	C_m	Equation (9.5)
28	0		0		1.000				
		0.20		0.13		1.017	0.98	1.38	$\frac{0.13}{1.38 + 0.10} = 0.088$
50	0.20		0.13		1.035				
		0.28		0.26		1.060	0.95	1.35	$\frac{1.35 - 0.14}{1.35 + 0.14} 0.088 + \frac{0.26}{1.35 + 0.14} = 0.246$
100	0.48		0.39		1.085				
		0.29		0.30		1.090	0.92	1.32	$\frac{1.32 - 0.145}{1.32 + 0.145} 0.246 + \frac{0.30}{1.32 + 0.145} = 0.400$
200	0.77		0.69		1.095				
		0.33		0.26		1.106	0.91	1.31	$\frac{1.31 - 0.165}{1.31 + 0.165} 0.400 + \frac{0.26}{1.31 + 0.165} = 0.487$
500	1.10		0.95		1.118				
		0.48		0.05		1.123	0.89	1.29	$\frac{1.29 - 0.24}{1.29 + 0.24} 0.487 + \frac{0.05}{1.29 + 0.24} = 0.368$
2000	1.58		1.00		1.129				
		0.36		0.0		1.130	0.88	1.28	$\frac{1.28 - 0.18}{1.28 + 0.18} 0.368 = 0.277$
∞	1.94		1.00		1.132				

As the plotted results in Fig. 9.3 show, the maximum value occurs after about 500 days and measures $0.49S_{r\infty}^{(e)}$. The ultimate value drops to $0.28S_{r\infty}^{(e)}$. Similarly, the examples cited in Section 9.3.1 were computed for $2A/u = 10$ and 80 cm, and the results are listed in Table 9.2 (page 142).

9.3.6. Auxiliary Means Developed in Accordance with Section 9.3.5 which Permit Obviating the Need for Coupling the Development Over Time of Settlement and Creep

It is not logical to mathematically couple the time-dependent development of settlement and creep since they are functions of entirely different factors. For the settlement curve, it seems appropriate to use an exponential equation of the form

$$a = (1 - e^{-\lambda ct})a_{\infty}. \tag{9.6}$$

However, it is more effective not to define the rate of settlement by the coefficient λc, which lacks any visual meaning, but rather by the half-life $T = 0.693(\lambda c)$. In his graphs, Dimel included half-times from 3 to 120 months. The T values considered by Trost, in his tables, range from a few days to a half-year for thin structural elements, but from a few days to 5 years for thick ones. However, the T values which are dependent only on the nature of the soil and foundations should always be chosen in regular intervals. The following half times appear appropriate:

$$T = 0, 3, 15, 30, 180, 360, \text{ and } 1800 \text{ days.}$$

For a structure subjected to settlement, other parameters must be considered:

The effective age when the constraint begins.

The effective thickness of the structural elements subjected to creep, which determines the development of creep over time.

The flow coefficient which depends on relative humidity.

The development over time of delayed elasticity which, for small values of T, has a strong influence on the maximum action effects.

The influence of concrete age on the modulus of elasticity of the concrete.

The numerical evaluations which take into account this multiplicity of influences were carried out by computers, based on Eq. (2.31), but with a very large number of steps ($\Delta\varphi_f < = 0.05$).

The modulus of elasticity E_c, which varies between steps, was taken into account by the factor β_e in accordance with Fig. 1.1. By contrast with the approximation in Appendix II, Section 3.2.4 (where delayed elasticity was fully applied in the first step), the actual course of delayed elasticity was taken into account here. A detailed description of the program is on file with the Institute of Reinforced Concrete Structures of the Munich Institute of Technology. The results are plotted in Figs. 9.4–9.12. These illustrations apply to the following input data:

relative humidity in %: 100, 90, 70, 40;
$2A/u$ in cm: 10, 40, 160.

Intermediate values must be determined by interpolation. The coordinates used are the half-life T, from 0 to 2000 days, and the age t_1 when the constraint becomes effective (0–1000 days). These values can thus be entered in continuous form.

The consequences of creep are represented by families of curves. The following parameters are given

The maximum value $\max \gamma = \max X_{rt}/X_{r\infty}^{(e)}$ (cf., Figs. 9.4a–9.12a). (9.7)
The ultimate value $\gamma_\infty = X_{r\infty}/X_{r\infty}^{(e)}$ (cf., Figs. 9.4b–9.12b). (9.8)
The time max t when the value $\max \gamma$ occurs, in days (cf., Figs. 9.4c–9.12c). (9.9)

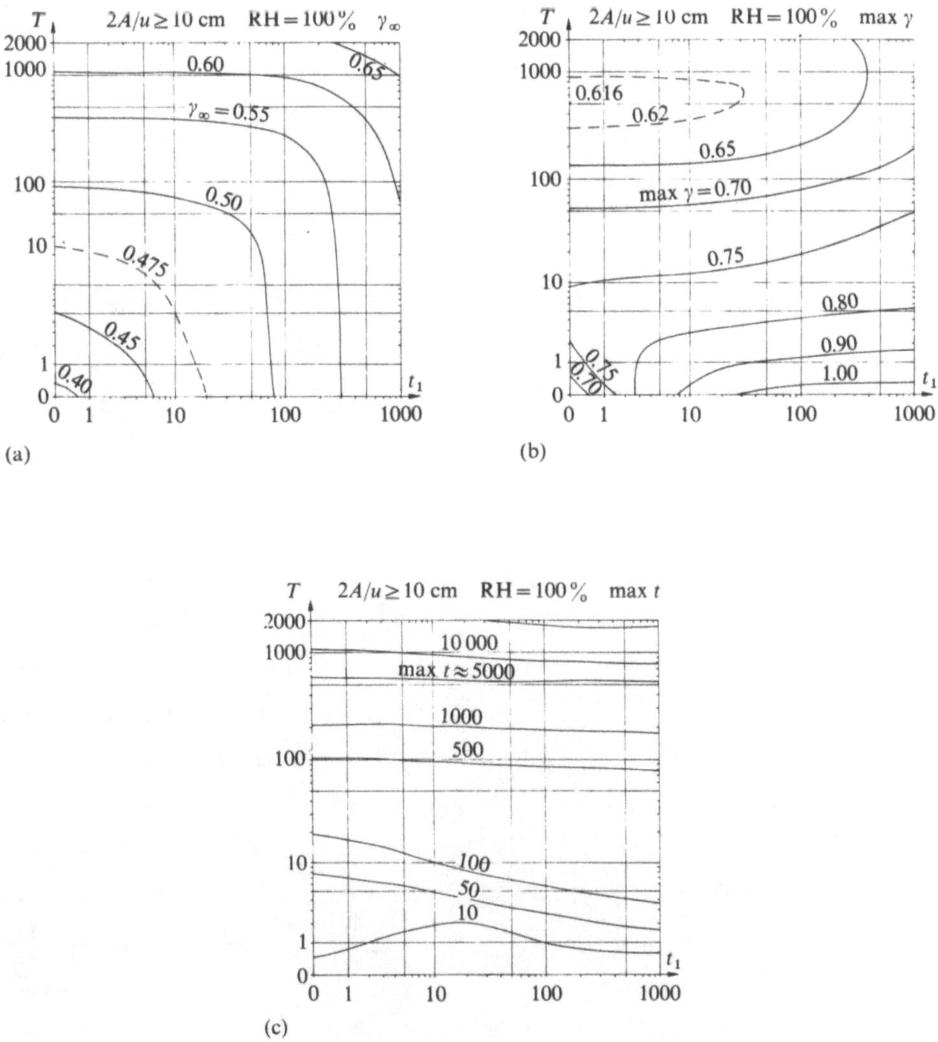

Figure 9.4.

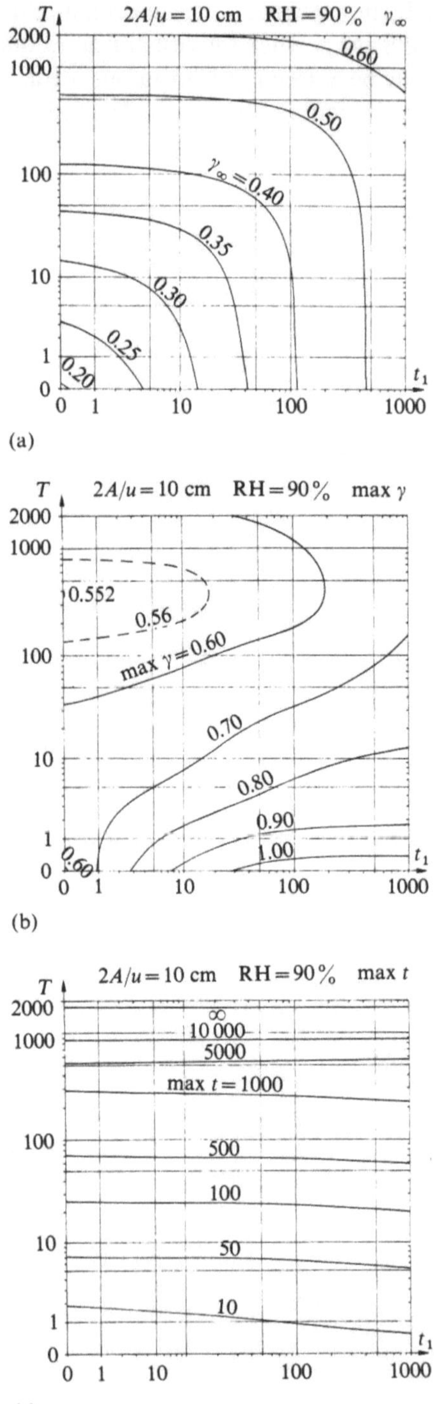

Figure 9.5.

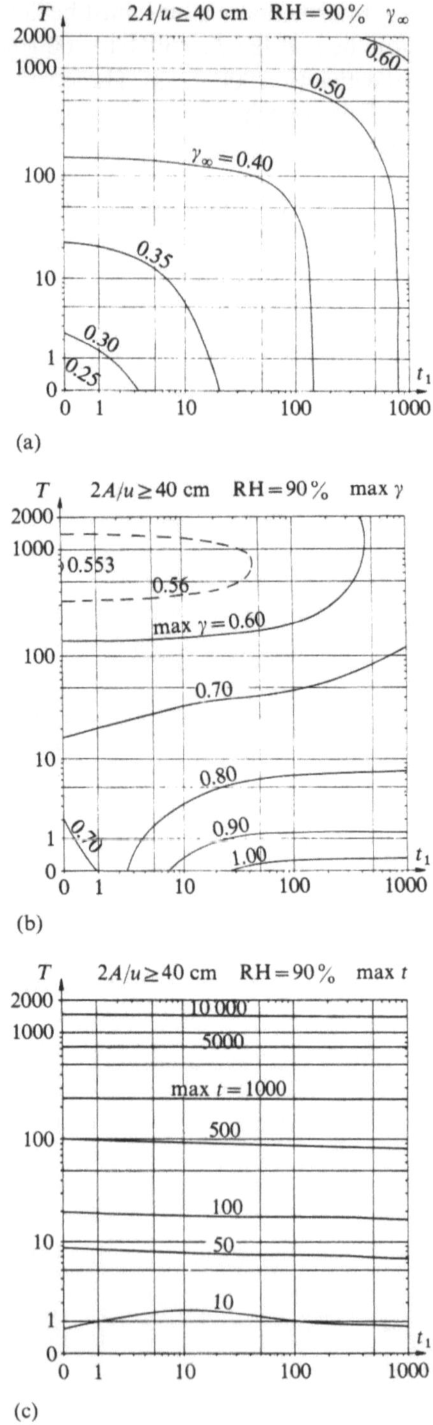

Figure 9.6.

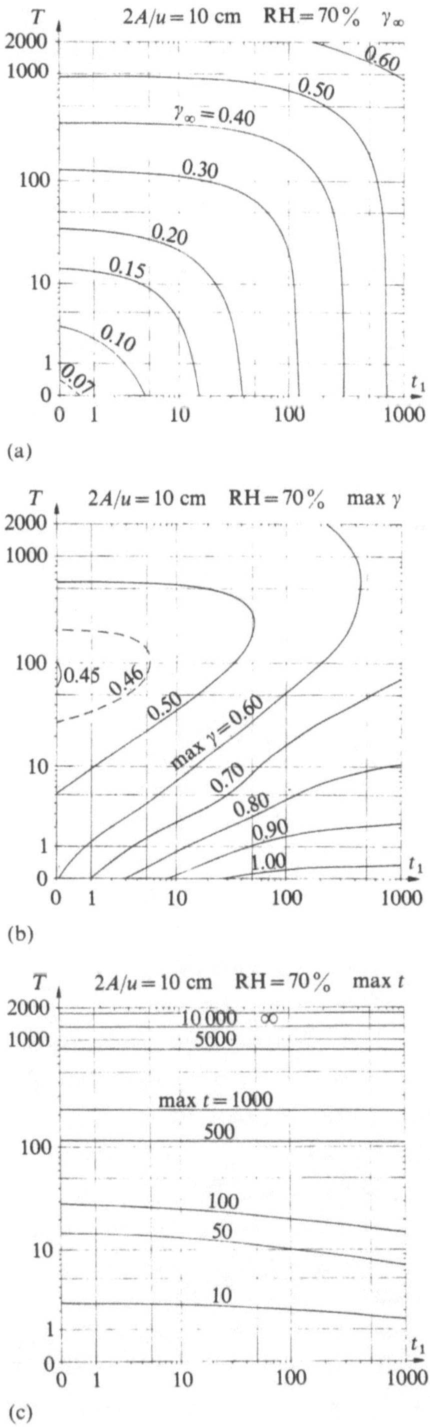

Figure 9.7.

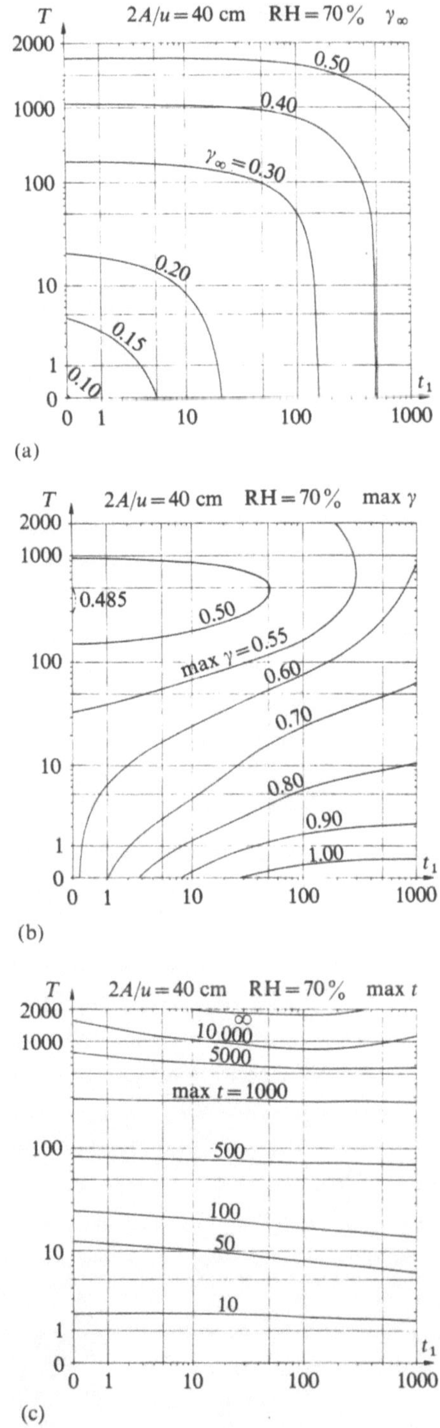

Figure 9.8.

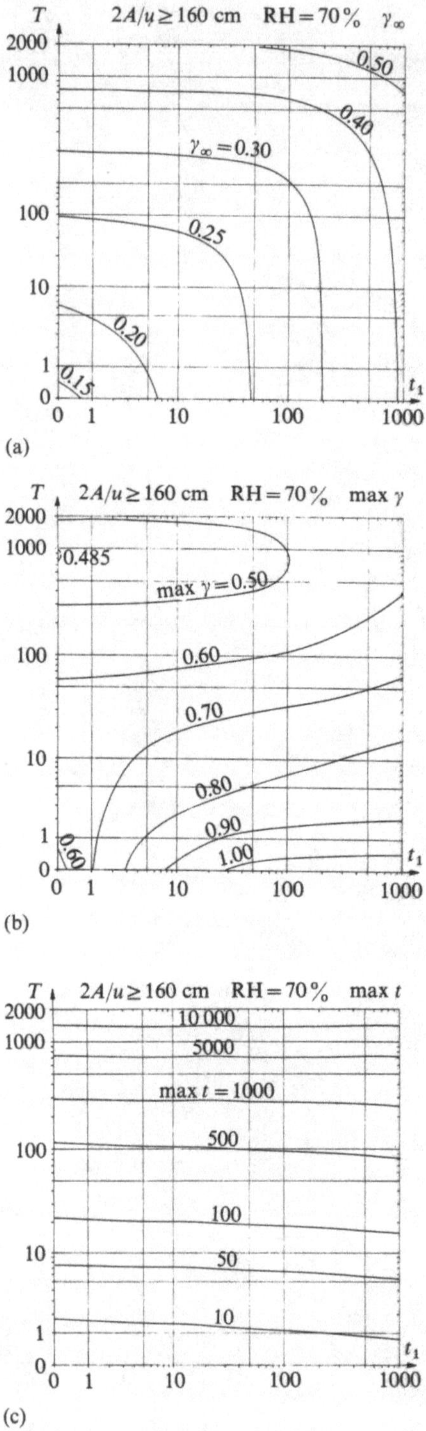

Figure 9.9.

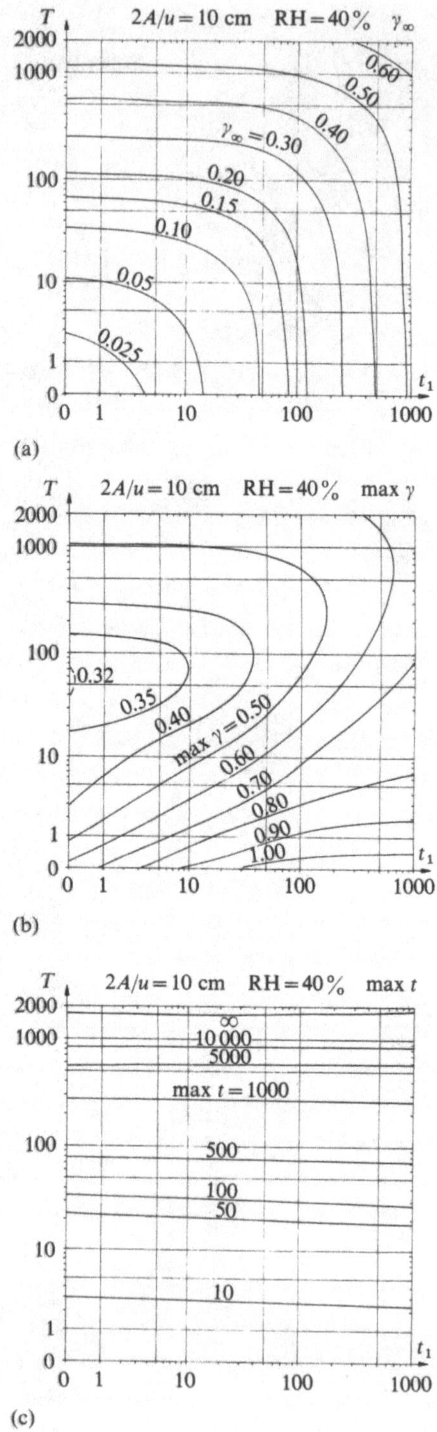

Figure 9.10.

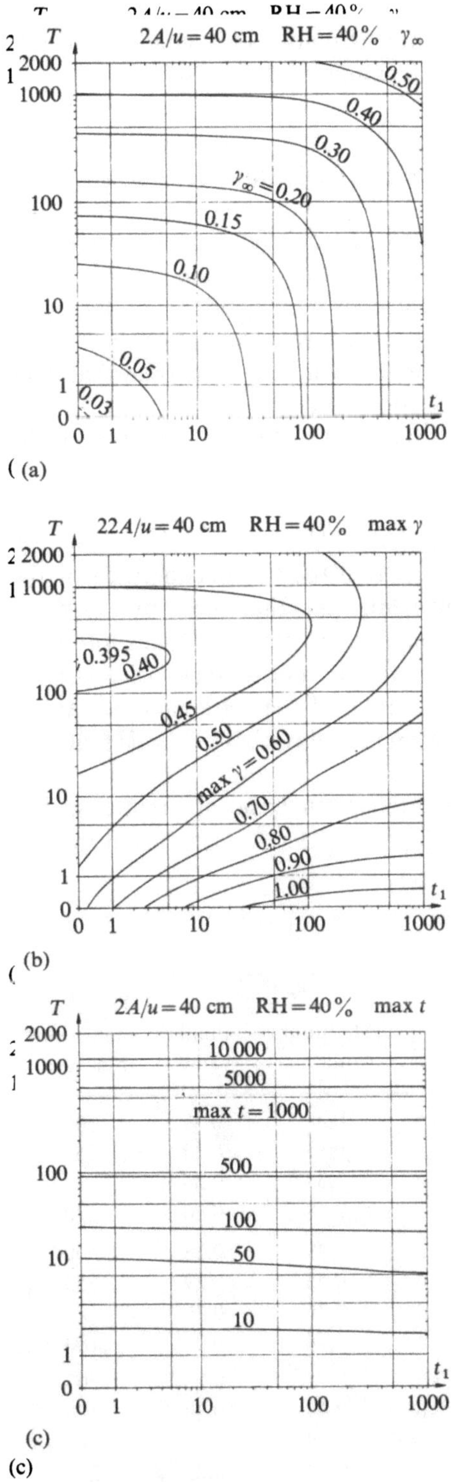

Figure 9.11.

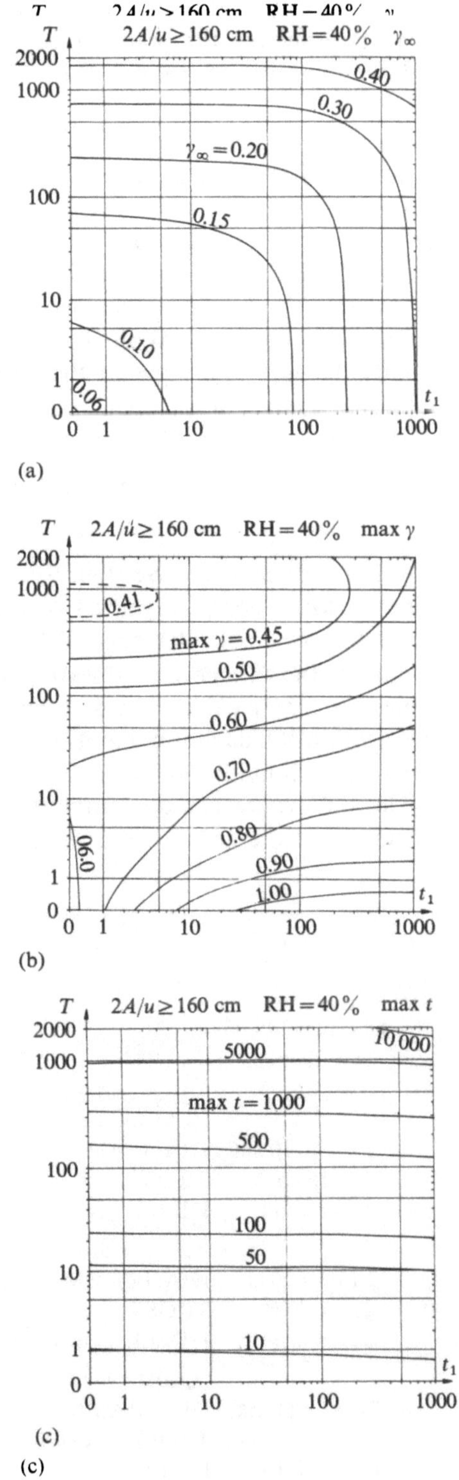

Figure 9.12.

In the form used, these results are valid only for $\alpha = 1$ and $\beta_{ce} = 1$ (Type I cement). However, they can also be used as approximations for smaller values of α, down to about 0.5, provided the following formula is adhered to:

$$\beta_{\alpha \neq 1} = \beta_{\alpha = 1}/\sqrt{\alpha}. \tag{9.10}$$

When other types of cement are used, one must take into consideration the fact that the effective concrete age is $t\beta_{ce}$. Thus, constraint starts at $t_1\beta_{ce}$ and creep develops β_{ce} times as fast. In order to compensate for this fictitious change in the time scale, one must also multiply the half-life for the constraint by β_{ce}. With $t_1\beta_{ce}$ and $T\beta_{ce}$, the data used yield correct values for max γ and γ_∞; on the other hand, $t_{\text{eff}} = \max t/\beta_{ce}$. With the choice of an appropriate half-life, this method can also be used to investigate the influence of the shrinkage constraint that does not develop over time similar to creep (cf., Section 10.4).

It should be noted that the value of $X_{r\infty}^{(e)}$ may also include the effect of elastic deformations of the soil (cf., also [17]).

9.3.7. Comparison of Results

It was already stated in Sections 9.3.3 and 9.3.4 that the values computed by the Dimel method apply to thick structural elements, and that those by the Trost method apply to slender ones. Table 9.2, therefore, particularly takes into consideration the influence of thickness ($2A/u = 10$, 20, and 80 cm). All tabulated values are predicted on the assumption that 50% of the settlement has taken place within 100 days and that its time development follows a hyperbolic function. The relative humidity is 70%.

The differences are due primarily to differences in basic assumptions. The advantage of the new analysis method developed in Section 9.3.6 lies in the fact

Table 9.2. Comparison of results.

RH = 70% Loading age = 28 days Support settlement $a_1 = a_\infty\,(1 - e^{-0.0069t})$		Section	$S_{rt}/S_{r\infty}^{(e)}$ Maximum value	Final value	Time for max. value in days
	f_m-method	9.3.5	0.46	0.30	500
$2A/u = 10$ cm	DIAGRAMM	9.3.6	0.53	0.30	550
$d = 15$ cm	TROST, $\lambda \approx 1.5$	9.3.4	0.34	0.34	∞
$\varphi\infty = 2.44$	DIMEL $\lambda \approx 6$	9.3.3	0.68	0.20	120
	f_m-method	9.3.5	0.49	0.28	500
$2A/u = 20$ cm	DIAGRAMM	9.3.6	0.55	0.29	550
$d_{\text{eff}} = 30$ cm	TROST, $\lambda \approx 1.5$	9.3.4	0.36	0.36	∞
$\varphi\infty = 2.34$	DIMEL, $\lambda \approx 6$	9.3.3	0.70	0.23	180
	f_m-method	9.3.5	0.55	0.26	500
$2A/u = 80$ cm	DIAGRAMM	9.3.6	0.59	0.28	550
$d_{\text{eff}} = 120$ cm	TROST, $\lambda \approx 1.5$	9.3.4	0.38	0.38	∞
$\varphi\infty = 2.10$	DIMEL, $\lambda = 6$	9.3.3	0.72	0.25	240

that the basis for the analysis can be better adapted to reality. The results are in good agreement with the f_m-method. Owing to the introduction of the true development of delayed elastic strain over time the maximum values are somewhat larger, since they correspond to reality.

9.4. Example Showing Coupling of Prefabricated Parts of Equal Age

The following example is intended to demonstrate that creep problems that appear very complicated can be reduced to a constraint problem by the method of load rearrangement.

Two prefabricated beams of a reinforced concrete floor system are installed as freely supported beams (time t_0) and, in this condition, must carry, in addition to their own weight, the weight of the concrete cast in situ. After hardening (time t_1), the latter transforms the system into a two-span slab whose center support moment at time t_1, however, still is $M_B = 0$. However, over the course of time, a center support moment will develop owing to the creep deformation induced by the sustained load.

The following type of computation is based on the assumption that the beams do not crack. If pronounced cracking is anticipated, its influence on the deformation parameters must be taken into account at least approximately, e.g., in accordance with Section 14.2.3 or with Appendix II, Section 8.2.5.

The problem due to the fact that during hardening of the concrete cast in situ the static system is modified can be solved in simple fashion by the method of load rearrangement. In accordance with Fig. 9.13, the moment distribution (line *a*) existing at time t_1 may be conceived as having occured in a two-span beam in which a settlement of the center support, i.e., a constraint, has caused the additional moments corresponding to line *c*. The imposed action effects thus appear as a difference between the transient condition and those action effects occuring in a continuous beam resting on a scaffold.

The moments represented by line *b* are not modified by creep, in accordance with Part A, Section 3.1.2. Section 9.2 applies with regard to reduction of the imposed moment line *c*. The support moment for dead weight at time $t = \infty$ then is

$$M_{B\infty} = M_{Bb} + \gamma_r M_{Bc} = \frac{dl^2}{8}(-1 + \gamma_r). \tag{9.11a}$$

On the basis of this formulation, action effects can be given for the ultimate state using a safety factor $\nu = 1.75$ for the load-dependent action effects and $\nu = 1.0$ for the action effects caused by constraint:

$$M_{Bu} = 1.75\ M_{Bb} + 1.0\gamma_r M_{Bc}. \tag{9.11b}$$

Assuming $\varphi = 2$, we get, in accordance with Fig. 2.2, $\gamma_r = 0.23$ and $M_{B\infty} = -0.77(dl^2/8)$ or $MB_{u\infty} = [-1.75(1) + 1.0(0.23)](dl^2/8) = -1.52(dl^2/8)$.

As line *d* in Fig. 9.13 shows, the new action effects are very close to line *b* for the load-dependent moment in a two-span continuous beam.

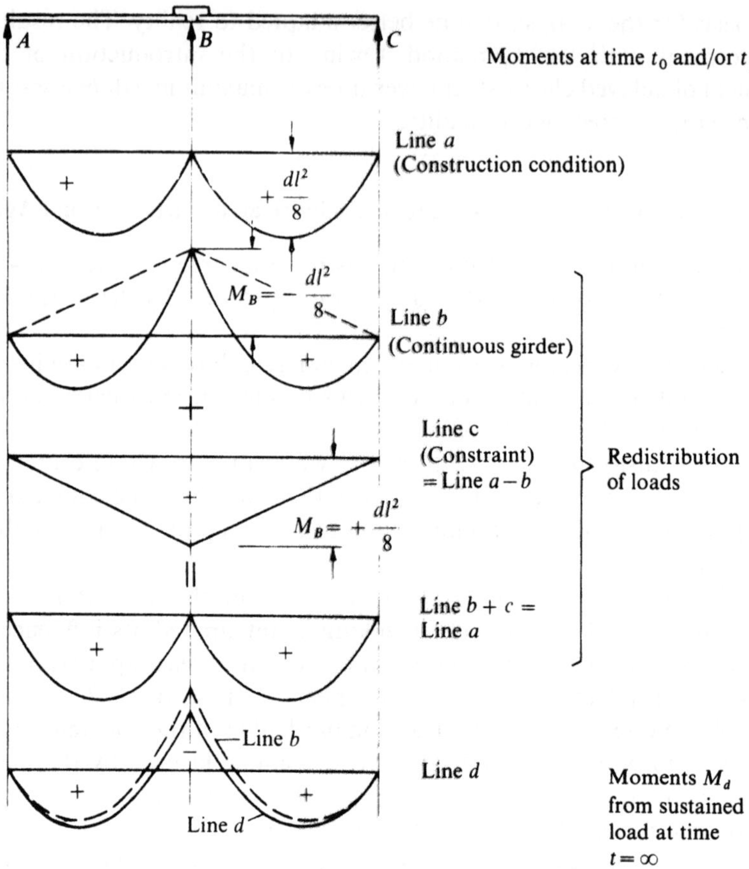

Figure 9.13.

In such a case, at mid-span the beams must be dimensioned for the conditions during construction and for the time at which exposure to life loads begins. At this time, only slight creep redistribution has occured. The sections at the center support have to be designed for the condition $t = \infty$ with full creep redistribution.

However, in actual practice, conditions are not always so unfavorable, since it is seldom that the entire dead-weight moment must be rearranged. This is shown by the following example.

The roof structure of a hangar is to be built with prefabricated girders which are connected rigidly above the supports to a continuous three-span beam before installation of the roof membrane made of lightweight concrete.

Design loads
girder weight $d_1 = 0.4d$
Roof membrane $d_2 = 0.6d$
$d_1 + d_2 = d$

Snow load $w = 0.3d$

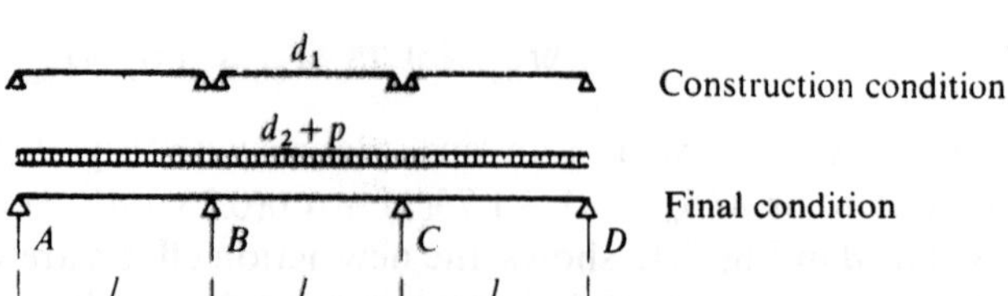

Load rearrangement is required only for d_1. It corresponds to a lowering against rigid supports; therefore, $\alpha = 1$. The creep coefficient is assumed to correspond to $\varphi = 2$ for the period from installing the rigid connection ($t = t_1$) to time $t = \infty$.

Constraint moment for $t = t_1$:

$$M_r = M_{\text{Constr}} - M_{d_1} = 0.100(0.4)dl^2 = +0.04dl^2.$$

Center support moment for $t = t_1$:

$$M_{t_1} = M_{\text{sus}} + M_r + M_w = -0.100(0.4d + 0.6d)l^2 + 0.04dl^2 - 0.100(0.3)dl^2 = -0.090dl^2.$$

Center support moment after termination of creep at $t = \infty$:

$$M_{t\infty} = M_{\text{sus}} + M_r\gamma_r + M_w = -0.100(0.4d + 0.6d)l^2 + 0.04dl^2 0.23 - 0.100(0.3)dl^2 = -0.121dl^2.$$

10. Constraints Induced by Shrinkage

10.1. General Considerations

Up to now the basis for the solution of these problems has been almost exclusively the engineer's experience and instinct. This approach is entirely acceptable for normal structures that are divided by joints into sufficiently short segments. In the domain of engineering design, however, there are frequent cases where the validity of such rules of thumb must be checked by a special analysis. The following is intended to show how one proceeds under such circumstances.

10.2. What Fraction of Total Shrinkage Can Be Compensated for by Keeping Shrinkage Joints Open?

There are widespread exaggerated conceptions regarding the effectiveness of shrinkage joints. That part of the effect of heat of hydration and, above all, that part of the effect of shrinkage which are compensated for by such joints are frequently so small that they cannot make up for the disadvantages that go with the creation of shrinkage joints: costly form work and an increased tendency to form faults in the concrete. This is particularly true for thick structural elements. More is achieved, in such cases, by trying to minimize the non-stress-induced deformations, e.g., by using cement with a low heat of hydration, minimizing cement content, using water reducing admixtures, maintaining a low fresh concrete temperature, and through appropriate curing.

In general, the requirements of construction operations do not permit keeping so-called shrinkage joints open for a long time. How little is achieved, therefore, particularly in thick structural elements, is convincingly shown by Table 10.1, whose values were derived with the help of Eq. (1.4), Table 1.2, and Fig. 1.7.

Thus, customary shrinkage joints can measurably reduce shrinkage even under uninhibited deformation only in slender structural members and at low relative

Table 10.1. Reduction of shrinkage constraint in percent in terms of relative humidity, of thickness $2A/u$, and the length of time in months during which the shrinkage joint is kept open.

Thickness $2A/u$ in cm at a relative humidity of				Reduction of shrinkage, in percent, as the shrinkage joint is kept open for the following number of months						
100%	80%	60%	40%	$\frac{1}{2}$	1	2	3	6	12	24
	3	4	5	41	53	66	74	83	90	95
	5	8	10	29	41	52	60	73	83	90
	10	17	20	15	26	33	40	53	67	80
	20	33	40	8	13	18	22	31	43	62
3	40	67	80	4	5	8	10	15	21	35
5	80	133	160	0	1	2	3	6	10	19

humidities. In many cases it will, therefore, not be possible to forego either reducing shrinkage by appropriate means or to compute the occuring constraint and, under certain circumstances, to make up for it with reinforcement.

10.3. Calculation of the Shrinkage Constraint Occuring in Uncracked Structural Members under the Simplifying Assumption that the Development of Shrinkage and Creep Is Similar over Time

10.3.1. Shrinkage of the Structural Member Is Constrained Elastically only at Its Ends

In such a case, Eq. (2.1) reads

$$S_{est} = \gamma_s S_{est}^{(e)}, \tag{10.1}$$

or, in accordance with Eq. (2.12b),

$$f_{cst} = \gamma_s \alpha |\varepsilon_{st}| E_c, \tag{10.2}$$

where $S_{est}^{(e)}$ designates those action effects in the shrinking member which would occur at the point of restraint. Both restraint and the member are assumed to be elastic. Hence, the notation $S^{(e)}$. The coefficient γ_s, a function of α and φ, characterizes the reduction in shrinkage constraint induced by creep. For totally rigid supports, $\alpha = 1$.

The following two examples are intended to show the normal range for such constraint.

Let us first examine a case of considerable shrinkage. A structural element with a mean thickness of $2A/u = 10$ cm is stored in very dry air under a relative humidity of 40%. In accordance with Section 1.3, we then have $d_{eff} = 10$ cm and $\varepsilon_{s\infty}$ becomes $40 \times 10^{-5}(1.05) = 42 \times 10^{-5}$. The accompanying creep coefficient, in accordance with Section 1.2, is $\varphi_\infty = 3.0(1.7) + 0.4 = 5.5$. The greatest constraint occurs, as shown in Fig. 2.3, at time $t = \infty$, and we get $\gamma_s = 0.19$. Therefore:

$$f_{cs\infty} = 0.19(42 \times 10^{-5})E_c = 8 \times 10^{-5}E_c. \tag{a}$$

The other extreme is provided by a structural element exposed to a relative humidity of 90%. Table 1.2 then yields $\beta_{\text{eff}} = 5$, and even structural elements that are only 32 cm thick result in $d_{\text{eff}} = 5(32) = 160$ cm. This leads to $\varepsilon_{s\infty} = 10 \times 10^{-5}(0.7) = 7 \times 10^{-5}$ and $\varphi_{\infty} = 0.4 + 1.3(1.12) = 1.85$. From Fig. 2.3, we then get $\gamma_s = 0.44$ and $f_{cs\infty} = 0.44(7 \times 10^{-5})E_c = 3.1 \times 10^{-5}E_c$.

It can thus be seen that even shrinkage strains ranging from 42×10^{-5} to 7×10^{-5} cause shrinkage constraints that differ very little since large values of ε_s always coincide with large φ values.

In the past, it was often customary to equate shrinkage with a temperature drop of 15–20°C. This is on the safe side. However, today we can readily obtain more reliable values with Eqs. (10.1) and (10.2) and also take into consideration the case where restraint of the shrinkage process occurs only at a later time.

A further example of the use of Eq. (10.1) is found in Section 3.1.

10.3.2. Continuous Restraint of Shrinkage by Elastic Elements

Examples of this type of constraint are provided primarily by large foundation slabs as well as by long runways and highway pavement slabs. The type of constraint occurring here is closely related to the problem of the effective width of T-beams as Fig. 10.1 illustrates. For this particular case one need only conceive of the compression region of the T-beam as being replaced by the restrained element which, however, connects only on one side. The foundation slab (or the runway) then takes the place of the web of the T-beam. It must also be taken into account that web and foundation slab or compression region of T-beam and foundation soil consist of materials with different moduli of elasticity.

In particular, Fig. 10.1 demonstrates how this analogy can be used. In both cases, a normal force N is introduced through shear stresses v into the elements which are separated. In freely supported T-beams under uniform loading, v is distributed approximately linearly. The same can be assumed for the shrinkage of the foundation slab. The differing values of E can be equalized by reducing the width b_S of the soil.

$$b_s = \frac{E_S}{E_c} b = nb \tag{10.3}$$

The effective width b_m (equal to the depth of the effective soil) can be determined for T-beams in terms of the ratios $d/d_0 = b/b_S = n$ and $b_i/l_0 = t_i/l_0 = \infty$. In this range, the tables for T-beams yield widths from $b_m = 0.20l$ to $0.30l$. However, in the case of soil, it is more appropriate to speak of effective depth t_m.

There are two more influences that must be taken into account, although they compensate each other in large measure. One of these is due to the fact that it is no longer a symmetrical T-beam that is involved. Since the foundation may then camber, t_m is reduced somewhat. The second factor is that the stresses in the foundation can radiate sideways. This makes b_S larger, and t_m grows. Normally, the following assumption is sufficient:

$$t_m \sim 0.25l. \tag{10.4}$$

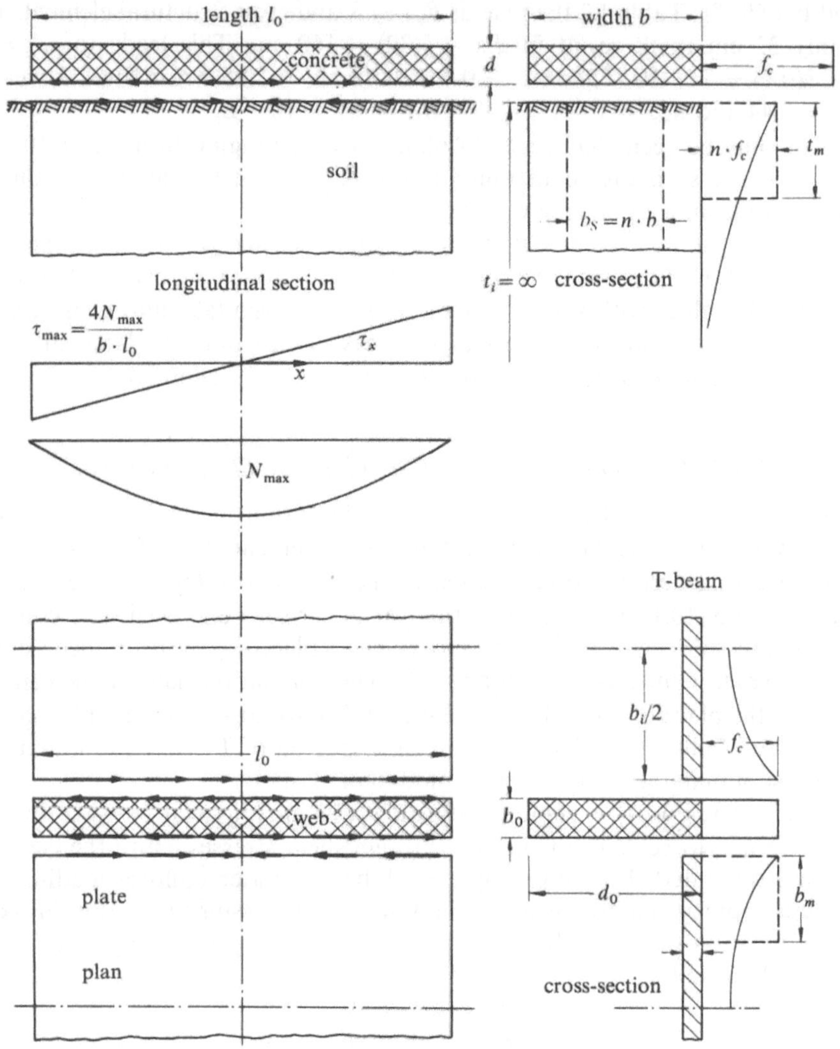

Figure 10.1.

In order to determine the normal force $N_{es}^{(e)}$, it is sufficient to apply the compatibility conditions for the beam center:

$$\varepsilon_s E_S - N_{es}^{(e)}\left(\frac{1}{t_m} + \frac{n}{d}\right) = 0,$$
$$N_{es}^{(e)} = \varepsilon_s E_S \frac{dl}{4d + nl}. \tag{10.5}$$

Since the concrete body and the soil are bonded, the stiffness ratio α is determined with Eq. (2.13) for $y_{\text{id}\,e} = 0$:

$$\alpha = n\frac{A_e}{A_{\text{id}}} = \frac{nt_m}{nt_m + d} = \frac{nl}{nl + 4d}, \tag{10.6}$$

and the uniform tensile stress acting in the concrete at time t is

$$f_{cst} = \gamma_s \frac{N_{es}^{(e)}}{d} = -\gamma_s \varepsilon_s E_c \frac{nl}{nl + 4d} = -\gamma_s \alpha \varepsilon_s E_c. \tag{10.7}$$

As the length l of the concrete body increases, α approaches 1 and we get

$$\max f_{cst} = \gamma_s^{(\alpha=1)} \varepsilon_s E_c. \tag{10.8}$$

10.3.3. Shrinkage Is Restrained by Structural Elements which Are Themselves Subject to Creep

Examples of this kind of constraint are tunnel linings which are prestressed against the surrounding rocks[1] and the interaction between a concrete frame and the time-dependent deformation of cohesive soils. Included in this category are also those cases where structural elements are interacting which, owing to differences in age or in concrete composition, present differential creep and shrinkage characteristics.

As long as one can assume, in such cases, that creep and shrinkage develop similarly over time, they can be solved by using the Dischinger–Kupfer method described in Section 2.1.3. Examples are described in Section 11.

10.3.4. Shrinkage of a Floor System Is Restrained by Supports

In the type of construction represented by Fig. 10.2, one must distinguish between steel and reinforced concrete columns. In the case of steel columns, the boundary between those structural elements that are subject to creep and those which behave elastically, lies at the head of the column; for reinforced concrete columns, on the other hand, it lies at the foot of the shrinkage-restraining foundations. In most cases, however, the deformation of the columns and the slab will be so preponderant that the deformation of the foundation can be disregarded.

When the number of columns is small, this problem can be solved taking into account multiple degrees of static indeterminacy, creep deformation being handled in accordance with Section 8. For a sufficiently high number of columns, one can try to reach a satisfactory approximate solution by proceeding in a similar fashion to Section 10.3.2. However, Fig. 10.3 and the following analysis show that the deformations along L no longer develop similarly over time. Therefore, as an approximation, the overall and maximum deformations may be made compatible. On this basis, the stiffness ratio may be determined.

[1] Regarding the deformability of rocks, see also Part B, Section 1.1.

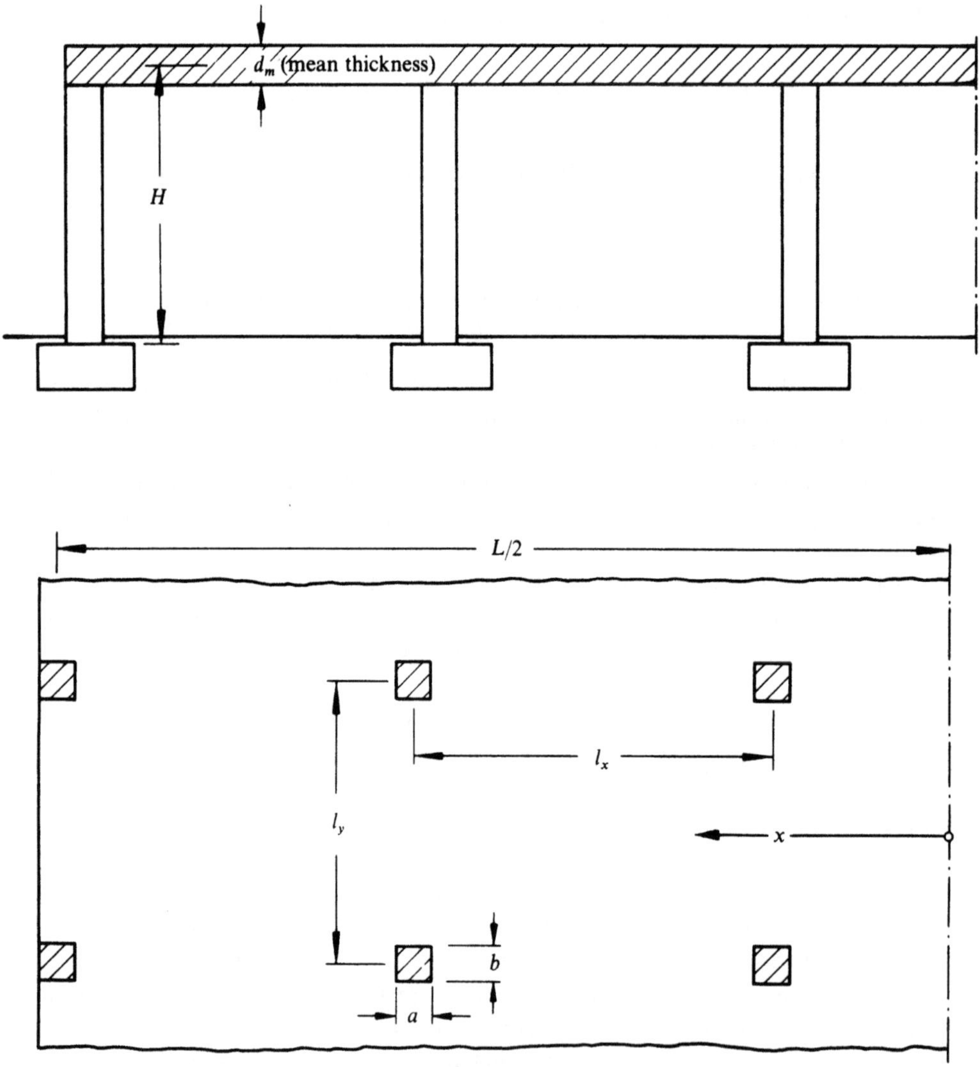

Figure 10.2.

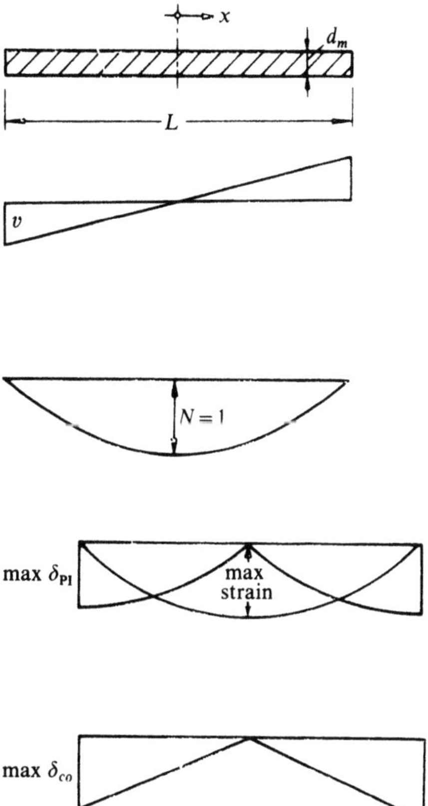

The cross section of floor system is

$$A_c = bd_m$$

The comparative distribution of the shear force acting between the creeping and the elastic structural elements is:

$$\frac{v_x}{l_y} = \frac{8}{L^2} x \text{ (kN/m}^2\text{)}.$$

The corresponding normal force is

$$\frac{N}{l_y} = 1 - \frac{4x^2}{L^2}$$

The strain and displacement max δ_{Pl} of floor system is

$$\max \delta_{Pl} = \frac{2}{3}\frac{L}{2}\frac{1}{d_m E_c} = \frac{L}{3d_m E_c} \tag{10.9}$$

The corresponding relative displacements of columns is

$$\delta_{co} = \frac{\beta(v_x/l_y)H^3}{E_c I_{co}}; \; I_{co} = \frac{a^3 b}{12 l_x l_y}$$

Figure 10.3. Structural system.

The coefficient β depends on the degree to which the end of the support is fixed (see Fig. 10.4).

EXAMPLES:

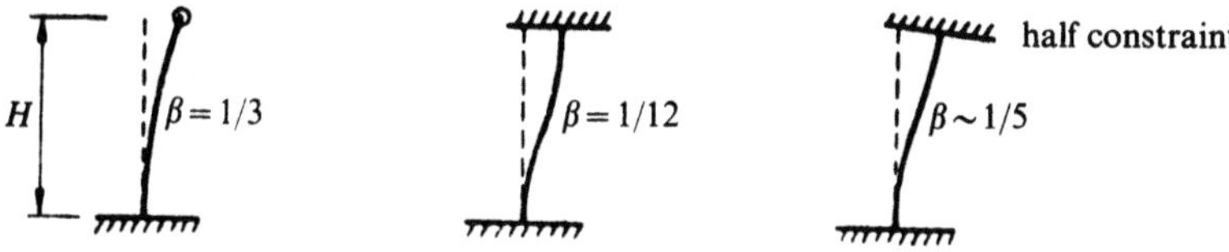

Figure 10.4. β coefficients.

$$\delta_{co} = \frac{\beta H^3}{E_{co}} \frac{8}{L^2} x \frac{12 l_x l_y}{a^3 b} = \frac{96 \beta H^3 l_x l_y}{a^3 b L^2 E_{co}} x.$$

$$x = \frac{L}{2}: \max \delta_{co} = \frac{48 \beta H^3 l_x l_y}{a^3 b L E_{co}}. \tag{10.10}$$

Equation (9.2) gives the normal force generated in the slab:

$$\max N^{(e)} = -\max\left(\frac{\delta_s}{\delta_{11}}\right) = \frac{\varepsilon_s L/2}{\max \delta_{\mathrm{Pl}} + \max \delta_{co}}, \tag{10.11}$$

$$\max N_c^{(e)} = \max N_c^{(e)} = \frac{\varepsilon_s L/2}{\dfrac{L}{3d_m E_c} + \dfrac{48\beta H^3 l_x l_y}{a^3 bLE_{co}}}.$$

Assumption I. The columns are made of steel; foundation deformation is disregarded. The displacement of the structural elements subjected to creep is

$$\max \delta_c = \max \delta_{\mathrm{Pl}} = \frac{L}{3\, d_m E_c}.$$

The displacement of the elastic structural elements is

$$\max \delta_e = \max \delta_{co} = \frac{48\beta H^3 l_x l_y}{a^3 bLE_e}.$$

Unfortunately, the deformations develop quite differently in this case, as shown by Fig. 10.3.

$$\alpha \approx \frac{1}{1 + \delta_e/\delta_c} = \frac{1}{1 + 144\beta m_1^3 m_2 m_3/n}, \tag{10.12}$$

where $m_1 = H/a$, $m_2 = l_x l_y/L^2$, $m_3 = d_m/b$, $n = E_e/E_c$.

$$\max N_e^{(e)} = -\frac{\varepsilon_s L/2}{\max(\delta_c + \delta_e)} = \max N_c^{(e)}, \tag{10.13}$$

$$\max f_{cst} = -\gamma_s \alpha N_c^{(e)}/d_m. \tag{10.14}$$

Assumption II. The columns, and the slab, are made of concrete. In that case,

$$\max \delta_c = \max(\delta_{\mathrm{Pl}} + \delta_{co}) = \frac{1}{E_c}\left(\frac{L}{3\, d_m} + \frac{48\beta H^3 l_x l_y}{a^3 bL}\right),$$

$$\max \delta_e = 0,$$

$$\alpha = \frac{1}{1 + \delta_e/\delta_c} = 1, \tag{10.15}$$

$$\max N_e^{(e)} = -\frac{\varepsilon_s L/2}{\max \delta_c} = \max N_c^{(e)}, \tag{10.16}$$

$$\max f_{cst} = -\gamma_s^{(\alpha=1)} \frac{\varepsilon_s L}{2 \max \delta_c d_m} \quad \text{(in the slab).} \tag{10.17}$$

10.3.5. Evaluation of the Risk of Shrinkage Cracks under Pure Shrinkage Constraint

As concrete ages, there is a simultaneous increase in the modulus of elasticity and in the tensile strength of the concrete as well as in its shrinkage. When shrinkage is restrained externally, tensile stresses are induced in the structural member under study. The question of whether these can cause a shrinkage crack depends on which dominates: the growing tensile strength or the shrinkage stresses which are a function of the modulus of elasticity and of creep. Additional tensile stresses that may be induced by such effects as temperature are disregarded in the following analysis.

If one makes do with the simplified assumption that ε_s and φ_f develop similarly over time, this process can be described as follows:

Their affinity is

$$\varepsilon_{st} = -\varphi_{ft}\frac{\varepsilon_{s\infty}}{\varphi_{f\infty}} = -(\varphi_t - 0.4\beta_d)\frac{\varepsilon_{s\infty}}{\varphi_{f\infty}}. \tag{10.18}$$

E_c from Eq. (1.1) is

$$E_{ct} = \beta_{et}E_{c28}, \tag{10.19}$$

From Eq. (3.9), Part B. $$\beta_{et} = \sqrt[3]{f'_{ct}/f'_{c28}} \tag{10.20}$$

The tensile strength in accordance with Section 2.2, Part B

$$f_{tu28} = C\sqrt[3]{(f'_{c28})^2} \tag{10.21}$$

Values for C are given in Section 2.2. Part B

$$f_{tut} = C\cdot\sqrt[3]{(\beta_t\cdot f'_{c28})^2} = C\cdot\beta_e\sqrt[3]{(f'_{c28})^2} \tag{10.22}$$

Assuming similarity in the development of shrinkage and creep, and disregarding the influence of the variation of E on the coefficient γ_s, one can write

$$f_{cst} = \gamma_s\alpha\varepsilon_{st}E_{ct}\ \text{(tension)}. \tag{10.7}$$

Using Eqs. (10.18) and (1.1),

$$f_{cst} = \gamma_s\alpha(\varphi_t - 0.4\beta_d)\frac{\varepsilon_{s\infty}}{\varphi_{f\infty}}\beta_{et}E_{c28}.$$

Applying Eq. (2.25) in Appendix II, we have

$$\varepsilon_s\alpha(\varphi_t - 0.4\beta_d) = \alpha\frac{1 - e^{-\alpha[(\varphi t - 0.4\beta_d)/(1 + 0.4\beta_d\alpha)]}}{\alpha(\varphi_t - 0.4\beta_d)}(\varphi_t - 0.4\beta_d)$$

$$= 1 - e^{-\alpha[(\varphi\ - 0.4\beta_d)/(1 + 0.4\beta_d\alpha)]} \tag{10.23}$$

and, thus,

$$\beta_{cst} = \beta_{et}(1 - e^{-\alpha\varphi_{ft}/(1 + \alpha\varphi_d)})\frac{\varepsilon_{s\infty}E_{c28}}{\varphi_{f\infty}}. \tag{10.24}$$

The safety against the development of shrinkage cracks is determined by the ratio between tensile strength and shrinkage stresses. Applying Eq. (10.24) and Eq. (2.5), Part B, we can write

$$\gamma_{C,s} = \frac{f_{tut}}{f_{cs,t}} = \frac{1}{A_t B} \tag{10.25}$$

$$A_t = [1 - e^{-(\alpha\varphi_{ft})/(1+\alpha\varphi_d)}]\beta_{et} \tag{10.26}$$

$$B = \frac{\varepsilon_{s\infty} E_{c28}}{\varphi_{f\infty} f_{tut}} \tag{10.27}$$

These equations were used for the two following examples with full shrinkage restraint ($\alpha = 1$) (cf., Fig. 10.5). The first one involves slender structural elements in a dry climate ($2A/u = 10$ cm, 40% relative humidity, $f'_c = 25$ N/mm^2; Type II cement); the second one pertains to thick structural members in a humid climate ($2A/u \geq 65$ cm, 80% relative humidity, $f'_c = 25$ N/mm^2; Type II cement). In both cases, measures were taken against the possible occurrence of cracks both for the mean as well as for the characteristic tensile strength (5% fractile). With the rapid drying rate on which the first example is predicated, a crack may develop in as little as 1 week; this possiblity can be counteracted by keeping the concrete moist

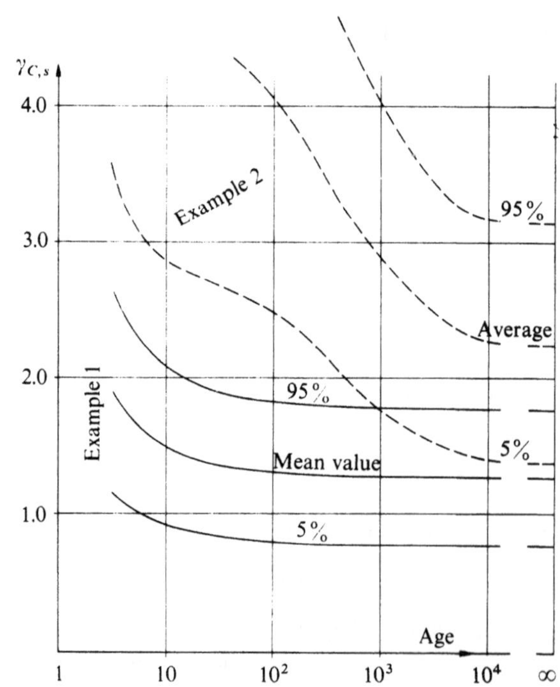

Figure 10.5. Safety factor $\gamma_{C,s}$ for the development of shrinkage cracks for full shrinkage restraint ($\alpha = 1$). It depends upon concrete age, relative humidity and the effective thickness, $2A/u$. This relationship shows considerable scatter.

until the growth in strength can combat the drying process to a sufficient degree. There is a 70% probability that the concrete will not develop cracks.

The effectiveness of moist curing is demonstrated by the second example, where one can, at worst, expect surface cracks from shrinkage, as long as other effects—such as temperature stresses—are not added.

In both cases, however, safety against crack development decreases monotonically over time. At the same time, the probability increases that other effects may accidentally induce large tensile stresses. This explains why, in reinforced concrete construction, the cracking anticipated during design often occurs only after half a year, or later.

10.4. When, to What Extent and with Which Consequences May Deviations from the Assumed Similarity of Shrinkage and Creep over Time Occur?

10.4.1. Magnitude of Deviation

Whether there is similarity of shrinkage and creep can best be seen in the form of the graphic approach chosen for Figs. 10.6 and 10.7 which show the interrelation between the simultaneously occurring ratios $\varepsilon_{st}/\varepsilon_{s\infty}$ and $\varphi_{ft}/\varphi_{f\infty}$. There would be complete similarity if the curves thus obtained were to form diagonals inclined at 45°. Naturally, this is never exactly the case. Considering the uncertainty that characterizes the values β_s and β_f, one must not be too demanding in evaluating such deviations, and attention should be paid only to significant and consistent differences.

It should be noted that this analysis is based on the concepts on creep and shrinkage as given in CEB–FIP 1978 and DIN 4227.

Curves running above the dotted-line diagonal signify that the shrinkage constraint is developing more slowly than creep. The consequences of this were shown in Section 9. The shrinkage stresses then tend monotonically to an ultimate value that exceeds the value computed for similarity. For curves that stay below the diagonal, the opposite occurs for the ultimate value—it becomes smaller. However, since shrinkage now develops faster than creep, there occurs an intermittent maximum value which exceeds even the ultimate value that applies for similarity of development over time.

Figures 10.6 and 10.7 represent extreme conditions. The conditions that occur in actual practice lie mostly between these extremes. Figure 10.6 shows, in particular, the influence of the type of cement and of the member thickness, while Fig. 10.7 illustrates the influence of climatic conditions and of the age of concrete at the beginning of the shrinkage constraint.

10.4.2. Consequences of Deviation

The following conclusions can be drawn from Figs. 10.6 and 10.7:

Larger ultimate values than for the case of similarity of development in time are to be expected primarily for moist curing and for thick structural elements in which shrinkage develops slower than creep.

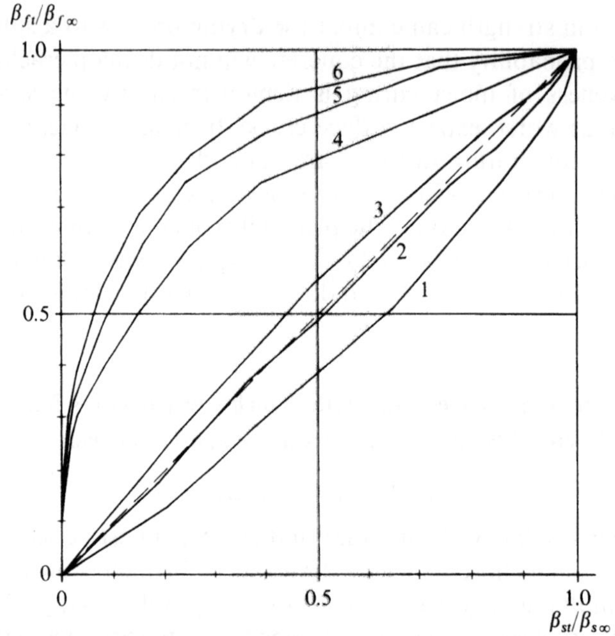

Figure 10.6. Effect of type of cement and of member size on the time development of flow and of shrinkage. The various relations are described in the following table.

Curve	Relative humidity	Thickness $2A/u$	Type of cement (cf., Section 1.2)	Initial age of concrete
1	40%	5 cm (1.96 in.)	Type II	$t_0 = 0$
2	40%	5 cm	Type I	$t_0 = 0$
3	40%	5 cm	Type III	$t_0 = 0$
4	40%	160 cm (63 in.)	Type II	$t_0 = 0$
5	40%	160 cm	Type I	$t_0 = 0$
6	40%	160 cm	Type III	$t_0 = 0$

Intermittent maximum values occur especially after dry storage and are largest for slender structural elements. However, the deviations from the values obtained for similarity between creep and shrinkage are not very significant.

To answer the question whether there is approximate similarity, one need not know exactly curves similar to those given in Figs. 10.6 or 10.7. It is enough to know within what period the half-life T for shrinkage occurs. If $\varphi_{ft}/\varphi_{f\infty}$ is also equal to 0.5 at that time, we get $T_f = T_s$, and the assumption of similarity represents at least a good approximation.

Under what conditions $T_f = T_s$ is shown by Fig. 10.8. There, existence of similarity is given as a function of the age of concrete at the start of the shrinkage

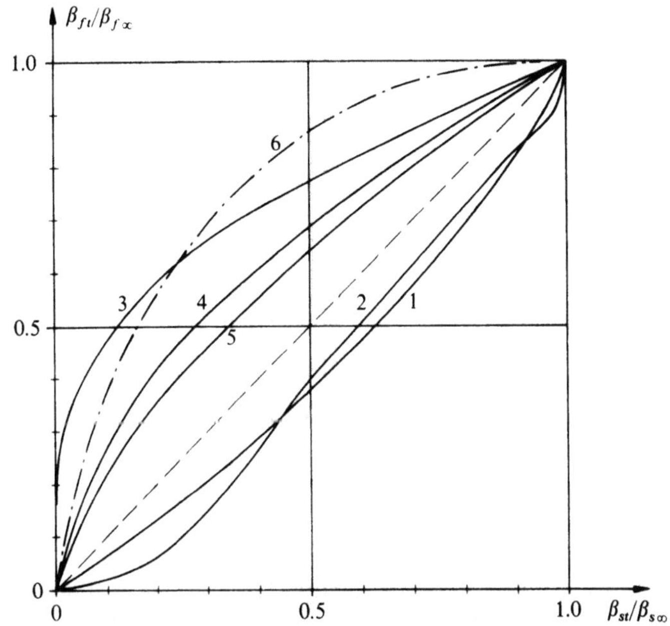

Figure 10.7. Effect of relative humidity and of concrete age at loading on the development of shrinkage. The various relations are described in the following table.

Curve	Type of cement (cf., Section 1.2)	Relative humidity	Thickness $2A/u$		d_{eff}		Initial age of concrete in days
1	Type II	40%	5 cm	(1.96 in.)	5 cm	(1.96 in.)	0
2	Type II	40%	5 cm		5 cm		28
3	Type II	100%	5 cm		150 cm	(59 in.)	0
4	Type II	100%	5 cm		150 cm		56
5	Type II	100%	5 cm		150 cm		365
6	Type III	100%	5 cm		150 cm		365

constraint, of the effective member thickness, and of the type of cement. The region above the curves indicates that shrinkage constraint develops more slowly than creep and where accurate computation yields larger ultimate values. In the region below the curves, intermittent maximum values are to be anticipated. The values on the curves indicate the age of the concrete in days when the shrinkage constraint becomes effective.

It must be stressed again that no exaggerated importance should be ascribed to deviations from similarity; one must always keep in mind that a "precise" computation, which can only be based on imprecisely known initial values, is not very meaningful. Thus, one will do well to make allowances for the shortcomings

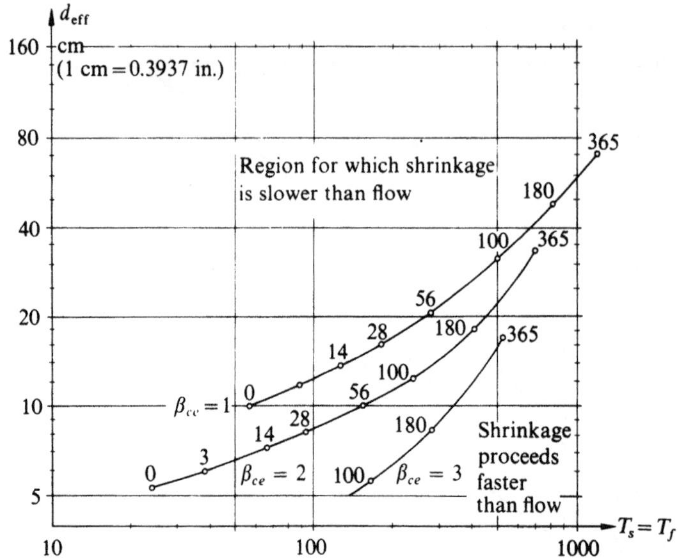

Figure 10.8. Shrinkage and flow are approximately similar if both relations have equal half time values. The relations valid for three types of cement show when this is true. The numbers correspond to the concrete age t_1 at which shrinkage restraint becomes effective. The values for the effective thickness d_{eff} and the half-life $T_s = T_f$ are given along the ordinates.

observed in a simple computation method based on similarity of the two types of deformations, and should pay closer attention only when thick structural elements are involved and in cases where shrinkage stresses play a particularly significant role.

10.4.3. Estimate of the Consequences of Large Deviations

In such cases, satisfactory values can be obtained rapidly by means of an analogy. One need only think of the shrinkage constraint as being replaced by the imposed stress described in Section 9.3.6 with the same half-time. Figure 10.9 shows which half-time values correspond to shrinkage. The following two examples illustrate the application of this method.

EXAMPLE 1 (With the aid of Fig. 10.9)

$2A/u = 32$ cm; 90% relative humidity; $\beta_{eff} = 5$; $d_{eff} = 160$ cm; Type I cement;

$t_1 = 3$ days; $\beta_{ce} = 2$; $t_{1f} = 6$ days.

$$\varphi_\infty = 1.3(0.92) + 0.4 = 1.60,$$

$$\varepsilon_{s\infty} = (-10)0.70 \times 10^{-5} = -7 \times 10^{-5},$$

$$\alpha = 1.0 \text{ or } 0.5.$$

Figure 10.8 shows that for $t_1 = 3$ days, $d_{eff} = 160$ cm, and $\beta_{ce} = 2$, shrinkage develops much more slowly than creep. Figure 10.9 yields $T_s = 2300$ days. In

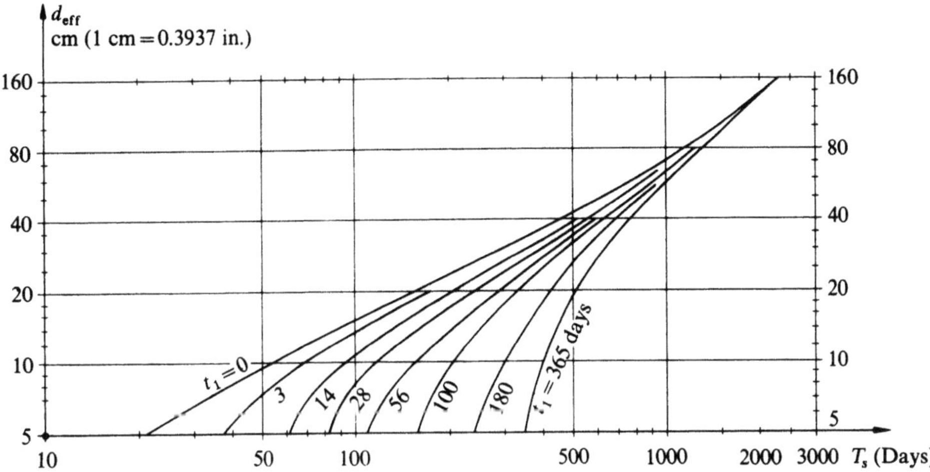

Figure 10.9. Half-life values for shrinkage T_s start at initiation of restraint at time $t = t_1$. They are given as functions of the effective thickness d_{eff} and of the concrete age at initiation of shrinkage restraint.

accordance with the explanations given at the end of Section 9.3.6, one must utilize the information provided there with the initial values $t_1 = 3\beta_{ce} = 6$ days and $T_s = 2300\beta_{ce} = 4600$ days. One finds from Fig. 9.6a, for 90 % relative humidity and $2A/u = 32$ cm, the approximate value $\gamma_\infty \sim 0.57$. That gives

$$f_{cs\infty} = \gamma_\infty \alpha E_c \varepsilon_{s\infty} = 0.57(1)3 \times 10^4(7 \times 10^{-5})$$
$$= 1.2 \text{ N/mm}^2 \text{ (172 psi)}.$$

Under the assumption of similar development over time and $E_c = E_{c28} =$ const and with the help of Fig. 2.3, one gets for $\alpha = 1$, the reference value $\gamma_s = 0.47$, and, consequently,

$$f_{cs\infty} = 0.47(1)3 \times 10^4(7 \times 10^{-5}) = 0.99 \text{ N/mm}^2 \text{ (14.2 psi)}.$$

Thus, the simplified method underestimates the final value by 16 %. However, despite this, the stress difference of 0.21 N/mm² is not of decisive importance in most cases.

When the supports are not rigid, i.e., when $\alpha < 1$, one can assume that γ will vary proportionally to γ_s:

$$\frac{\gamma_{\infty,\alpha\neq 1}}{\gamma_{\infty,\alpha=1}} = \frac{\gamma_{s,\alpha\neq 1}}{\gamma_{s,\alpha=1}}.$$

With $\gamma_{s,\alpha=0.5} = 0.65$, Fig. 2.3 yields as an approximation

$$\gamma_{\infty,\alpha=0.5} = 0.57\frac{0.65}{0.47} = 0.78$$

and

$$f_{cs\infty} = 1.65 \text{ N/mm}^2 \text{ (237 psi)}.$$

Table 10.2[a]

With similar time development				With time development as in Fig. 10.10			
$\Delta\varphi_f$	$\Delta\varepsilon_s/\varepsilon_{s\infty}$	Eq. (2.31)		$\Delta\varphi_f$	$\Delta\varepsilon_s/\varepsilon_{s\infty}$	Eq. (2.31)	
1	2	3		4	5	6	
0.71	0.20	$0 + \frac{0.20}{1.4 + 0.5(0.71)} =$	0.114	0.71	0.215	$0 + \frac{0.215}{1.4 + 0.5(0.71)} =$	0.122
0.71	0.20	$\frac{1.4 - 0.5(0.71)}{1.4 + 0.5(0.71)} 0.114 + 0.14$	0.181	0.71	0.240	$\frac{1.4 - 0.5(0.71)}{1.4 + 0.5(1.71)} 0.122 + \frac{0.240}{1.755} =$	0.209
0.71	0.20	$0.595(0.181) + 0.114 =$	0.222	0.71	0.205	$0.595(0.209) + \frac{0.205}{1.755} =$	0.241
0.71	0.20	$0.595(0.222) + 0.114 =$	0.247	0.71	0.200	$0.595(0.241) + \frac{0.200}{1.755} =$	0.257
0.71	0.20	$0.595(0.247) + 0.114 =$	0.262	0.71	0.140	$0.595(0.257) + \frac{0.140}{1.755} =$	0.230
3.55	1.00			3.55	1.00		

[a] The numerical value of the numerator in the fraction of the first line and the numerical value of the fraction in the second line are the same for all the following lines.

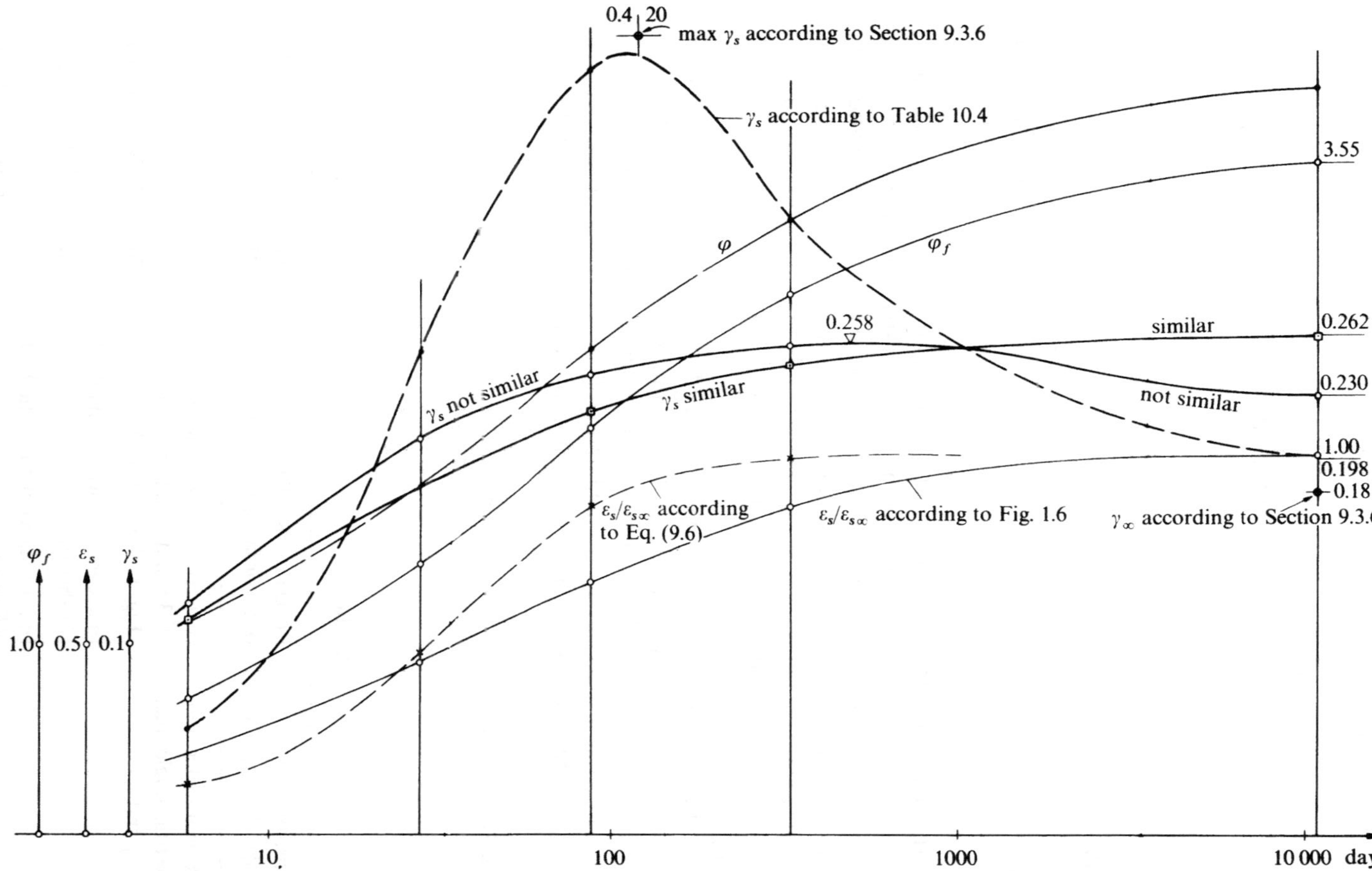

Figure 10.10. Development of φ_f, $\varepsilon_s/\varepsilon_{s\infty}$ the γ-values given in Table 10.2 with and without similarity and the more accurate values γ_s from Table 10.4 (Example 2).

The approximation obtained in accordance with Section 9.3.6,

$$\gamma_{\infty,\alpha\neq 1} = \gamma_{\infty,\alpha=1}(1/\sqrt{\alpha}),$$

yields the same result.

EXAMPLE 2 (With the aid of the f_m-method).

$$2A/u = 5 \text{ cm}; 70\% \text{ relative humidity};$$

$$\beta_{\text{eff}} = 1.5; d_{\text{eff}} = 1.5(5.0) = 7.5 \text{ cm}; \text{Type II cement}; t_1 = 0.$$

$$\varphi_f = 2(1.77) = 3.54; \varphi = 3.54 + 0.40 = 3.94;$$

$$\varepsilon_{s\infty} = -25(1.1 \times 10^{-5}) = -27.5 \times 10^{-5}; \alpha = 1.$$

Under the assumption of similar development over time and $E_c = E_{c28} =$ const, Fig. 2.3 yields the value $\gamma_s = 0.258$. Figure 10.8 shows that for $t_1 = 0$ and $d_{\text{eff}} = 7.5$ cm, shrinkage develops faster than creep. From Fig. 10.9, we find that $T \sim 1$ month. With this information, the data in Section 9.3.6 (approximations from Figs. 9.7a–9.7c) yield max $\gamma \sim 0.42$ after about 4 months and $\gamma_\infty \sim 0.18$.

For the sake of completeness, the purpose of Example 2 is to show that reliable results can also be obtained rapidly by means of step-by-step integration with the f_m-method (cf., Section 2.3 and the example in Section 9.3.5). The left-hand side of Table 10.2 is predicated on similar development of shrinkage and creep (five steps with $\Delta\varphi = \varphi_\infty/5$ and $\Delta\varepsilon_s/\varepsilon_{s\infty} = \varepsilon_s/5\varepsilon_{s\infty}$), while the right-hand side uses, for the same steps of $\Delta\varphi$, the corresponding $\Delta\varepsilon_s$ values from Fig. 1.6 (cf., Fig. 10.10, $E_c = E_{c28}$).

The values obtained in the same manner as for Example 1, and the characteristic figures for the values plotted in Fig. 10.10 are summarized in Table 10.3.

The first two lines clearly demonstrate the reliability of the f_m-method, since values identical to Fig. 2.3 were obtained in only five steps. High stresses occurring in young concrete (Example 2), however, are not accounted for accurately since

Table 10.3. Comparison of γ-coefficients.

		Example 1	Example 2	
Method		γ_∞	max γ	γ_∞
Approximation with similar development	In accordance with Fig. 2.3	0.470	—	0.258
	In accordance with f_m-method	0.460	—	0.262
Taking into account the true development of φ_f and ε_s	In accordance with Section 9.3.6 and T_s taken from Fig. 10.9 with variable E_c	0.570	0.420	0.180
	In accordance with f_m-method (Table 10.2)	0.580	0.258	0.230
	In accordance with f_m-method (Table 10.4)		0.411	0.198

the f_m-method applies the delayed elastic strain in full in the first step. In addition, the penultimate line disregards the variability of E_c.

Utilization of the information presented in Section 9.3.6 leads more rapidly to the solution and is normally applicable for all ranges. Variations of E_c are automatically included. The development of shrinkage can be determined by means of a hyperbolic function and the half-times with sufficient accuracy.

For thin bodies and early load application, the e-function deviates a great deal, from the true course of shrinkage in the region between the half-life and the extreme points. The maximum developing in Example 2 is therefore too high. The real value is between the value obtained through the f_m-method, max $\gamma = 0.258$, and 0.420.

For the sake of completeness, we want to show in Example 2 that by allowing for the true development of φ_d over time, the f_m-method according to Section 2.3 yields results which, for the most part, agree with the basic data developed in Section 9.3.6. In both cases, time development of shrinkage is described by a hyperbolic function with a half-life of 1 month. With a correspondingly high number of steps, both solutions merge.

First of all, we expand the detailed deduction in Appendix II, Section 3.2.

The following are compatibility conditions for Step i:

Change of restraint:

$$\varepsilon_{i+1} - \varepsilon_i = \Delta\varepsilon_i.$$

Flow:

$$\frac{S_i + S_{i+1}}{2}\delta_{c11}^{28}\Delta\varphi_{fi}.$$

Then we consider the elastic and delayed elastic recovery.

Each partial step is analyzed separately, and the results of all steps are superimposed. Step 1 takes into consideration the full amount of elastic recovery, but only a partial amount of the delayed elastic recovery $-S_1\delta_{c11}^{28}C_{1m}$. The remaining $\Delta\varphi_d$ values appear in the following steps as additional deformation. In Step 2, we then have

$$[-S_1C_{d_12} - (S_1 - S_2)C_{2m}]\delta_{c11}^{28}$$

and in Step 3,

$$[-S_1C_{d_13} - (S_1 - S_2)C_{d_23} - (S_3 - S_2)C_{3m}]\delta_{c11}^{28}.$$

Then

$$C_{d_12} = \Delta\varphi_{d_12}\left[1 + \frac{1-\alpha}{\alpha(1/\beta_{em} + \Delta\varphi_{d_12})}\right] \rightarrow \text{for } \alpha = 1\colon C_{d_12} \approx \Delta\varphi_{d_12};$$

$$C_{1m} = \frac{1}{\alpha} + \frac{1}{\beta_{em}} + \Delta\varphi_{d_11} - 1 \rightarrow \text{for } \alpha = 1\colon C_{1m} = \frac{1}{\beta_{em}} + \Delta\varphi_{d_11};$$

$$C_{2m} = \frac{1}{\alpha} + \frac{1}{\beta_{em}} + \Delta\varphi_{d_22} - 1 \rightarrow \text{for } \alpha = 1\colon C_{2m} = \frac{1}{\beta_{em}} + \Delta\varphi_{d_22}.$$

Table 10.4

Time (days)	φ_{ft}	$\Delta\varphi_{ft}$	$10^5\varepsilon_{st}$	$\frac{\Delta\varepsilon_{st}}{\varepsilon_{s\infty}}$	β_{et}	$\left(\frac{1}{\beta_{et}}\right)_m$	β_d for the steps 1	2	3	4	5	$C_{d_xi} = \Delta\varphi_{d_xi}$ for the 1	2	3	4
1	2	3	4	5	6	7	8					9			
0	0		0		0.422										
		0.71		0.13		1.779						0.148			
6	0.71		0.13		0.842		0.37								
		0.71		0.35		1.094						0.052	0.192		
28	1.42		0.48		1.000		0.50	0.48							
		0.71		0.39		0.968						0.072	0.076	0.240	
88	2.13		0.87		1.067		0.68	0.67	0.60						
		0.71		0.12		0.920						0.092	0.096	0.120	0.352
360	2.84		0.99		1.108		0.91	0.91	0.90	0.88					
		0.71		0.01		0.894						0.036	0.036	0.040	0.048
∞	3.55		1.00		1.131		1.00	1.00	1.00	1.00	1.00				
	Σ	3.55		1.00							Σ	0.400	0.400	0.400	0.400

The compatibility conditions for step i result in:

$$S_{i+1} = \frac{S_i(C_{im} - \Delta\varphi_i/2) + \sum \Delta S_{xi} C_{xi} + \Delta\varepsilon_i \delta_{b_1 1}^{28}}{C_{im} + \Delta\varphi_i/2} \tag{10.28}$$

The hyperbolic function for the imposed action in Example 2, is then, in accordance with Eq. (9.6), Section 9.3.6. (half-life 1 month),

$$\frac{\varepsilon_{st}}{\varepsilon_{s\infty}} = 1 - e^{-0.0231t} \quad \text{(see column 5, Table 10.4).}$$

The calculation is done in Table 10.4. The results are plotted in Fig. 10.10 (see, also, Table 10.3).

10.5. The Effect of Shrinkage Reinforcement in Structural Elements which Are Primarily Subject to Shrinkage Stresses

10.5.1. General

It is often assumed that more reinforcement will prevent the development of cracks caused by temperature change and shrinkage. However, there is little chance of that. As we can see from the following formula, the concentric tensile stress is reduced only slightly by such measures.

$$f_{ct} = \frac{N}{A_i} = \frac{N}{A_c + nA_s} = \frac{1}{1 + n\rho}\frac{N}{A_c}. \tag{10.29}$$

Since, on the average, $n = 6$, f_{ct} is reduced by only about 6% after increasing the amount of reinforcement by $\Delta\rho = 1\%$.

Under these circumstances, the measures listed in Section 10.2, which aim at reducing the shrinkage stresses, promise to be more successful because they have

steps 5	C_{xm}	According to Eq. (10.28)	
	10	11	
	1.927	$\frac{0.13}{1.927 + 0.71/2}$	$= 0.057$
	1.286	$\frac{1.286 - 0.71/2}{1.286 + 0.71/2}0.057 + \frac{0.052(0.057) + 0.35}{1.286 + 0.71/2}$	$= 0.253$
	1.208	$\frac{1.208 - 0.71/2}{1.208 + 0.71/2}0.253 + \frac{0.072(0.253) + 0.076(0.196) + 0.39}{1.208 + 0.71/2}$	$= 0.400$
	1.272	$\frac{1.272 - 0.71/2}{1.272 + 0.71/2}0.400 + \frac{0.092(0.400) + 0.096(0.196) + 0.120(0.147) + 0.12}{1.272 + 0.71/2}$	$= 0.325$
0.400	1.294	$\frac{1.294 - 0.71/2}{1.294 + 0.71/2}0.325 + \frac{0.036(0.325) + 0.036(0.196) + 0.040(0.147) + 0.048(0.075) + 0.01}{1.294 + 0.71/2}$	$= 0.198$
0.400			

favorable effects in two respects. This is particularly true for suitable concrete curing.

Equation (10.29) takes into account only the influence of the action effects on the concrete tensile stress. Internal stresses alone (see Part A, Figure 2), which develop as a result of the nonlinear distribution of the thermal strains and shrinkage, can, under unfavorable circumstances exceed the local tensile strength. This is particularly true if these internal stresses develop early due to rapid drying or as a result of temperature differences which occur as a consequence of the diminishing heat of hydration.

This does not mean, however, that shrinkage reinforcement is useless. It accomplishes two important tasks:

(a) It is designed to prevent internal stress peaks (see Fig. 10.11)—which always develop near the exterior surface and cover only a limited part of the cross section—from causing continuous or bothersome cracks on the surface. To prevent this is the task of the reinforcement near the surface.

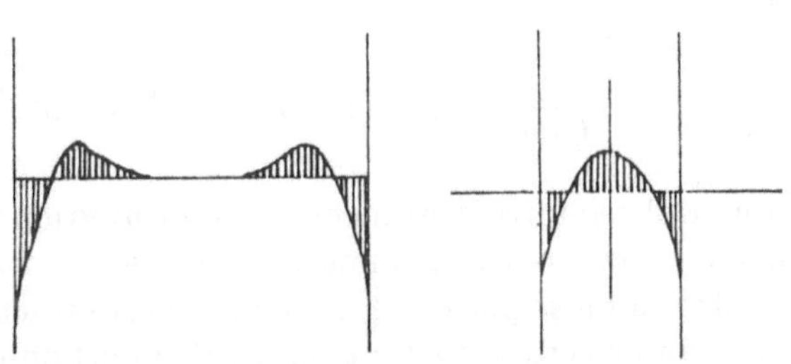

Figure 10.11. Distribution of internal shrinkage stresses in thick and in thin members.

(b) It is also designed to control the width of cracks in the tensile zone, which can never be avoided completely without prestressing. To achieve this, the amount of reinforcement has to be above a threshold value.

Problems connected with the calculation of such threshold values have been dealt with previously [10], [11], [12]. However, they deal only with short-term influences though, shrinkage causes sustained stresses. The following explanations are designed to complement the previous work by taking into consideration the reduction of shrinkage stresses through creep. They are limited to cases where the shrinkage stresses prevail, such as is frequently the case with interior walls and underground construction.

10.5.2. Reinforcement Near the Surface

It makes little sense to calculate the internal stresses mentioned in (a) since they depend to a great degree on local conditions and requirements. Pertinent regulations thus specify minimum values for surface reinforcement (e.g., ACI-318, DIN 4227, or CEB/FIP 1978).

10.5.3. The Amount of Reinforcement Needed for Crack Control

10.5.3.1. Very Long Structural Components without Contraction Joints

We have to say in advance that such structural components are seldom used in practice. Very long structures which are primarily subject to shrinkage stresses will always need supports which, in turn, impede the shrinkage strain and thus shorten the free length. Thus, the following explanations are, above all, an introduction to the subsequent section.

The tensile force which causes the first crack is

$$F = f_{tu} A_{\text{id}}. \tag{10.30}$$

According to Section 10.3.5, the existing tensile strength is estimated to be

$$f_{tu28} = C \cdot \sqrt[3]{(f'_{c28})^2} \quad \text{and} \quad f_{tut} = C \cdot \sqrt[3]{(\beta_t \cdot f'_{c28})^2}$$

In very long structural components, the additional tensile strain which develops at the first crack causes only a negligible reduction of the tensile force. If the reinforcement is not strong enough to take up the tension which causes cracking without yielding, then shrinkage will cause a single, wide crack. This leads to the following requirement:

$$\min \rho = \frac{A_s}{A_c} \geq \frac{f_{tu}(A_c + n \cdot A_s)}{f_s \cdot A_c} = (1 + n \cdot \rho) \cdot \frac{f_{tu}}{f_s} \sim 1.05 \frac{f_{tu}}{f_s} \tag{10.31}$$

The necessary amount of reinforcement increases in proportion to the tensile strength of the concrete. How do we estimate it? First, we can presume that cracks will always develop in those places which have the lowest tensile strength. On the other hand, we can be certain that a considerable portion of this tensile strength will be used up by the above-mentioned internal stresses (see Fig. 10.11).

Section 10.5.3.2 underscores the finding, already developed in Section 10.3.5, that cracks which are the result of shrinkage alone will appear relatively late; that is, after 6 months to 1 year. During that period, many load conditions can occur which result in unfavorable internal stresses. It is therefore presumably safer to insert for C a value between C_m and $C_{5\%}$ (C_m is the mean value, $C_{5\%} = 5\%$ fractile). This also applies to thick structural elements, since according to [8], the tensile strength decreases substantially with increasing member thickness.

Example for Concentric Stresses
Assumptions:

$$2A/u = 8\text{ cm};\ 80\%\text{ relative humidity};\ \beta_{\text{eff}} = 2.4;\ d_{\text{eff}} = 2.4(8) = 20\text{ cm};$$

$$\text{Cement Type II, } f'_{c28} = 25\text{ N/mm}^2$$

$$E_c = 30\,000\text{ N/mm}^2;\ t_1 = 0;\ t = 365\text{ days};$$

$$\varphi = 1.55(1.7) + 0.4 = 3.0;\ \varepsilon_s = 0.90(20)10^{-5} = 18.0 \times 10^{-5};\ \alpha = 1;$$

$$f_y = 420.0\text{ N/mm}^2\text{ (yield stress of steel).}$$

Assuming similarity, we obtain from Eq. (2.1) and Fig. 2.3 for $t = \infty$,

$$f_{ts} = 0.32(18.0 \times 10^{-5})30\,000 = 1.69\text{ N/mm}^2\text{ (242 psi).}$$

The tensile strength for $t = \infty$ is

$$f_{tu} = \frac{0.22 + 0.36}{2}\,1.13^2\sqrt[3]{25^2} = 2.77\text{ N/mm}^2\text{ (397 psi).}$$

Thus, the crack will develop only if there are additional internal stresses. According to Eq. (10.31), we then obtain

$$\min \rho = 1.05(1.69/420) = 0.42\%.$$

However, this requirement itself is not sufficient. By choosing the correct bar diameter Φ, we also have to make sure that the crack width—when the crack develops—stays within tolerable limits. For this example, we select $w = 0.3$ mm as a maximum width. For the crack spacing to be expected, the crack theory, as described in [9], Section 21.7, applies for crack width control. It results in the following equation:

$$\Phi \leq \frac{3v_1 wE_s}{f_{sd}^2}. \tag{10.32}$$

According to [9], Figs. 6.15 and 6.16, v_1 can be assumed to be $0.15f'_c$ for bars which are placed horizontally in the concrete [f'_c (as the average value) = 25.0 + 5.0 N/mm²).

$$\Phi \leq \frac{3(0.15)(25.0 + 5.0)0.03(210\,000)}{420.0^2} = 0.48\text{ cm} \sim 5\text{ mm (0.189 in.).}$$

If we want to increase the bar diameter to 10 mm, we have to reduce the steel stress to $f_s = 420.0\sqrt{4.8/10} = 290.0\ \text{N/mm}^2$. We then have

$$\min \rho = 1.05(1.69/290.0) = 0.61\,\%.$$

10.5.3.2. Short Structural Components

As has been shown in the preceding section, very long structural components require high amounts of reinforcement in order to prevent the development of wide cracks. In practice, we almost always use shorter components, for which the conditions are more favorable as shown in [11], because the additional strain caused by each crack noticeably reduces the shrinkage stress.

In the following, we will explain this by using the same example. The dimensions and material properties are therefore the same. The shrinkage strain of the 10-cm-thick wall, however, is to be impeded every 10 m by rigid supports.

To begin with, we examine (in Table 10.5) the development of shrinkage stresses in the uncracked cross section, in accordance with Section 2.3. As is shown in Fig. 10.8 for $d_{\text{eff}} = 20$ cm, the use of the simplified similarity assumption is not quite correct here because shrinkage develops more slowly than flow. In addition, the time dependence of the modulus of elasticity is taken into consideration.

The values computed in Table 10.5 and plotted in Fig. 10.12 lead to the following conclusions:

(a) The shrinkage stresses alone will cause cracks only if the tensile strength does not exceed the characteristic value $f_{tu5\%}$.
(b) Since there are always internal stresses, the first crack will presumably develop no later than 1 year after construction. (The assumption of a later date leads to the highest amount of reinforcement.)

For short structures, the additional strain at the crack will lead immediately to a decrease in the steel stress. The additional elongation Δl can be computed with the following formula (see [9], Section 21.7):

$$\Delta l = w\,\frac{\max f_s - n \max f_c}{\max f_s} \tag{10.33}$$

In this case, w stands for the permissible crack width which has to be secured by an appropriate choice of the bar diameter. For $w = 0.03$ cm and $f_s = 420\ \text{N/mm}^2$, we obtain with $\max f_c \triangleq f_{tu}$ after 1 year $\sim 2.4\ \text{N/mm}^2$.

$$\Delta l = w\,\frac{420 - 7(2.4)}{420} \sim 0.95w = 0.0285\ \text{cm}.$$

Thus, we now have $\Delta f_c = \Delta l \beta_e E_{c28} = 0.95(0.03)1.09(30\,000) = 0.93\ \text{N/mm}^2$. According to Fig. 10.12, the tensile stress caused by shrinkage alone will amount to $1.71\ \text{N/mm}^2$ after 1 year. As had to be expected, the more exact calculation resulted in a higher value than the one assuming similarity ($1.69\ \text{N/mm}^2$ according to Section 10.5.3.1).

Table 10.5. Example of a 10-cm-thick wall between rigid supports at a distance of 10 m; $2A/u = 8$ cm; 80% relative humidity; $\beta_{eff} = 2.4$; $d_{eff} = \beta_{eff} 2A/u = 20$ cm; Type II cement; concrete: $f'_c = 25$ N/mm²; $E_{c28} = 30\,000$ N/mm²; $t_1 = 0$; $\varphi_{f0} = 1.70$; $\varphi_{f\infty} = 1.70(1.55) \approx 2.60$; $\varepsilon_{s0} = 20 \times 10^{-5}$; $\varepsilon_{s\infty} = 20(0.90 \times 10^{-5}) = 18 \times 10^{-5}$; $f_{tut} = C\beta_c^2 \sqrt[3]{f_c'^2}$ (values for C from Part B, Section 2.2); C_m values for Eq. (2.31) according to Eq. (2.26).

Day	$\Delta\varphi_f$	$10^5\Delta\varepsilon_s$	β_e	β_{em}	C_m	$f_{cst} = \frac{C_m - 0.5\Delta\varphi_f}{C_m + 0.5\Delta\varphi_f} f_{cst-1} + \frac{\Delta\varepsilon_s E_{c28}}{C_m + 0.5\Delta\varphi_f}$	f_{tu} 5%	f_{tu} m	f_{tu} 95%	f_{cst}/f_{tu} 5%	f_{cst}/f_{tu} m	f_{cst}/f_{tu} 95%
0												
	0.52	1.3			~2.00							
5			0.82			$0 + \frac{0.3(1.3)}{2.00 + 0.26} = 0 + 0.173 = 0.173$	1.12	1.84	2.60	0.15	0.09	0.07
	0.52	2.9		0.915	1.49							
31			1.01			$\frac{1.49 - 0.26}{1.49 + 0.26} 0.173 + \frac{0.3(2.9)}{1.49 + 0.26} = 0.122 + 0.499 = 0.621$	1.70	2.78	3.87	0.36	0.22	0.16
	0.52	3.7		1.040	1.36							
110			1.07			$\frac{1.36 - 0.26}{1.36 + 0.26} 0.621 + \frac{0.3(3.7)}{1.36 + 0.26} = 0.425 + 0.681 = 1.112$	1.91	3.12	4.35	0.58	0.36	0.26
	0.52	5.7		1.090	1.32							
500			1.11			$\frac{1.32 - 0.26}{1.32 + 0.26} 1.112 + \frac{0.3(5.7)}{1.32 + 0.26} = 0.752 + 1.086 = 1.838$	2.05	3.36	4.68	0.90	0.55	0.39
	0.52	4.4		1.113	1.30							
∞			1.115			$\frac{1.30 - 0.26}{1.30 + 0.26} 1.838 + \frac{0.3(4.4)}{1.30 + 0.26} = 1.235 + 0.849 = 2.084$	2.07	3.40	4.78	1.01	0.61	0.44
	2.60	18										

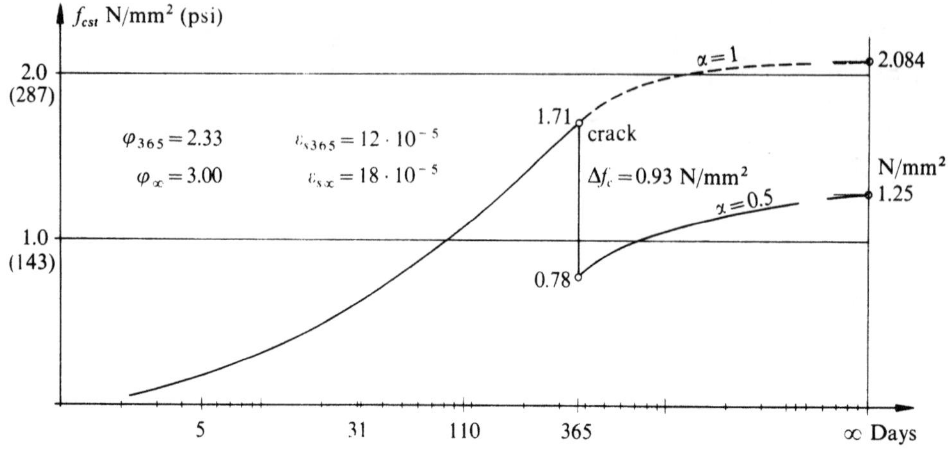

Figure 10.12.

Thus, we obtain

$$\rho \geq 1.05\,\frac{1.71 - 0.93}{420} = 0.20\,\%.$$

As has been shown in Section 10.5.3.1, the steel stress has to be reduced if 10-mm bars are used [see Eq. (10.32)].

$$\rho \geq 1.05\,\frac{1.71 - 0.93}{290} = 0.28\,\%.$$

This is only 46% of the threshold value calculated for extremely long structural components with the favorable assumption of similarity and discounting changes in E_c.

Further shrinkage leads again to an increase in tensile stress which, however, is limited through creep. When calculating these values, we have to take into consideration that the crack acts somewhat like a spring. This is the reason why $\alpha < 1$. We obtain:

$$\alpha = \frac{f_{cs} - \Delta f_c}{f_{cs}} = \frac{1.71 - 0.93}{1.71} = 0.46.$$

The final value $f_{cs\infty}$ is influenced by three factors (see Fig. 10.12):

The initial value $f_{cs} = +1.71$ N/mm².

The stress increase which is caused by the remaining shrinkage $\Delta\varepsilon_s = 6 \times 10^{-5}$ with $\alpha = 0.46$.

The constraint $\Delta f_c = 0.93$ N/mm², which is caused by the crack and changed by creep.

The calculation is done analogous to Table 10.5, whereby $\alpha = 0.46$ [according to Eq. (2.31)]:

$$f_{cs\infty} = \frac{(1/\alpha + 1/\beta_{em} - 0.6 - 0.5\Delta\varphi_f)f_{cs365} + \Delta\varepsilon_s E_{c28}}{1/\alpha + 1/\beta_{em} - 0.6 + 0.5\Delta\varphi_f}$$

$$= \frac{[1/0.46 + 1/1.113 - 0.6 - 0.5(0.67)](1.71 - 0.93) + 6(0.3)}{1/0.46 + 1/1.113 - 0.6 + 0.5(0.67)}$$

$$= 1.25 \text{ N/mm}^2 < 1.71.$$

If we have primarily shrinkage stresses, a second crack is not very likely to develop.

It can also be proven that all subsequent cracks are narrower. Deliberations concerning minimum reinforcement of structural components subjected to tension [Eq. (10.31)] can also be applied to those subjected to bending.

10.5.4 Special Cases: Watertight Concrete and/or Crack-Free Concrete

For the sake of completeness, this section deals with a question which is important in engineering practice. It deals only indirectly with creep and shrinkage.

The calculations demonstrated so far become questionable if watertight and/or crack-free concretes are required. In such cases, specifications for reinforced concrete normally require permissible tensile stresses which are lower than the tensile strength of the concrete. Since imposed stresses have to be considered here, this criterion reaches intolerable proportions with increasing thickness of the structure. Without prestressing, this condition frequently can only be met by uneconomical expenditures.

However, for many applications of watertight concrete, it is not necessary to stay below the tensile strength of the concrete.

As long as the cross section is subjected to bending and the thickness of the compression region >15 cm (>5.9 in.) [13], watertightness is guaranteed, and we can forego expensive additional water proofing. Frequently, it is cheaper to reduce the steel stress in order to enlarge the compression region sufficiently. In the area of small moments, concentric tension develops because of restrained shrinkage. In such cases, the stresses should be less than the tensile strength for concrete, taking into account appropriate safety factors.

Otherwise, we have to proceed according to Section 10.5.3 with respect to the minimum reinforcement and crack width.

11. Coupling of Structural Components with Different Creep and Shrinkage Properties

11.1. Coupling through One Statically Indeterminate Action Effect

11.1.1. General

For such cases, the *Dischinger/Kupfer* method is particularly suited. However, this method presupposes a similar development over time for the various creep values. Since there are only small deviations, this simplification has no marked consequences. Only a coupling of very thin structural components with very thick components will lead to more prominent deviations from similarity. We will demonstrate in Section 12 how to proceed in case the deviations are prominent.

The following examples do not deal with prestressing separately. With good approximation, we can assume the effects of different displacements due to prestressing δ_p for small values of α to be similar to the effect of a sustained load, if we take into account half of the loss of prestress; that is, if we proceed from an average value. We can thereby avoid taking into account the additional statically indeterminate action effect which develops as a consequence of the particular pattern of prestressing tendons.

11.1.2. Two-Span Continuous Girder with Successively Cast and Demolded Spans

The following example is calculated as usual with the simplifying assumption that creep in both spans of different ages develops similarly over time.

As demonstrated in earlier examples (see Section 9.4, for example), it is useful in such cases to divide the moments, which developed after the second span was finished, by the load redistribution method into imposed action effects and load effects (Fig. 11.1).

In order to simplify the example, $l_1 = l_2$ and $E_{cU} = E_{cW}$. The total moment for time $t = 0$ is $M_B = -dl^2/16$ and is divided into a load effect: $M_B = dl^2/8$ (both spans built at the same time) and an imposed action effect $M_B = +dl^2/16$ (result of subsequent construction).

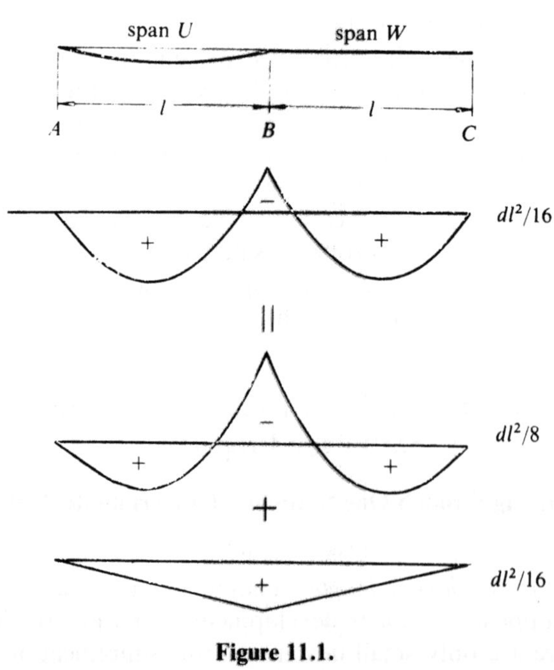

Figure 11.1.

In order to make a comparison with the example in Section 12.1 possible, the following creep values are used: $\varphi_U = 2.18$; $\varphi_W = 2.62$. Since $l_U = l_W$, $\delta_{U_11} = \delta_{W_11}$. $= \delta_{W_11}$.

According to Eq. (2.30) in Appendix II:[1]

$$\varphi'_{UW} = \frac{\delta_{U_11}(\varphi_U - 0.4) + \delta_{W_11}(\varphi_W - 0.4)}{1.4(\delta_{U_11} + \delta_{W_11})} = \frac{\varphi_U + \varphi_W - 0.8}{2.8} = 1.43,$$

$$e^{-\varphi_{UW}} = 0.239;\ \alpha_U = 1.4\frac{1.43}{1.78} = 1.12;\ \alpha_W = 1.4\frac{1.43}{2.22} = 0.90.$$

Thus, we obtain with, Eq. (2.30) in Appendix II,

$$-M_{Bt} = \left(\frac{1 - (1 - 1.12)0.239}{1.12}\right)\frac{dl^2}{16} + \left(\frac{1 - (1 - 0.90)0.239}{0.90}\right)$$

$$\times \frac{dl^2}{16} - \frac{0.239}{1.4}\frac{dl^2}{16} = (0.92 + 1.08 - 0.17)\frac{dl^2}{16} = 0.91\frac{dl^2}{8}.$$

The surprising thing about these results is that the difference in creep behavior does not influence the sustained load effect at all. That is, the first two terms of the preceding equation result in $(0.46 + 0.54)dl^2/8 = 1.00dl^2/8$. The cause can be found in the symmetry of the structure. Since $\delta_{UO} = \delta_{WO} = \delta_{10}/2$ and $\delta_{U_11} = \delta_{W_11} = \delta_{11}/2$, we obtain

$$X_t = \frac{\delta_{UO}\varphi_U + \delta_{WO}\varphi_w}{\delta_{U_11}\varphi_U + \delta_{W_11}\varphi_W} = \frac{\delta_{10}(\varphi_U + \varphi_W)}{\delta_{11}(\varphi_U + \varphi_W)} = \frac{\delta_{10}}{\delta_{11}} = X_0.$$

11.1.3. Construction of a Rigid Frame with Two Hinged Supports Consisting of a Prefabricated Truss and Cast in situ Concrete Columns

A similar example will be described in Section 7.2 of Appendix II where the basic outlines of methods suited for such cases are explained. The following calculation is above all designed to show the extent of calculations required.

The rigid frame with two hinged supports in Fig. 11.2 consists of cast in situ concrete columns and a prefabricated truss. We assume that the truss will creep less than the columns.

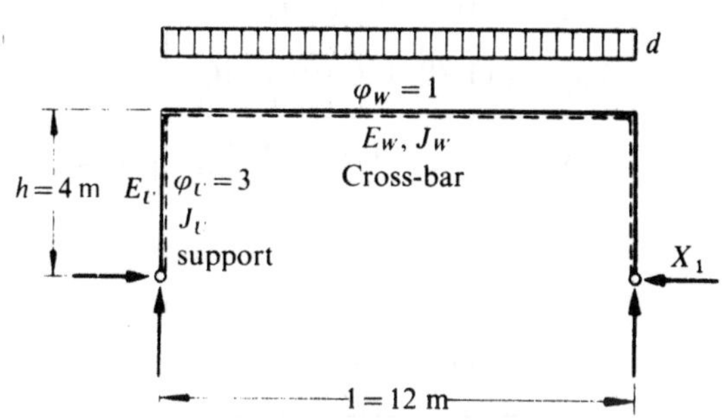

Figure 11.2.

[1] Similar to Section 11.1.3 we can also use the transformed relations of Section 2.1.3.

Assumptions

Truss: $l = 12.0$ m; $I_W = 0.0105$ m^4; $\varphi_W = 1$; $\varepsilon_{sW\infty} = 10 \times 10^{-5}$;

$$E_{cW} = 3.0 \times 10^6 \text{ N/cm}^2.$$

Column: $h = 4.0$ m; $I_U = 0.0072$ m^4; $\varphi_U = 3$; $E_{cU} = 3 \times 10^6$ N/cm^2.

The truss is prefabricated, installed without auxilliary supports and subsequently connected to the columns. Over the course of time, a negative moment develops due to the dead weight of the truss which, however, is partially reduced through shrinkage.

The moments developing under the dead weight of the truss are subdivided into a load dependent and an imposed action dependent part, in accordance with example 11.1.2:

Load condition under $X = 1$ kN

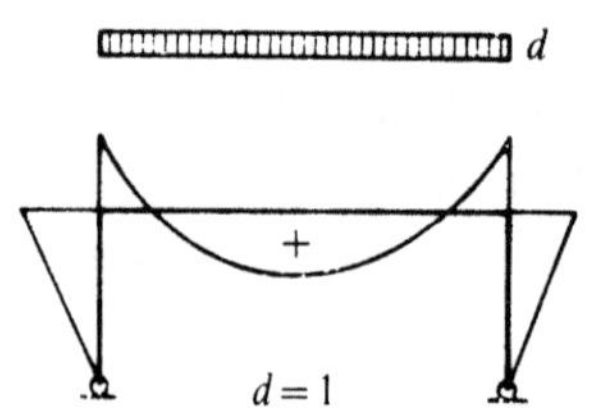

$$\delta_{W_1d} = -\frac{2}{3}\frac{dl^2}{8}\frac{hl}{E_W I_W} = \frac{2(1)12^2(4)12}{3(8)3.0 \times 10^6(0.0105)},$$

$$= -18.3 \times 10^{-3}$$

$$\delta_{U_11} = \frac{2}{3}\frac{h^3}{E_U I_U} = \frac{2(4^3)}{3(3.0 \times 10^6)0.0072} = 1.98 \times 10^{-3},$$

$$\delta_{W_11} = \frac{lh^2}{E_W I_W} = \frac{12(4^2)}{3.0 \times 10^6(0.0105)} = 6.10 \times 10^{-3},$$

$$\delta_{11} = \delta_{U_11} + \delta_{W_11} = (1.98 + 6.10) \times 10^{-3} = 8.08 \times 10^{-3},$$

$$X_d = X_{dW} = -\frac{\delta_{W_1d}}{\delta_{11}} = \frac{18.3}{0.08} = +2.26.$$

Constraint $X_t = -2.26$ kN

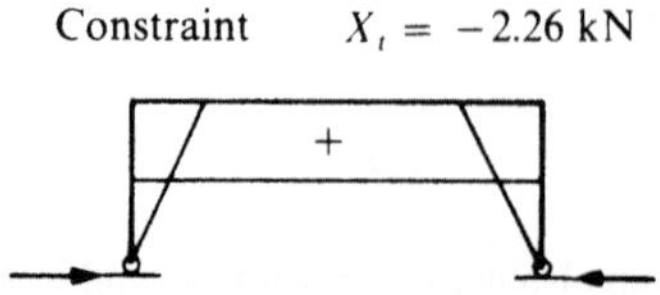

The sum of both cases provides the condition at time $t = 0$,

$$X_t = X_{dt=0} + X_{rt=0} = 0.$$

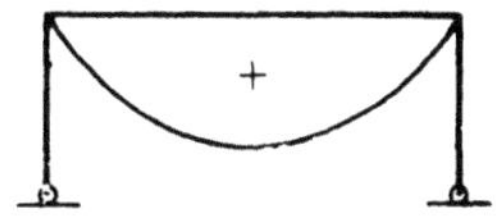

For time $t = \infty$ we obtain, from Eq. (2.19),

$$X_{t=\infty} = \gamma_{\mathrm{sus}\,U} X_{dU} \frac{1 + 0.4\alpha_U}{1.4} + \gamma_{\mathrm{sus}\,W} X_{dW} \frac{1 + 0.4\alpha_W}{1.4} + \gamma_r X_{r0} + \gamma_s X_{s\infty}^{(e)}.$$

Thus,

$$X_{dU} = 0 \text{ because } \delta_{U_1 d} = 0,$$

$$X_{s\infty}^{(e)} = -\frac{\delta_s}{\delta_{11}} = \frac{\varepsilon_s l}{\delta_{11}} = \frac{-10 \times 10^{-5}(12)}{10^{-3}(1.98 + 6.10)} = -1.48 \text{ kN},$$

$$\varphi_{UW} = \frac{\delta_{U_1 1}\varphi_U + \delta_{W_1 1}\varphi_W}{\delta_{U_1 1} + \delta_{W_1 1}} = \frac{1.98(3) + 6.10(1)}{1.98 + 6.10} = 1.49.$$

The γ coefficients we obtain from Table 2.1 and Eq. (2.21). $\gamma_{\mathrm{sus}\,W} = 0.51$ in accordance with Fig. 2.1, with

$$\alpha_W = \frac{\varphi_{UW} - 0.4}{1.4\varphi_W - 0.4\varphi_{UW} - 0.4} = \frac{1.49 - 0.4}{1.4(1) - 0.4(1.49) - 0.4} = 2.7$$

and $\varphi_W = 1$; $\gamma_r = 0.33$ according to Fig. 2.2, with $\alpha = 1$ and $\varphi = \varphi_{UW} = 1.49$; $\gamma_s = 0.50$ according to Fig. 2.3, with $\alpha = 1$ and $\varphi = \varphi_{UW} = 1.49$.

Thus, we have

$$X_{t=\infty} = 0.51 \frac{1 + 0.4(2.7)}{1.4} 2.26 - 0.33(2.26) - 0.50(1.48)$$

$$= 1.72 - 0.75 - 0.74 = 0.23 \text{ kN (51 lbs)}.$$

This means that over the course of time, the horizontal thrust increases from 0 to 0.23 kN. If we disregard shrinkage, we obtain $X_{t\infty} = 0.97$ kN, as opposed to $X_d = 2.26$ kN for a truss cast in situ. That is, $0.97/2.26 \triangleq 43\%$ disregarding the effect of shrinkage. Note that shrinkage would occur in both cases.

The sustained loads, which develop after the prefabricated truss and the cast in situ columns are joined, are also redistributed through the varied creep behavior. Here the horizontal thrust decreases at a ratio of 1.72/2.26 to 76% because the columns, which creep more, are more flexible than the truss.

11.2. Multiple-Degree Statically Indeterminate Coupling of Structural Components with Different Creep and Shrinkage Properties

11.2.1. General

It is possible to treat this problem rigorously from a mathematical point of view by using the relations in Section 8 and Knittel's work ([19], Appendix II), but for practical use this is too involved. It is better to use the Busemann method as described in Section 5. Occasionally, we can find a satisfactory approximation by neglecting individual coupling forces, as will be shown in Section 11.2.3.

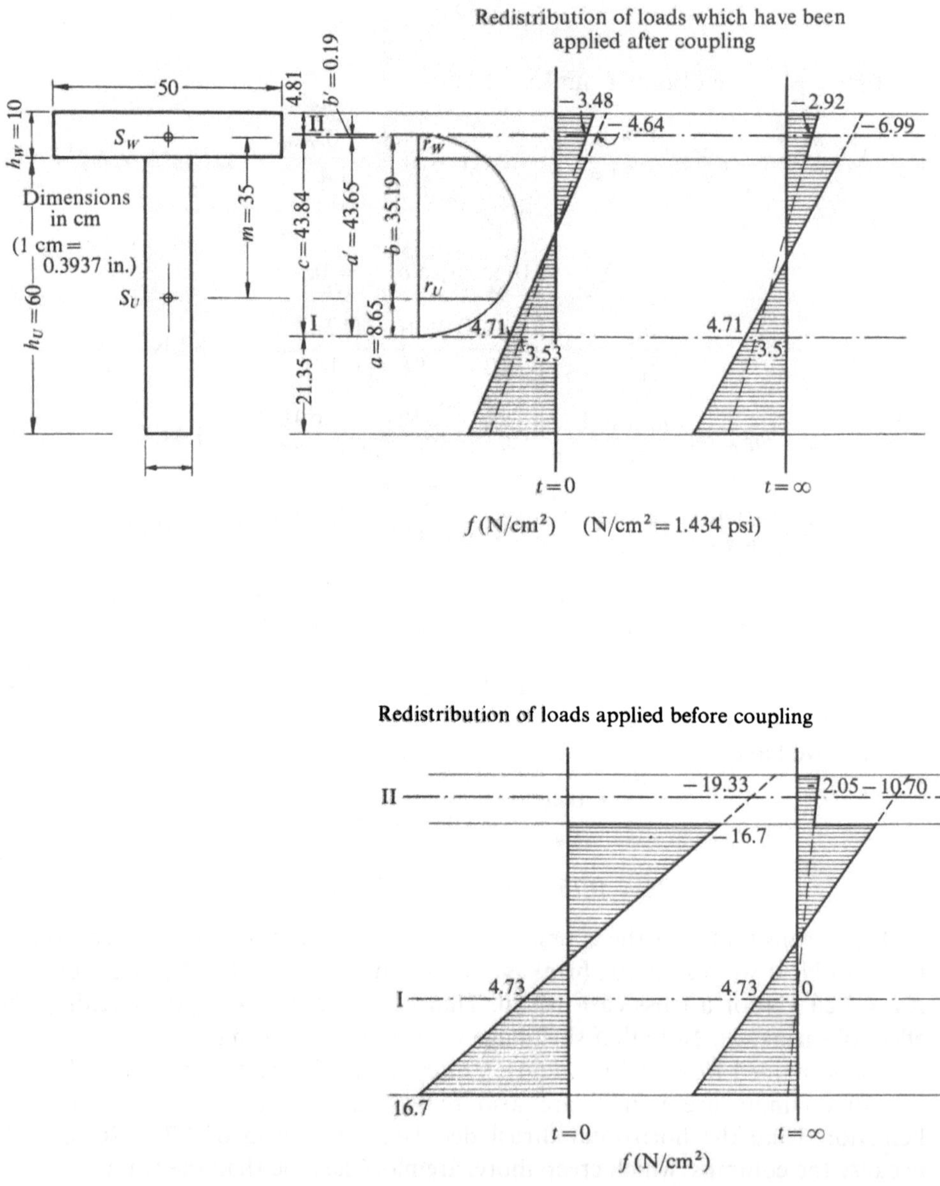

Figure 11.3.

11.2.2. Consideration of Flexural Rigidity of a Cast in situ Slab, Interacting with a Prefabricated Web (Busemann Method)

11.2.2.1. Assumptions

Here we have two coupling conditions which determine the shear stresses v and the normal stresses f, effective at the joint.

With the aid of Busemann's creep fiber method (see Appendix II, Section 6), the composite cross section is divided into concentrically stressed elements in place of the creep fibers, as shown in the example in Section 5. Each of these elements consists of two segments derived from the prefabricated part W and the cast in situ concrete U. Because these segments exhibit different creep and shrinkage properties, a stress redistribution takes place over the course of time. It is solved as follows, in accordance with the Dischinger/Kupfer method.

Figure 11.3 shows the dimensions of the cross section.

Cross section characteristics for the Busemann method [see Eqs. (6.1)–(6.8) in Appendix II].

U For the prefabricated part	W For the cast in situ concrete
$A = 60(10) = 600\ \text{cm}^2$	$A = 50\,(10) = 500\ \text{cm}^2$
$I = 10(60^3/12) = 180\,000\ \text{cm}^4$	$I = 50(10^3/12) = 4170\ \text{cm}^4$
$r_U = \sqrt{180\,000/600} = 17.3\ \text{cm}$	$r_W = \sqrt{4170/500} = 2.88\ \text{cm}$

$m = 35\ \text{cm}$

$\varkappa = 17.3^2 + 35^2 - 2.88^2 = 1512$

$$b' = \frac{\sqrt{1512^2 + 4(35^2)2.88^2} - 1512}{2(35)} = 0.19\ \text{cm}$$

$$a' = \frac{r_W^2}{b'} = \frac{2.88^2}{0.19} = 43.65\ \text{cm}$$

$c = a' + b' = 43.65 + 0.19 = 43.84\ \text{cm}$

$a = a' - m = 43.65 - 35.00 = 8.65\ \text{cm}$

$b = c - a = 43.84 - 8.65 = 35.19\ \text{cm}$

Area CI

$$A_{UI} = \frac{A_U b}{c} = \frac{600(35.19)}{43.84} = 482\ \text{cm}^2 \qquad A_{WI} = \frac{500(0.19)}{43.84} = 2\ \text{cm}^2$$

Area CII

$$A_{UII} = \frac{A_U a}{c} = \frac{600(8.65)}{43.84} = 118\ \text{cm}^2 \qquad A_{WII} = \frac{500(43.65)}{43.84} = 498\ \text{cm}^2$$

$E_c = 39\,000\ \text{N/mm}^2$	$E_c = 30\,000\ \text{N/mm}^2$
$\varphi_U = 1.5$	$\varphi_W = 3.0$
$\varepsilon_{sU} = 10 \times 10^{-5}$	$\varepsilon_{sW} = 20 \times 10^{-5}$

Differences in the values of E for the prefabricated part and for the concrete cast in situ can, for the elastic case, be taken into account through a fictitious reduction of the cross section of the concrete cast in situ.

$$n = \frac{E_W}{E_U} = \frac{300\,000}{390\,000} \sim 0.75.$$

The cross section characteristics corrected by n are

for creep fiber I:

$$A_{UI} + nA_{WI} = 482 + 0.75(2) = 482 + 2 = 484 \text{ cm}^2;$$

for creep fiber II:

$$A_{UI} + nA_{WII} = 118 + 0.75(498) = 118 + 374 = 492 \text{ cm}^2.$$

Calculation is done with the expanded creep relations in Appendix II, Section 2.2.4 in order to take into consideration the delayed elasticity effect. For comparison, the original relations (without delayed elasticity, Section 2.1.4, Appendix II) are given in parentheses.

Creep fiber I	Creep fiber II
$\varphi'_{UWI} = \dfrac{(1.5-0.4)/482 + (3.0-0.4)/2}{1.4(1/482 + 1/2)} = 1.86(2.99)$	$\varphi'_{UWII} = \dfrac{(1.5-0.4)/118 + (3.0-0.4)/374}{1.4(1/118 + 1/374)}$ $= 1.04(1.86)$
$\alpha_{UI} = 1.4(1.86/1.1) = 2.36(1.99)$	$\alpha_{UII} = 1.4(1.04/1.1) = 1.32(1.24)$
$\alpha_{WI} = 1.4(1.86/2.6) = 1.00(1.00)$	$\alpha_{WII} = 1.4(1.04/2.7) = 0.56(0.62)$
$e^{-\varphi'_{UWI}} = 0.156(0.05)$	$e^{-\varphi'_{UWII}} = 0.353(0.156)$

As a rule, the prefabricated web receives part of the load d_1 prior to coupling, while the sustained load parts d_2 become effective later on. Section 11.2.2.2 initially examines the effect of creep for part d_2. The more complicated solution of the first problem is described in Section 11.2.2.3.

11.2.2.2. Redistribution of Action Effects which Develop after Coupling

Action Effect: $M = 1$ kNm;

Couple of forces: $N_I = -N_{II} = \dfrac{1}{0.4384} = 2.281$ kN.

In the coupled cross sections, this force is distributed at time $t = 0$ as follows ($1 \text{ N/cm}^2 = 1.434$ psi):

Creep Fiber I (Tension)
Prefabricated part U:

$$N_{UI} = N_I A_{UI}/(A_{UI} + nA_{WI}) = 2281(482/484) = 2272 \text{ kN}.$$

Cast in situ slab W:

$$N_{WI} = N_I nA_{WI}/(A_{UI} + nA_{WI}) = 2281 - 2272 = +9 \text{ N};$$

$$f_{UIO} = N_{UI}/A_{UI} = N_{WI}/(nA_{WI}) = 4.71 \text{ N/cm}^2;$$

$$f_{WIO} = nf_{UIO} = 0.75(4.71) = 3.53 \text{ N/cm}^2.$$

Creep Fiber II (Compression)
Prefabricated part U:

$$N_{UII} = N_{II} A_{UII}/(A_{UII} + nA_{WII}) = -2281(118/492) = -547 \text{ N}.$$

Cast in situ slab W:

$$N_{WII} = N_{II} - N_{UII} = -2281 + 547 = -1734 \text{ N};$$

$$f_{UIIO} = N_{UII}/A_{UII} = N_{WII}/(nA_{WII}) = -4.64 \text{ N/cm}^2;$$

$$f_{WIIO} = 0.75(-4.64) = -3.48 \text{ N/cm}^2.$$

Coupling Forces in Creep Fibers

Creep Fiber I
The insignificant influence of the cast in situ concrete on the cross section characteristics immediately suggests that creep redistribution can be neglected. Thus, after redistribution, we have

$$f_{UIt} \approx f_{UIO} = 4.71 \text{ N/cm}^2,$$

$$f_{WIt} \approx f_{WiO} = 3.53 \text{ N/cm}^2.$$

Creep Fiber II
Creep fiber II is composed of two cross sections. They are

from the prefabricated part U:

$$N_{UIIO} = -547 \text{ N};\ A_{UII} = 118 \text{ cm}^2;\ f_{UIIO} = -4.64 \text{ N/cm}^2.$$

from the cast in situ concrete W:

$$N_{WIIO} = -1734 \text{ N};\ nA_{WII} = 374 \text{ cm}^2;\ f_{WIIO} = nf_{UIIO} = 3.48 \text{ N/cm}^2.$$

We have demonstrated above that the strains in both cross sections are compatible at time $t = 0$ because $N_{UII}/A_{UII} = N_{WII}/(nA_{WII})$. However, the two cross sections exhibit varied creep and shrinkage properties which result in a redistribution of action effects. To use Eq. (2.30), Appendix II, we need the following initial values:

$$X_{dU} = -\frac{\delta_{UId}}{\delta_{II}} = -\frac{N_U(A_U E_{cU})}{1/(A_U E_{cU}) + 1/(A_W E_{cW})} = -\frac{N_U n A_W}{A_U + nA_W},$$

$$X_{dW} = -\frac{\delta_{WId}}{\delta_{II}} = -\frac{N_W/(A_W E_{cW})}{1/(A_U E_{cU}) + 1/(A_W E_{cW})} = -\frac{N_W A_U}{A_U + nA_W},$$

$$X_{\Delta s}^{(e)} = -\frac{\delta_{Is}}{\delta_{II}} = -\frac{\Delta\varepsilon_s}{1/(A_U E_{cU}) + 1/(A_W E_{cW})} = -\frac{\Delta\varepsilon_s A_U n A_W E_{cU}}{A_U + nA_W}.$$

If applied, we then obtain:

$$X_{dU} = -\frac{-547(374)}{118 + 374} = +416 \text{ N},$$

$$-X_{dW} = -\frac{-1734(118)}{18 + 374} = +416 \text{ N},$$

$$X_{\Delta s}^{(e)} = -\frac{-10 \times 10^{-5}(118)374(3 \times 10^6)}{118 + 374} = 26.90 \text{ kN}.$$

According to Eq. (2.30), Appendix II, we obtain for $M = 1$ kNm:

$$X_t = \frac{1 - (1 - [1.4\varphi'_{UW}/(\varphi_U - 0.4)])e^{-\varphi'_{UW}}}{1.4\varphi'_{UW}/(\varphi_U - 0.4)} X_{dU0}$$
$$+ \frac{1 - (1 - [1.4\varphi'_{UW}/(\varphi_W - 0.4)])e^{-\varphi'_{UW}}}{1.4\varphi'_{UW}/(\varphi_U - 0.4)} X_{dW0}.$$

We then obtain for the coupling force:

$$X_t = \frac{1 - (1 - 1.32)0.353}{1.32} 416 - \frac{1 - (1 - 0.56)0.353}{0.56} 416$$
$$= 351 - 628 = -277 \text{ N } (-283 \text{ N}).$$

As had to be expected, part of the sustained load is shifted to the prefabricated part which creeps less.

$$N_{UIIt} = N_{UII0} + X_{dt} = (-547 - 277) = -824 \text{ N};$$

$$N_{WIIt} = N_{WII0} - X_{dt} = (-1734 + 277) = -1457 \text{ N}.$$

This leads us to

$$f_{UIIt} = \frac{-824}{118} = -6.99 \text{ N/cm}^2;$$

$$f_{wIIt} = \frac{-1457}{498} = -2.92 \text{ N/cm}^2.$$

The stresses which are effective during both points in time were plotted in Fig. 11.3.

The higher shrinkage rate of the cast in situ concrete slab is effective in the same direction as the sustained load redistribution. We obtain from Eq. (2.30), Appendix II:

$$X_{\Delta st} = \frac{1}{1.4}\frac{(1 - e^{-\varphi'_{UW}})}{\varphi'_{UW_t}} X^{(e)}_{\Delta s} = \frac{1 - 0.353}{1.4(1.04)} 26.90 = 11.95 \text{ kN } (12.20 \text{ kN});$$

$$N_{UII_t} = 11.95 \text{ kN}, f_{UII_t} = \frac{11\,950}{118} = 101 \text{ N/cm}^2;$$

$$N_{WII_t} = -11.95 \text{ kN}, f_{WII_t} = \frac{11\,950}{488} = 24 \text{ N/cm}^2.$$

11.2.2.3. Redistribution of Loads which Developed Prior to Coupling

Assumption: The total moment $M = 1$ kNm is effective at time $t = 0$ in the prefabricated part U. The accompanying extreme fiber stress at the place of coupling is

$$f_U = My/I_U = 100\,000 \frac{30}{180\,000} = 16.7 \text{ N/cm}^2.$$

In this case too, it is most practical to proceed from the principle of stress redistribution as with the two previous examples. First, moment $M = 1$ is permitted to act upon the homogeneous total cross section at time $t = 0$, and then a constraint is superimposed which after superposition results in stresses in the prefabricated web only.

At time $t = 0$, the following forces are effective in the creep fibers:

For the homogeneous cross section (see Section 11.2.2.2):

$$N_{\text{I}} = -N_{\text{II}} = \frac{M}{c} = 2281 \text{ N}; N_{U\text{I}0} = +2272 \text{ N}: N_{W\text{I}0} = +9 \text{ N}.$$

$$N_{U\text{II}0} = -547 \text{ N}; N_{W\text{II}0} = -1734 \text{ N}.$$

As a result of the constraint:

$$N_{U\text{I}0} = +9 \text{ N}; N_{W\text{I}0} = -9 \text{ N}.$$

$$N_{U\text{II}0} = -1734 \text{ N}; N_{W\text{II}0} = +1734 \text{ N}.$$

Following superposition:

$$N_{U\text{I}0} = +2272 + 9 = +2281 \text{ N}: N_{W\text{I}0} = +9 - 9 = 0.$$

$$N_{U\text{II}0} = -547 - 1734 = -2281 \text{ N}; N_{W\text{II}0} = -1734 + 1734 = 0.$$

$$f_{U\text{I}0} = \frac{2281}{482} = 4.73; f_{U\text{II}0} = -\frac{2281}{118} = -19.33 \text{ N/cm}^2.$$

For time $t = \infty$, the influence of creep on the homogeneous condition has already been calculated in Section 11.2.2.2. Also, the constraint of creep fiber I is so small that it is useless to estimate changes caused by creep.

For creep fiber II, the following applies according to Eq. (2.30) in Appendix II:

$$X_{rt} = \frac{1}{1.4} e^{-\varphi'_{UW}} X_{r0} = \frac{0.353}{1.4} 1734 = 437 \text{ N } (270 \text{ N}).$$

Superposition then results in

$$N_{U\text{II}t} = N_{U\text{II}dt} + X_{rt} = -824 - 437 = -1261 \text{ N},$$

$$N_{W\text{II}t} = N_{W\text{II}dt} + X_{rt} = -1457 + 437 = -1020 \text{ N}.$$

Thus, we have

$$f_{U\text{II}t} = \frac{-1261}{118} = 10.70 \text{ N/cm}^2 \ (9.29 \text{ N/cm}^2),$$

$$f_{W\text{II}t} = \frac{-1020}{498} = 2.05 \text{ N/cm}^2 \ (2.37 \text{ N/cm}^2).$$

Originally, the compressive force of 2281 N in creep fiber II had to be taken up by the web alone. As a result of creep 1020 N of this force is redistributed to the cast in situ concrete slab. This represents 70% of $N_{W\text{II}t} = 1457$ N, which develops if the load is not effective until after coupling, and/or 59% of $N_{W\text{II}0} = 1734$ N.

11.2.3. Approximation by Neglecting the Flexural Rigidity of the Slab (Dischinger/Kupfer Method)

If the cast in situ concrete slab is, by comparison, thin (Fig. 11.4), we can frequently neglect its flexural rigidity compared to that of the prefabricated part. We can then immediately write the solution with the aid of the formulas in Section 2.1.3 or Appendix II, Section 2.1.4.

In the following, this method is used for the example mentioned in the previous section.

Displacements for the Dowel Force X = 1

$$E_U \delta_{U_11} = \frac{1}{A_U} + \frac{h_U/2}{I_U}\frac{h_U}{2} = \frac{1}{A_U} + \frac{h_U^2}{4I_U} = \frac{1}{600} + \frac{60^2}{4(180\,000)} = \frac{1}{150},$$

$$E_U \delta_{W_11} = \frac{1}{nA_W} = \frac{1}{0.75(500)} = \frac{1}{375}.$$

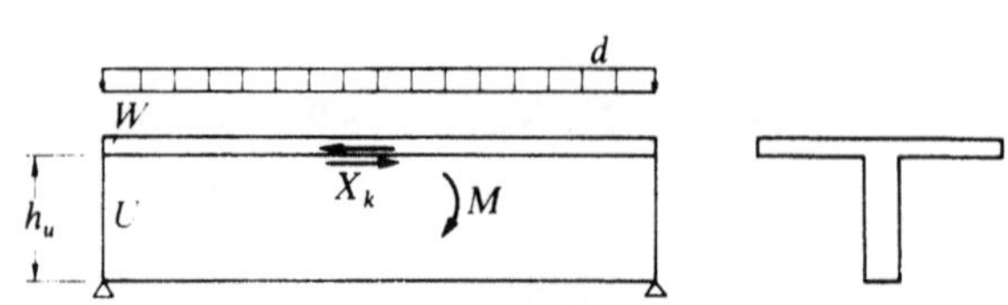

Figure 11.4.

According to Appendix II, Eq. (2.30):

$$\varphi'_{UW} = \frac{(1.5 - 0.4)/150 + (3.0 - 0.4)/375}{1.4(1/150 + 1/375)} = 1.10,$$

$$\alpha_W = 1.4[1.1/(3.0 - 0.4)] = 0.59; e^{-1.1} = 0.333.$$

In this case, too, we have to divide the loads, in accordance with the following diagram, into a sustained load *a* and a constraint portion *b*.

Using the creep and shrinkage coefficients from the previous example, we obtain the following for coupling after the prefabricated part is loaded:

$$X_{d0}^{(e)} = \frac{QS}{I}.$$

At time $t = 0$:

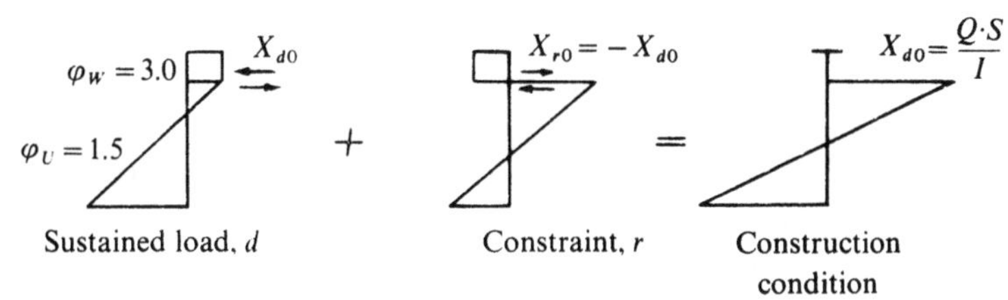

At time $t = \infty$, according to Eq. (2.30), Appendix II:

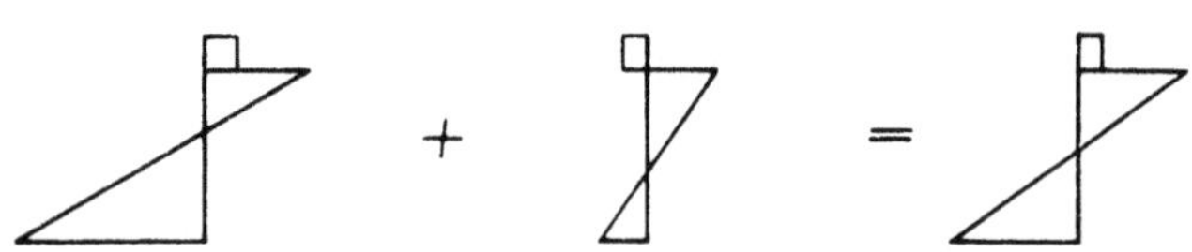

$$X_i = \frac{1 - (1 - 1.4)0.333}{1.4} X_{d0} - \frac{0.333}{1.4} X_{d0} = (0.809 - 0.238) X_{d0} = 0.571 X_{d0}.$$

Initially, the prefabricated part is subjected to 100% of the weight of the cast in situ slab. By the end of the creep process, 57.1% of the load has shifted to the composite cross section.

12. Use of Step-by-Step Integration for Cases which Cannot, or Only with Difficulty, Be Solved in a Closed Mathematical Form

12.1. Redistribution of Action Effects for a Two-Span Continuous Girder Constructed in Two Parts, Taking into Consideration the Dissimilar Development of Creep over Time of Both Spans

A similar problem appeared in Section 11.12. However, for the sake of simplicity, it was assumed that the time-dependent deformations in both spans were not the same, but that their development over time was similar. In the case at hand, this assumption is avoided while everything else remains the same.

The two-span support is constructed one span at a time. The form work for the first span is to be removed after 15 days, and the one for the second span after 30 days. At the time of removal of the form work, the concrete in each case is 0.5 months old (see Fig. 12.1).

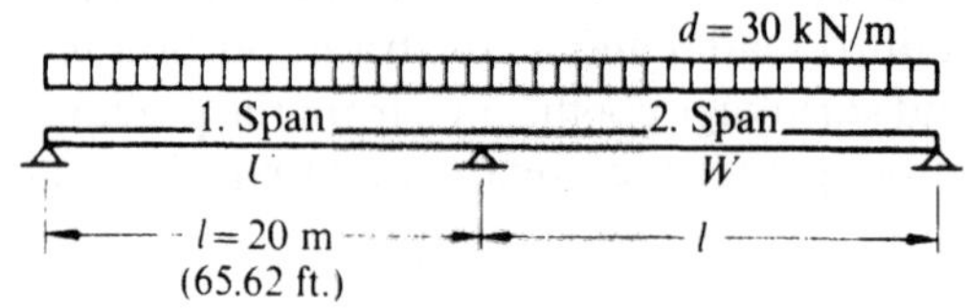

Figure 12.1.

After removal of the form work for the second span, the negative moment at the center support is initially only half of that for a two span continuously cast girder. However, when creep subsequently increases the deformation, the moment at the center support increases again.

For the example at hand, creep is subdivided into three time intervals only in order to limit the scope of the calculation. In a normal case, one would use four to five intervals. For practical reasons, they are chosen such that almost equal steps of $\Delta\varphi$ develop. A general description of the finite difference method appears in Section 3, Appendix II.

The additional moment at the center support ΔX_n, which develops at each time interval, acts as a load on the following interval. The individual creep stages of beams 1 and 2 are offset by 1 month. The corresponding values $\Delta\varphi$ are determined in Table 12.1 for 70% relative humidity, $d_{\text{eff}} = 20$ cm, and Type II cement.

The course of the calculation follows:

(a) Removal of form work from span U at time $t = 15$ days; no center support moment.

(b) Removal of form work from span W at time $t_0 = 45$ days; for $d = 30$ kN/m; the center support moment at this point in time becomes

$$X_0 = -\frac{EI\delta_{1dW}}{EI\delta_{11}} = \frac{-1000}{0.667 + 0.667} = \frac{dl^2}{16} = -750 \text{ kNm}. \qquad (12.1)$$

(c) The first time interval: $\Delta t^{\text{I}} = 45$–105 days.

$$\delta_{U_1d}\Delta\varphi_{U1}^{\text{I}*} + \delta_{W_1d}\Delta\varphi_{W1}^{\text{I}} + X_0(\delta_{U_11}\Delta\varphi_{U1}^{\text{I}} + \delta_{W_11}\Delta\varphi_{W1}^{\text{I}}) + \Delta X_1(\delta_{U_11} + \delta_{W_11})$$
$$+\Delta X_1[\delta_{U_11}(\Delta\varphi_{dU1}^{\text{I}} + \tfrac{1}{2}\Delta\varphi_{fU1}^{\text{I}}) + \delta_{W_11}(\Delta\varphi_{dW1}^{\text{I}} + \tfrac{1}{2}\Delta\varphi_{fW1}^{\text{I}})] = 0. \qquad (12.2)$$

The fact that the moment ΔX, which develops within a time interval, corresponds to a higher age at loading than moment X which is effective from the beginning can be neglected in view of the degree of accuracy achieved by selecting β_f.

The sub- and superscripts used in the formulas above, e.g., in $\varphi^{\mathrm{I}}_{W3}$, have the following meaning:

I stands for the time interval when the action effect begins to be effective;
W is the right span;
3 stands for the time interval during which the action effect is effective.

(d) The second time interval is $\Delta t^{\mathrm{II}} = 105\text{–}445$ days.

$$\begin{aligned}
&\delta_{U_1d}\Delta\varphi_{U2} + \delta_{W_1d}\Delta\varphi^{\mathrm{I}}_{W2} + X_0(\delta_{U_11}\Delta\varphi^{\mathrm{I}'}_{U2} + \delta_{W_11}\Delta\varphi_{W2}) \\
&\quad + \Delta X_1(\delta_{U_11}\Delta\varphi^{\mathrm{I}}_{U2} + \delta_{W_11}\Delta\varphi^{\mathrm{I}}_{W2}) + \Delta X_2(\delta_{U_11} + \delta_{W_11}) \\
&\quad + \Delta X_2[\delta_{U_11}(\Delta\varphi^{\mathrm{II}}_{dU2} + \tfrac{1}{2}\Delta\varphi^{\mathrm{II}}_{fU2}) + \delta_{W_11}(\Delta\varphi^{\mathrm{II}}_{dW2} + \tfrac{1}{2}\Delta\varphi^{\mathrm{II}}_{fW2})] = 0. \qquad (12.3)
\end{aligned}$$

(e) third time interval: $\Delta t^{\mathrm{III}} = 445 \cdot$ day through ∞.

$$\begin{aligned}
&\delta_{U_1d}\Delta\varphi^{\mathrm{I}}_{U3} + \delta_{W_1d}\Delta\varphi^{\mathrm{I}}_{W3} + X_0(\delta_{U_11}\Delta\varphi^{\mathrm{I}}_{U3} + \delta_{W_11}\Delta\varphi^{\mathrm{I}}_{W3}) \\
&\quad + \Delta X_1(\delta_{U_11}\Delta\varphi^{\mathrm{I}}_{U3} + \delta_{W_11}\Delta\varphi^{\mathrm{I}}_{W3}) + \Delta X_2(\delta_{U_11}\Delta\varphi^{\mathrm{II}}_{U3} + \delta_{W_11}\Delta\varphi^{\mathrm{II}}_{W3}) \\
&\quad + \Delta X_3(\delta_{U_11}\Delta\varphi^{\mathrm{I}}_{U3} + \delta_{W_11}\Delta\varphi^{\mathrm{I}}_{W3}) + \Delta X_2(\delta_{U_11}\Delta\varphi^{\mathrm{II}}_{U3} + \delta_{W_11}\Delta\varphi^{\mathrm{II}}_{W3}) \\
&\quad + \Delta X_3(\delta_{U_11} + \delta_{W_11}) + \Delta X_3[\delta_{U_11}(\Delta\varphi^{\mathrm{III}}_{dU3} + \tfrac{1}{2}\Delta\varphi^{\mathrm{III}}_{fU3}) \\
&\quad + \delta_{W_11}(\Delta\varphi^{\mathrm{III}}_{dW3} + \tfrac{1}{2}\Delta\varphi^{\mathrm{III}}_{fW3})] = 0. \qquad (12.4)
\end{aligned}$$

The final action effect is

$$X = X_0 + \sum \Delta X_i. \qquad (12.5)$$

Solving Eq. (12.5) gives us (1 kNm = 87.49 × 10^2 in. lbs)

$$X_0 = -750 \text{ kNm}.$$

Creep interval 1:

$$\begin{aligned}
-\Delta X_1 &= \frac{1}{0.667[2 + 0.24 + \frac{1}{2}(0.40) + 0.24 + \frac{1}{2}(0.68)]}[1000(0.44 + 0.92) \\
&\quad -750(0.64 + 0.92)0.667] = 288 \text{ kNm},
\end{aligned}$$

$$X_1 = X_0 + \Delta X_1 = -750 - 288 = -1038 \text{ kNm}.$$

Creep interval 2:

$$\begin{aligned}
-\Delta X_2 &= \frac{1}{0.667[2 + 0.37 + \frac{1}{2}(0.64) + 0.37 + \frac{1}{2}(0.76)]}[1000(0.77 + 0.89) \\
&\quad - 750(0.77 + 0.89)0.667 - 288(0.77 + 0.89)0.667] = 223 \text{ kNm},
\end{aligned}$$

$$X_2 = X_1 + \Delta X_2 = 1038 - 223 = -1261 \text{ kNm}.$$

Table 12.1. Calculation of creep coefficients valid for the individual time intervals. [$\Delta\varphi = 0.4\beta_{d(t-a)} + 2(\beta_{ft} - \beta_{fa}); d_{eff} = 20$ cm].

Action effect develops in the time interval		Left span U: I	II	III	Right span W: I	II	III
Designation of action effect		$X_0, \Delta X_1$	ΔX_2	ΔX_3	$X_0, \Delta X_1$	ΔX_2	ΔX_3
Age of concrete at start of effect of X (days)		45	105	445	15	75	415
Creep in time interval I	Time period (days)	45–105			15–75		
	Delayed elastic strain	0.61 (0.40)			0.61 (0.40)		
	Flow	+2.0 (0.86 − 0.66)			+2.0 (0.78 − 0.44)		
	Sum	$\Delta\varphi^{I}_{U1} = 0.24^{a} + 0.40$			$\Delta\varphi^{I}_{W1} = 0.24 + 0.68$		
Creep in time interval II	Time period (days)	105–445	105–445		75–415	75–415	
	Delayed elastic strain	0.32 (0.40)	0.92 (0.40)		0.32 (0.40)	0.92 (0.40)	
	Flow	+2.0 (1.18 − 0.86)	+2.0 (1.18 − 0.86)		+2.0 (1.16 − 0.78)	+2.0 (1.16 − 0.78)	
	Sum	$\Delta\varphi^{I}_{U2} = 0.77$	$\Delta\varphi^{II}_{U2} = 0.37 + 0.64$		$\Delta\varphi^{I}_{W2} = 0.89$	$\Delta\varphi^{II}_{W2} = 0.37 + 0.76$	
Creep in time interval III	Time period (days)	445 − ∞	445 − ∞	445 − ∞	445 − ∞	445 − ∞	445 − ∞
	Delayed elastic strain	0.07 (0.40)	0.08 (0.40)	1.00 (0.40)	0.07 (0.40)	0.08 (0.40)	1.00 (0.40)
	Flow	+2.0 (1.55 − 1.18)	+2.0 (1.55 − 1.18)	+2.0 (1.55 − 1.18)	+2.0 (1.55 − 1.16)	+2.0 (1.55 − 1.16)	+2.0 (1.55 − 1.16)
	Sum	$\Delta\varphi^{I}_{U3} = 0.77$	$\Delta\varphi^{II}_{U3} = 0.77$	$\Delta\varphi^{III}_{U3} = 0.40 + 0.74$	$\Delta\varphi^{I}_{W3} = 0.81$	$\Delta\varphi^{II}_{W3} = 0.81$	$\Delta\varphi^{II}_{W3} = 0.40 + 0.78$
		$\sum \varphi^{I}_{U} = 2.18$			$\sum \varphi^{I}_{W} = 2.62$		

[a] A part of $\Delta\varphi_d$ has already developed for ΔX_1 between days 15 and 45; thus we have $\Delta\varphi^{I}_{U1\Delta X} = 0.11(0.4) + 2.0(0.86 - 0.66) = 0.44$.

Creep interval 3:

$$-\Delta X_3 = \frac{1}{0.667[2 + 0.40 + \frac{1}{2}(0.74) + 0.40 + \frac{1}{2}(0.78)]}[1000(0.77 + 0.81)$$

$$- 1038(0.77 + 0.81)0.667 - 223(0.77 + 0.81)0.667] = 106 \text{ kNm},$$

$$X_3 = X_2 + \Delta X_3 = 1261 - 106 = 1367 \text{ kNm}.$$

This value corresponds to 0.91 $(dl^2/8)$.

The example in Section 11.1.2, which proceeded from the assumption that φ_U and φ_W are similar over time, also resulted in $0.91(dl^2/8)$ under the same conditions. This affinity, however, does not exist, as we can see from a comparison of the ratio of φ_W/φ_U. Table 12.1 shows that it varies between 1.04 and 1.44 which, however, does not change the results. Also, we mentioned before that three creep intervals are often not sufficient. If we take into account five intervals, we have $0.90(dl^2/8)$. As we can see, the difference in the case at hand is not great enough to justify the larger number of intervals (see Fig. 12.2).

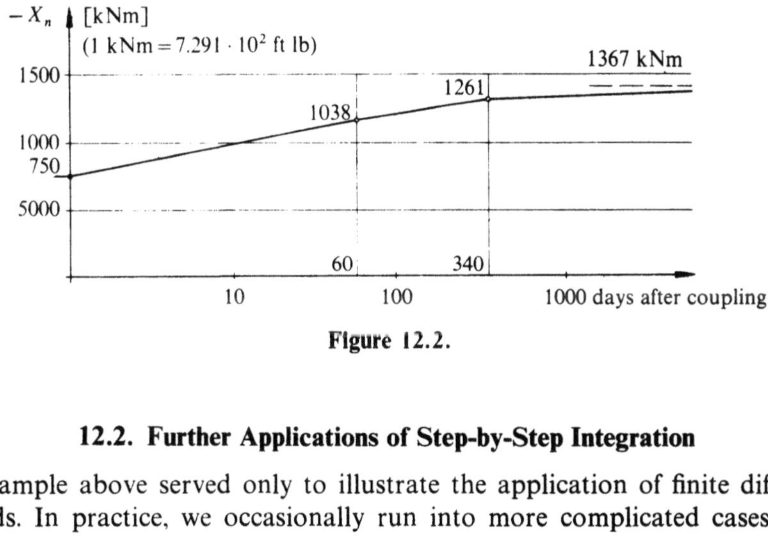

Figure 12.2.

12.2. Further Applications of Step-by-Step Integration

The example above served only to illustrate the application of finite difference methods. In practice, we occasionally run into more complicated cases which cannot be solved in a closed form. As an example, we can point to the box girder section of a wide-span concrete arch which rests on a free-span scaffolding and which is cast in three sections at three points in time (lower slab, webs, and upper slab).

Another example, mentioned in Section 10.3.5, is the development of internal shrinkage stresses in young concrete. We find similar conditions if we have to determine internal stresses resulting from the heat of hydration of young concrete. In such cases, it may be desirable to also consider those physical factors which change with increasing age, such as the coefficient of thermal expansion, the modulus of elasticity, and the changing temperature gradient.

13. Possible Ways of Reducing the Redistribution of Action Effects Caused by Creep

An engineer will always try to keep the redistribution of action effects as low as possible if the effects on the structural behavior are adverse. Sections 9.4, 11.1.2, and 12.1, for example, dealt with this redistribution caused by creep in structures consisting of prefabricated elements or of concrete cast in segments. In a two-span continuous girder, the extreme case can occur that the moment under sustained load in the span is $M_d = dl^2/8$, at time $t = 0$, and that the center support moment at time $t = \infty$ has almost the same value (see, also, Fig. 9.22). In the following, we will describe possible ways to avoid such an over-dimensioning.

13.1. Stage-by-Stage Construction of Continuous Girders

If multiple-span bridges are constructed in stages, in order to save scaffolding, we do not have to proceed as described in the examples in Sections 11.1.2 and 12.1. Those examples are designed to explain a calculation method. If we proceed as described in Fig. 13.1 for the span-by-span construction according to

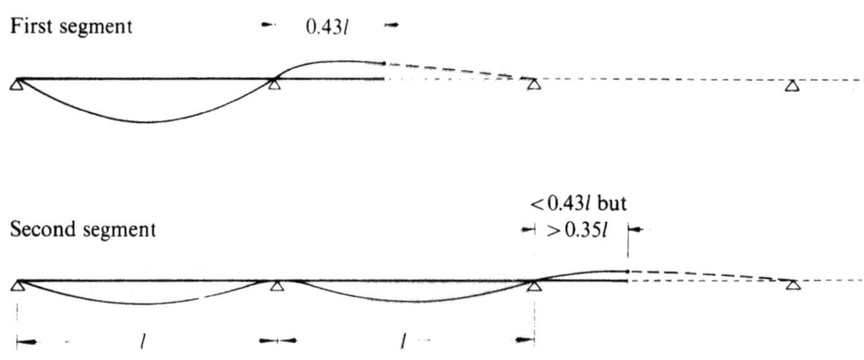

Figure 13.1. Step by step construction of continuous girders. Each span has a cantilever whose slope at the free end points towards the neighboring support.

[14], we can avoid a large redistribution of moments. Each new span projects far enough into the next span so that, following removal of the scaffold, the tangent at the end—or, for that matter, its imagined extension through a weightless and infinitely rigid element—points precisely to the next support. If, in the following construction segment, this imagined extension is replaced by concrete, no change of the static system occurs, and thus no imposed action, when the scaffold is removed from the next span. The imagined system receives only additional loads. With uniform loads, this condition is met when the cantilever length in the second span of a continuous girder with equal spans is $0.43l$ and with an infinite number of spans is $0.35l$.

13.2. Construction of Continuous Girders by Using a Series of Symmetrical Cantilevers

As shown in Fig. 13.2, continuous girders with wide spans are often built by joining separately constructed symmetrical free cantilevering girders. The moments that develop during construction of the individual spans are different from those

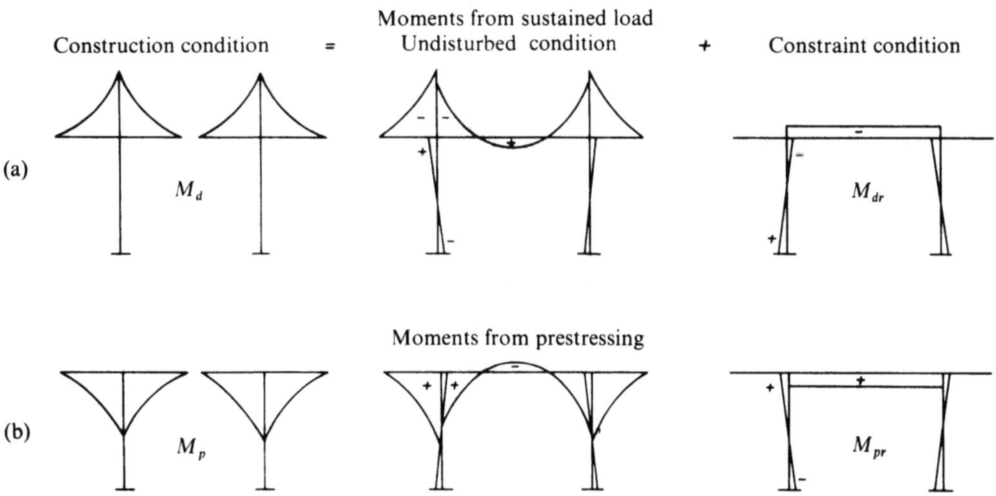

Figure 13.2. With later joining of symmetrical prestressed cantilever beams the redistribution of moments caused by creep can be reduced considerably through a suitable choice of prestressing. Redistribution moments from *d* and *p* are nearly inversely equal.

which would occur in the undisturbed condition, that is, in construction done with a continuous scaffold (see Fig. 13.2). The difference is the moment due to constraint M_{dr}. Thus, we can imagine that the construction condition develops through superposition of the undisturbed condition and the constraint condition. Creep does not change the moment diagram of the undisturbed condition, but it will greatly reduce the constraint moments. The distribution of moments therefore approaches that which would develop during construction in a single operation.

However, wide-span structures are almost always prestressed. Thus, according to Fig. 13.2b, we have the chance to choose the moments M_p which are imposed on the concrete through prestressing such that they will nearly neutralize the moments M_d developing during construction. This causes constraint moments M_{pr} to develop which also approximately neutralize M_{dr}, so that a creep redistribution is avoided.

Thus, cantilevered beams can be built practically without curvature with the aid of prestressing. However, this will never be completely accomplished because the prestress, and thus the counter-acting prestressing moments M_p, are partially reduced by creep.

14. Calculation of Deformations and Camber

14.1. General

Above all, deformation calculations serve two purposes. They permit an assessment of the dangers which deformation may cause, and they provide a base for calculating the required camber of the scaffolding. This is shown in Fig. 14.1, where *a* is counted from the stress-free condition, while the scaffolding has to be raised by *u* with respect to the required baseline.

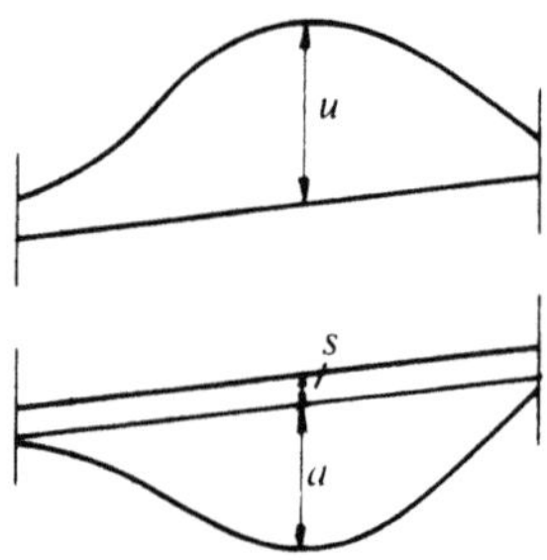

Camber *u* with respect to the required baseline

Deflection *a* with respect to the required baseline where *s* is caused by foundation settlement

Figure 14.1.

We can only estimate the deformations because the input variables E_c, φ_t, and ε_{st}, as well as the extent of cracking, vary considerably. Under these circumstances, one has to select values from the band of scatter which best serve the specific purpose of the deformation calculation.

Three groups have to be distinguished here:

Estimates of maximum deformations are needed where damage caused by too large deformations, such as cracks in partition walls or water traps on roofs are to be avoided. The same applies if the deformations endanger the stability of a structure (buckling).

Estimates of mean values of deformation are needed if the amount of camber is determined primarily from an esthetic point of view.

We proceed from estimates of minimum deformations if we want to determine the most adverse values of constraint.

The data given in Section 1 for the quantities E_c, φ, and ε_s are an attempt to describe mean behavior. In Appendix I, Section 3 we will show in this connection, that for φ and ε_s deviations in the order of magnitude of at least $\pm 20\%$ have to be expected.

For E_c, however, much greater deviations may occur if we consider only the influence of concrete compressive strength (up to $\pm 50\%$). The particulars given in Part B, Section 3.1 reduce this error considerably, especially if we can choose the coefficient β_a for the type of locally available aggregates on the basis of previous experience. Since it is very unlikely that E_c, φ, and ε_s will assume unfavorable values at the same time, we can normally expect an accuracy for the estimate of deflections at time $t = \infty$ of $a_{m\infty} = \pm 15\%$ and at time $t = 0$, which depends on E_c only, $a_{m0} = \pm 20\%$ provided E_c has been determined according to Part B, Section 3.1.

More accurate calculations taking into account upper and lower bound values are rarely needed, except for special cases such as very large spans. In most instances an estimate will be sufficient.

When calculating the deformation, we have to take into consideration that the creep coefficient is related to the elastic strain at an age of 28 days. This is why the total strain of a fiber has to be calculated in accordance with the following formula:

$$\varepsilon_t = \frac{f_{ct_1}}{E_{ct_1}} + \frac{f_{ct_1}}{E_{c28}} \varphi_t = \frac{f_{ct_1}}{E_{c28}} \left(\frac{E_{c28}}{E_{ct_1}} + \varphi_t \right) = \frac{f_{ct_1}}{E_{c28}} (\beta_\varepsilon + \varphi_t). \tag{14.1}$$

Here t_1 refers to the concrete age at the time of load application and β_ε is the inverse value, shown in Fig. 14.2, of the coefficient β_e which was deducted in Part B, Sections 2 and 3.1 (see, also, Fig. 1.1). If the sustained load is applied at points in time $t_1, t_2, \ldots$, we have to use analogous methods.

Of course, the deformation and/or camber values to be calculated have to include the foundation settlement which can be expected. For camber, deformation of the scaffolding and settlement of its foundation have to be added.

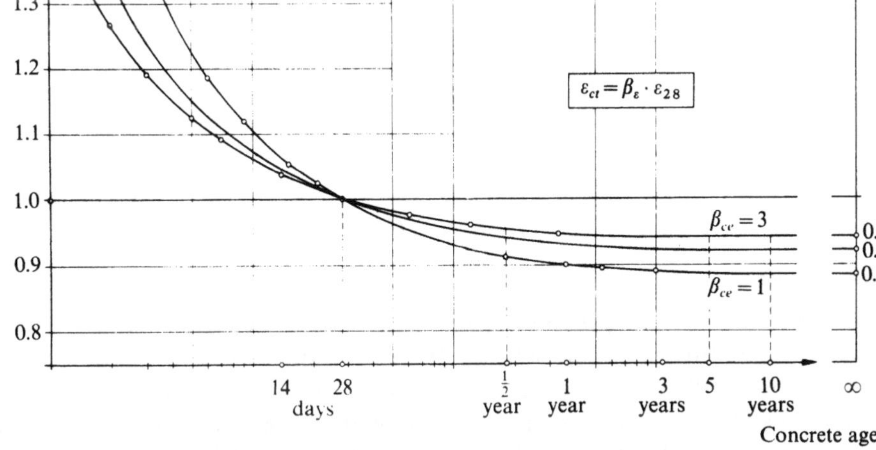

Figure 14.2. Effect of concrete age on elastic strain (β_e from Fig. 1.1).

14.2. The Effects of Creep and Shrinkage on the Deformation of Structures: Deductions and Aids

14.2.1. Introduction

Even if we take into consideration creep and shrinkage, it is necessary to make a reliable estimate of the deformation of structures, particularly in the following cases:

(a) In the case of *stability problems*, the load-carrying capacity can decrease dramatically under the influence of time-dependent deformations. Here, the expected maximum value of the additional deformation is the governing factor.

(b) Deformations can affect the *serviceability of the structure*. This is true, for example, if a water trap forms on a roof, or if a continuous girder assumes the shape of a garland. Such defects have to be avoided through proper selection of the camber of the form work. For a roof structure, this camber is usually chosen such that even with the highest possible deflections, no damage will occur. For a continuous girder, on the other hand, such a step may lead to a garland with opposite sign. We then have to proceed from the most likely value.

(c) Extensive deformations of a structure can adversely affect the serviceability of adjoining, non-load carrying elements. Examples are disturbing cracks in partitions and broken windowpanes.
When designing measures to prevent such damage, we have to proceed from the highest value of the deflection that can be expected. In simple cases, limitations of the slenderness of a structural member is sufficient.

The difference between the most probable deformation and its highest possible value is due to the scatter of the characteristic values used in the deformation calculation.

They are:

modulus of rupture of the concrete;
the influence of cracking on the effective cross sectional area;
the magnitude of the modulus of elasticity E_c;
the creep coefficient φ;
and the shrinkage strain ε_s.

14.2.2. Deformation Calculation

14.2.2.1. General

In Part A, Section 3.5.1 we provided a general view of the effects of cracking and creep on the deformation of structures. In the example therein, the deflection increased from 100% to 277% during the change from state I (uncracked) to state II (cracked). Due to subsequent creep with a creep coefficient $\varphi = 2.5$, another increase of 187% to a total at 464% of the initial value occurred. Thus, creep increased the initial deformation, which developed in state II, 464/277 = 1.67 times. This deliberation shows that deformations in state II are influenced at least as much by cracking as by creep.

In contrast, creep in state I causes an increase of (100 + 250)/100 = 3.5 times.

The deduction of realistic methods for deformation calculations is difficult, due to the effects and importance of cracking. Deformation calculations generally used to determine the action effects of statically indeterminate structures are not very helpful in this case because only relative values are needed here. It is considerably more difficult to determine absolute values. However, a number of methods have been published recently which seem suitable for the analysis of structures, considering the difficulties mentioned ([20], [22], [23], and [24]).

The following recommendation borrows extensively from the draft for the *Aids for Calculation and Dimensioning of Reinforced Concrete Elements*, and the pertinent proposals [28][1] by Thielen, Grasser, and Krämer which are based on [20] and [22]. On the basis of these data, deformation of a reinforced concrete member may be calculated for state I (uncracked) and state II (cracked), for the initial condition $t = 0$ as well as for the final condition $t = \infty$ which includes creep and shrinkage. The probable deflection is then estimated according to Section 14.2.2.5. The summary in Table 14.1 reviews the formulas thus developed, which are then explained and substantiated in the subsequent sections. We have to keep in mind, however, that these formulas cover only the effect of bending. The effect of normal forces rarely plays a role when calculating the deflection of structural elements. Generally, it is sufficient to consider the normal force effects only when differentiating between the cracked and uncracked areas (Section 14.2.2.5). Stability problems, on the other hand, are different. They are dealt with separately in Section 14.2.3.

With extremely compact beams or slender T-beams, it may be necessary to estimate deformations due to shear as well. Such cases are rare, though, and can be dealt with in accordance with [26] and/or the above-mentioned aids.

Shrinkage influences deformations in three different ways. It reduces an existing prestress; it can produce constraint in case of statically indeterminate structures; and it increases the curvature in the case of cracked sections because only the compression zone shortens due to shrinkage. The first effect is taken into account through the final value of the prestress force P_t; the second can be determined, like any other constraint, as being independent of E_c and φ, provided we can consider the supports to be rigid. Only the third effect requires special data (see Section 14.2.2.4).

In uncracked regions, the influence of reinforcement can be neglected completely or estimated roughly. In cracked regions, though, it becomes more important as the amount of reinforcement increases.

The formulas listed in the following sections are generally valid; but, for the time being, the diagrams for the pertinent coefficients $\varkappa$ apply only to rectangular cross sections.

14.2.2.2. Deformation of Uncracked Prestressed Concrete Structures (State I)

For the elastic deflection developing at time t_0, also called the basic value of deflection, the following formula applies:

$$a_{c0}^{\mathrm{I}} = \frac{\beta M l^2}{E_c I_c}. \tag{14.2}$$

[1] This work uses the mean creep-inducing stress method to determine the $\varkappa$-coefficients. It also contains data on deflection of T-beams.

Table 14.1. Summary of formulas proposed for deflection calculations; subscripts: c = concrete, d = permanent load, w = loads with short-term effect, $q = d + w$, r = constraint, p = prestress, φ = creep, s = shrinkage, 0 = time of initial load application; and t = final point in time. Coefficients: β_e from Fig. 1.1; $\varkappa_b$ from Fig. 14.3; $\varkappa_\varphi$ from Fig. 14.4; $\varkappa_s$ from Fig. 14.5; and $\varkappa_\alpha$ from Fig. 14.6.

Condition	Area of application	Section	Time $t = 0$	Time t
I uncracked cracked	full and limited prestressing	14.2.2.2	$a^{\text{I}}_{c0} = a^{\text{I}}_{c0d} + a^{\text{I}}_{c0w} + a^{\text{I}}_r + a^{\text{I}}_{c0p}$ $= a^{\text{I}}_{c0q} + a^{\text{I}}_r + a^{\text{I}}_{c0p}$	$a^{\text{I}}_{ct} = a^{\text{I}}_{c0q} + a^{\text{I}}_r + a^{\text{I}}_{ctp} + \beta_e\varphi[a^{\text{I}}_{c0d} + 0.5(a^{\text{I}}_{c0p} + a^{\text{I}}_{ctp})]$ $= a^{\text{I}}_{c0q} + a^{\text{I}}_r + a^{\text{I}}_{ctp} + a^{\text{I}}_{c\varphi d} + a^{\text{I}}_{c\varphi p}$
	reinforced concrete and partial prestressing	14.2.2.3	$a^{\text{I}}_0 = \varkappa^{\text{I}}_b a^{\text{I}}_{c0q} + a^{\text{I}}_r$ $= a^{\text{I}}_{0q} + a^{\text{I}}_r$	$a^{\text{I}}_t = a^{\text{I}}_{0q} + a^{\text{I}}_r + \varkappa^{\text{I}}_b \varkappa^{\text{I}}_\varphi \varphi\beta_e a^{\text{I}}_{c0d} + a^{\text{I}}_s$ $= a^{\text{I}}_{0q} + a^{\text{I}}_r + a^{\text{I}}_{d\varphi} + a^{\text{I}}_s$
II Fully cracked	reinforced concrete	14.2.2.4	$a^{\text{II}}_0 = \varkappa^{\text{II}}_b \left(\frac{h}{d}\right)^3 a^{\text{I}}_{c0q} + a^{\text{II}}_r$ $= a^{\text{II}}_{0q} + a^{\text{II}}_r$	$a^{\text{II}}_t = a^{\text{II}}_{0q} + a^{\text{II}}_r + \varkappa^{\text{II}}_b \left(\frac{h}{d}\right)^3 \varkappa^{\text{II}}_\varphi \varphi\beta_e a^{\text{I}}_{c0d} + \varkappa^{\text{II}}_s a^{(e)}_s \left(\frac{h}{d}\right)$ $= a^{\text{I}}_{0q} + a^{\text{II}}_r + a^{\text{II}}_{d\varphi} + a^{\text{II}}_s$
I and II partially cracked	reinforced concrete and partial prestressing	14.2.2.5	$a_0 = a^{\text{I}}_0 + \varkappa_{\alpha 0}(a^{\text{II}}_0 - a^{\text{I}}_0)$	$a_t = a^{\text{I}}_t + \varkappa_{\alpha t}(a^{\text{II}}_t - a^{\text{I}}_t)$

Table 14.2. System coefficients for constant flexural rigidity.

Load	M	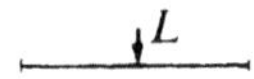 L			 $q = \frac{L}{l}$			 $q = \frac{2L}{l}$		
Static system	β	α	β	$10^3\,\gamma$	α	β	$10^3\,\gamma$	α	β	$10^3\,\gamma$
	0.50	1.000	0.333	333.3	0.5	0.25	125	0.333	0.20	66.6
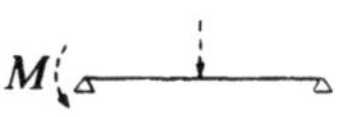	0.063	0.250	0.0833	20.8	0.125	0.104	13.0	0.125	0.104	13.0
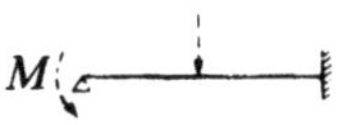	0.031	0.156	0.0584	9.32	0.0625	0.0834	5.24	0.0583	0.080	4.66
	-	0.125	0.0417	5.21	0.0417	0.0625	2.61	0.0417	0.0625	2.61

As shown in Table 14.2, β is a coefficient which depends on the shape of the moment diagram. If the modulus of elasticity deviates strongly from normal values, e.g., because of very early load application, it has to be determined in accordance with Part B, Section 3.1. If we denote the load which acts on the span under consideration by L, we then obtain the maximum moment with the aid of the moment coefficient α (for cantilever beams, the fixed end moment):

$$M = \alpha L l. \tag{14.3}$$

Thus, Eq. (14.2) may be written as follows for the time t_0 prior to the effect of creep and shrinkage:

$$a^{\mathrm{I}}_{c0} = \frac{\alpha\beta L l^3}{E_c I_c} = \frac{\gamma L l^3}{E_c I_c}. \tag{14.4}$$

Table 14.2 is a compilation of the coefficients α, β, and γ for some frequently occurring systems with constant flexural rigidity and types of load. The data always refer to the deflection at mid-span and/or at the end of a cantilever. If the data did refer to the point of maximum moment or maximum deflection, it would be impossible to obtain the maximum total deformation for different simultaneously occurring types of load by superimposing the individual loading conditions. Furthermore, the deflection at mid-span usually does not differ much from the maximum value.

With a highly varying moment of inertia along the length of the beam, it may become necessary, for statically indeterminate systems, to change the coefficient β as compared to Table 14.2. The new values are obtained from

$$\beta = \frac{1}{l^2}\int_0^l \frac{M_0}{M_c} M_1 \frac{I_c}{I}\,ds,$$

where M_0 is the distribution of moments resulting from the external load, M_c is the M_0 moment at mid-span, and I_c is the moment of inertia at the same location (see also the example in Section 14.3). M_1 corresponds to the distribution of moments caused by a concentrated load $L = 1$ at mid-span. Frequently, we can also use the aids [27] developed by Guldan.

Under the influence of creep, the deformation which is caused by sustained loads increases to φ times the initial value if we can neglect the effects of reinforcement. We have to take into consideration, however, that the creep coefficient $\varphi = \varepsilon_c/\varepsilon_{c28}$ relates to concrete deformation at an age of 28 days. If the concrete is loaded at a time t_0, which differs strongly from 28 days, we have to replace a_{c0}^{I} with $a_{c0}^{\mathrm{I}} E_{c0}/E_{28} = a_{c0}^{\mathrm{I}} \beta_e$ for creep deformation calculations. In that case, β_e can be taken from the diagram in Fig. 1.1.

The same is true for creep caused by prestress. However, we have to take into account that the prestress decreases from P_0 to P_t due to the effects of creep and shrinkage. In this case, the mean value $P = (P_0 + P_t)/2$ is relevant for the pertinent creep deformation.

Neglecting the effects of reinforcement and a possibly existing normal force, we arrive at the following formulas:

At time t_0:

$$a_{c0}^{\mathrm{I}} = a_{c0d}^{\mathrm{I}} + a_{c0w}^{\mathrm{I}} + a_r^{\mathrm{I}} + a_{c0p}^{\mathrm{I}} = a_{c0q}^{\mathrm{I}} + a_r^{\mathrm{I}} + a_{c0p}^{\mathrm{I}}. \tag{14.5}$$

At time t:

$$a_{ct}^{\mathrm{I}} = a_{c0q}^{\mathrm{I}} + a_r^{\mathrm{I}} + a_{ctp}^{\mathrm{I}} + \beta_e \varphi \left(a_{c0d}^{\mathrm{I}} + \frac{a_{c0p}^{\mathrm{I}} + a_{ctp}^{\mathrm{I}}}{2} \right)$$

$$= a_{c0q}^{\mathrm{I}} + a_r^{\mathrm{I}} + a_{ctp}^{\mathrm{I}} + a_{c\varphi d}^{\mathrm{I}} + a_{c\varphi p}^{\mathrm{I}}. \tag{14.6}$$

The individual values in these formulas are obtained as follows:

$$a_{c0d}^{\mathrm{I}} = \frac{\gamma_d L_d l^3}{E_{c0} I_c}; \quad a_{c0w} = \frac{\gamma_w L_w l^3}{E_{c0} I_c}. \tag{14.7}$$

γ_d, γ_w are coefficients which depend on the structural system and the type of load. In complicated cases, they are obtained for a constant moment of inertia by dividing the loads into components according to Table 14.2 and subsequent superposition.

E_{c0} is the modulus of elasticity for the concrete age at the time of initial load application. For earlier load application, $E_{c0} = \beta_e E_{c28}$.

β_e $= E_{c0}/E_{c28}$ is a coefficient taken from Fig. 1.1, which expresses the effect of concrete age on E_c. If at t_0 the concrete age does not differ substantially from $t = 28$ days, then $\beta_e = 1$.

P_0, P_t are prestress forces at time t_0 and time t, respectively, which has been reduced through creep and shrinkage.

a_r^{I} is the deformation caused by an imposed action, such as foundation settlement or shrinkage which acts against rigid supports. In principle, the extent of this deformation is changed neither by the duration of load application nor by the change from state I to state II.

14.2.2.3. *Deformation of Uncracked Reinforced Concrete and Partially Prestressed Structures (State I)*

With low amounts of reinforcement, reinforced concrete structures frequently have no cracks at all or only a few which develop after some time.

With partial prestressing, we can generally anticipate that under sustained loads cracks are compressed, meaning they are closed. However, for the maximum deformation which develops under maximum load, we have to proceed in accordance with Section 14.2.2.4.

Even, for lightly stressed structural elements, their behavior in an uncracked state is of interest because they are free of cracks, at least in some sections. In this case, the most likely deformation value is between both extremes describing the behavior in the uncracked and totally cracked condition.

The deformation of reinforced concrete elements is reduced because of the amount of reinforcement which is noticeably higher than that of prestressed concrete. This not only applies to t_0; at time t, concrete creep forces the tension reinforcement to participate more. These effects can be determined as a rough but satisfactory approximation through coefficients $\varkappa_c^{\mathrm{I}}$ and $\varkappa_\varphi^{\mathrm{I}}$.

With increasing amounts of reinforcement, however, we have to consider the effects of shrinkage as well. This can be done with the following formula:

$$a_s^{\mathrm{I}} = \frac{\varkappa_s^{\mathrm{I}} \beta_s \varepsilon_s l^2}{h} = \varkappa_s^{\mathrm{I}} a_s^{(e)}, \tag{14.8}$$

where β_s is 0.125 for single-span beams, 0.0625 for continuous beams, and 0.500 for cantilevers.

With this, we obtain, for time t_0 with the use of Eq. (14.5):

$$a_0^{\mathrm{I}} = \varkappa_b^{\mathrm{I}} a_{\mathrm{c}0q}^{\mathrm{I}} + a_r^{\mathrm{I}} = a_{0q}^{\mathrm{I}} + a_r^{\mathrm{I}}, \tag{14.9}$$

and for time t using the values from Eqs. (14.9), (14.6), and (14.8):

$$a_t^{\mathrm{I}} = a_{0q}^{\mathrm{I}} + a_r^{\mathrm{I}} + \varkappa_b^{\mathrm{I}} \varkappa_\varphi^{\mathrm{I}} \varphi \beta_e a_{\mathrm{c}0d}^{\mathrm{I}} + a_s^{\mathrm{I}} = a_{0q}^{\mathrm{I}} + a_r^{\mathrm{I}} + a_{d\varphi}^{\mathrm{I}} + a_s^{\mathrm{I}}. \tag{14.10}$$

As in Section 2.2, $\beta_e = 1$ can be used for normal cases. Different values are to be used only if the concrete is subjected to loads at a very early stage, so that $a_{\mathrm{c}0d}^{\mathrm{I}}$ was calculated with $E_c < E_{c28}$.

The new coefficients are:

- $\varkappa_b^{\mathrm{I}}$ the reduction factor taking into account effects of longitudinal reinforcement on the load-dependent deflection, in accordance with Fig. 14.3.
- $\varkappa_\varphi^{\mathrm{I}}$ the reduction factor taking into account the additional effect of reinforcement on creep deformation in accordance with Fig. 14.4.
- $\varkappa_s^{\mathrm{I}}$ the coefficient taking into account the effect of reinforcement on the deflection caused by shrinkage in accordance with Fig. 14.5.
- β_e $= E_{c0}/E_{c28}$, the coefficient taken from Fig. 1.1 which is used to take into account that all creep coefficients φ relate to E_{c28}.

The data given in Figs. 14.3–14.5 for coefficients $\varkappa_b^{\mathrm{I}}$, $\varkappa_\varphi^{\mathrm{I}}$, and $\varkappa_s^{\mathrm{I}}$ apply only to rectangular cross sections. For other cross sections, we can proceed by applying ρ

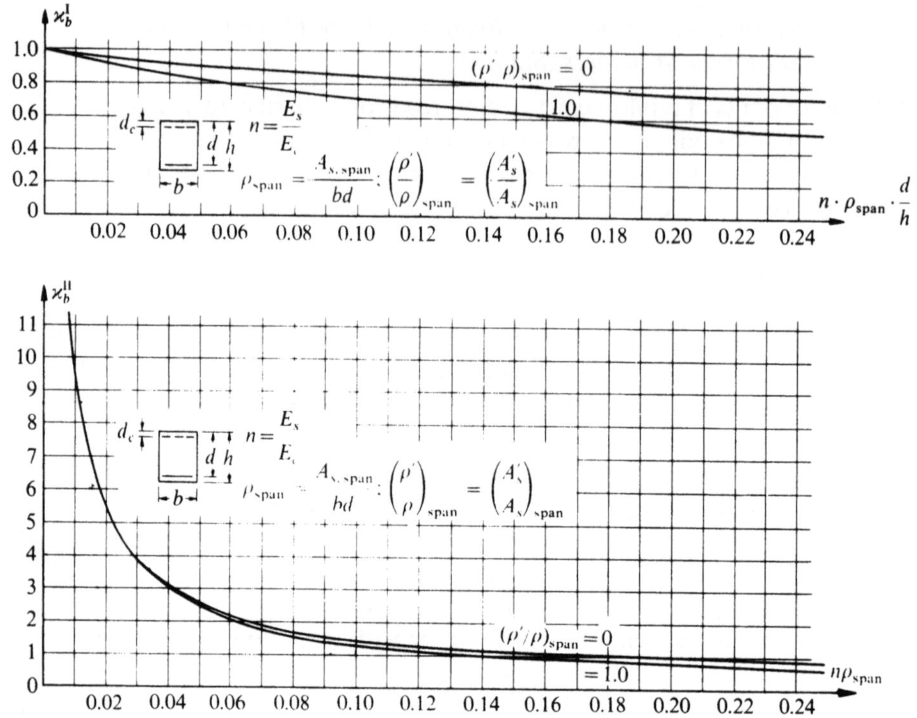

Figure 14.3. Coefficients $\varkappa_b^{I}$ and $\varkappa_b^{II}$ are used to take into account the effect of reinforcement in the calculation of upper and lower values of elastic deflections from the basic value a_c of a beam with rectangular cross section.

to the average width of the tension region and ρ' to the average width of the compression region.

14.2.2.4. Deformation of Fully Cracked Reinforced Concrete Elements (State II)

The following applies if we assume that a structural element is in its full length in a cracked state:

$$a^{II} = \beta \frac{\varepsilon_c + \varepsilon_s}{d} l^2. \tag{14.11}$$

In reality, however, the concrete between cracks is also effective. Thus for ε_s, we have to use the average steel stress $\varepsilon_{sm} = a_{sm}/E_s$. How much ε_{sm} differs from ε_s at a cracked section depends on the extent to which a_s surpasses the pertinent value for deflection, on the bond properties of the reinforcement and on the extent to which the tensile strength of the concrete is reduced through atmospheric influences.

In the following, it is suggested that the values a^{II} for the cracked state be deduced approximately with the aid of suitable coefficients from the values computed for the uncracked state in accordance with Section 14.2.2.3. The effects of staggering

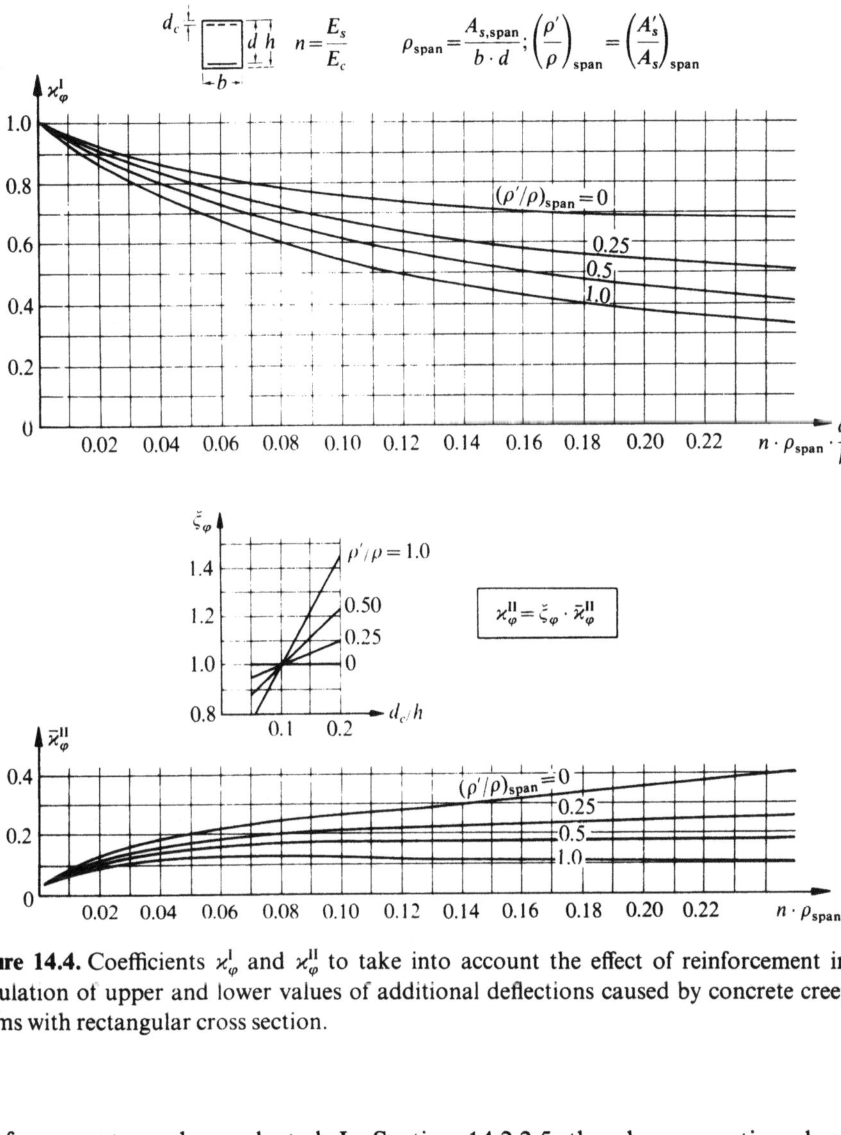

Figure 14.4. Coefficients $\varkappa_\varphi^I$ and $\varkappa_\varphi^{II}$ to take into account the effect of reinforcement in the calculation of upper and lower values of additional deflections caused by concrete creep for beams with rectangular cross section.

reinforcement can be neglected. In Section 14.2.2.5, the above-mentioned added effect of concrete between cracks is estimated.

We set

$$a_0^{II} = \varkappa_b^{II}\left(\frac{h}{d}\right)^3 a_{co}^I \qquad (a_{co}^I \text{ according to Eq. (14.4)}), \tag{14.12}$$

$$a_{d\varphi}^{II} = \varkappa_b^{II}\left(\frac{h}{d}\right)^3 \varkappa_\varphi^{II}\varphi\beta_e a_{cod}^I \qquad (a_{cod}^I \text{ according to Eq. (14.4)}), \tag{14.13}$$

$$a_s^{II} = \frac{\varkappa_s^{II}\beta_s\varepsilon_s l^2}{d} = \varkappa_s^{II} a_s^{(e)}\frac{h}{d} \qquad (a_s^{(e)} \text{ according to Eq. (14.8)}). \tag{14.14}$$

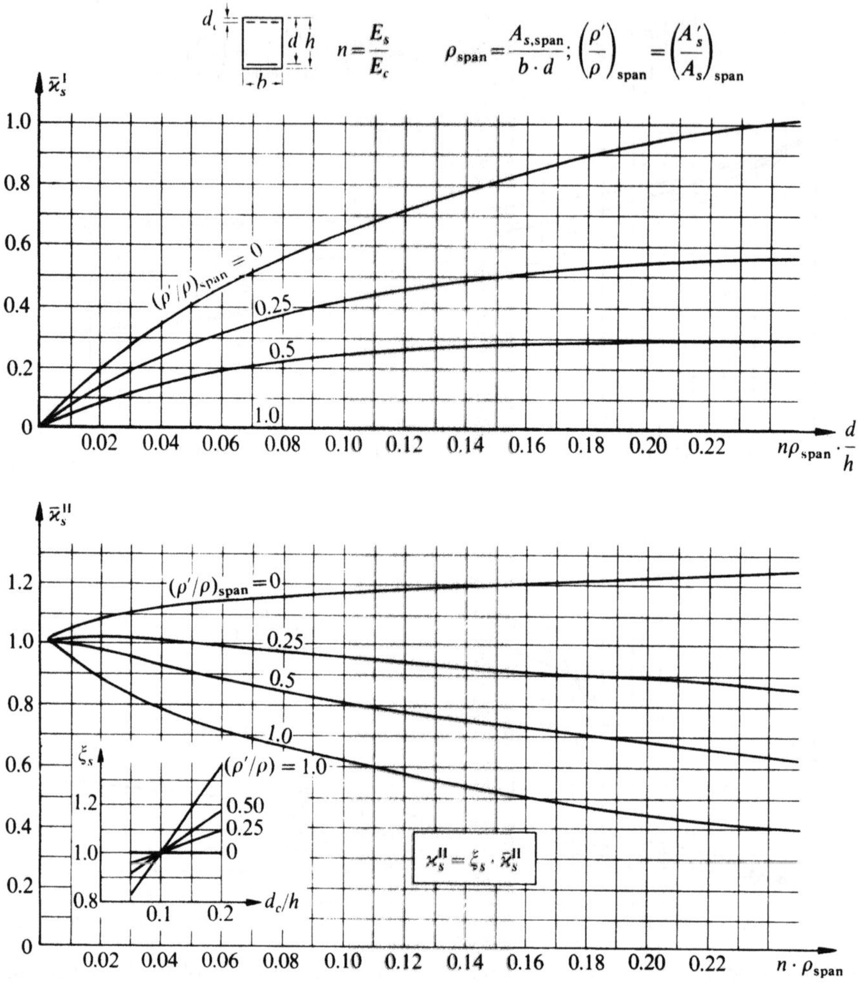

Figure 14.5. Coefficients $\varkappa_s^{I}$ and $\varkappa_s^{II}$ to consider upper and lower values of the additional deflections caused by concrete shrinkage in beams with rectangular cross sections.

We thus obtain the following at t_0 by using the results of Eq. (14.9):

$$a_0^{II} = \varkappa_b^{II}\left(\frac{h}{d}\right)^3 a_{coq}^{I} + a_r^{II} = a_{0q}^{II} + f_r^{II}, \tag{14.15}$$

and for t, by using the results of Eqs. (14.15), (14.6), and (14.8):

$$a_t^{II} = a_0^{II} + \varkappa_b^{II}\left(\frac{h}{d}\right)^3 \varkappa_\varphi^{II}\varphi\beta_e a_{cod} + a_r^{II} + \varkappa_s^{II} a_s^{(e)}\frac{h}{d} \tag{14.16}$$

$$= a_0^{II} + a_{d\varphi}^{II} + a_r^{II} + a_s^{II}.$$

The comments concerning Eq. (14.10), with respect to β_e, apply to this case as well.

The new coefficients are defined as follows:

$\varkappa_b^{II} = a_0^{II}/a_0^{I}$, the magnification coefficient according to Fig. 14.3, to consider the effects of cracking on deflection, whereby partial effectiveness of concrete between cracks is not taken into consideration.

β_e is the coefficient in accordance with Fig. 1.1 (see Section 14.2.2).

$\varkappa_\varphi^{II}$ is the reduction factor for the effects of φ which is intended to cover not only the consequences of the stress redistribution described in Part A, Section 3.5.1, but also the additional effects of a possible compression reinforcement. Figure 14.4 is valid for an average $d_c/d = 0.1$. The example in Section 14.3.13 illustrates the deduction of $\varkappa_\varphi^{II}$ for simple cases.

$\varkappa_s^{II}$ is the reduction factor for the effects of shrinkage, according to Fig. 14.5, which takes into account the effects of cracking as well as shrinkage. Deviations from the assumption $d_c/d = 0.1$ can be neglected.

The values to be used for $\varkappa_b^{II}$, $\varkappa_\varphi^{II}$, and $\varkappa_s^{II}$ can be taken from Figs. 14.3–14.5 for rectangular cross sections. They may also be used as approximations for other cross sections if ρ and ρ' are related to an equivalent rectangular cross section of the same height, the width b_m of which is selected such that about $\frac{4}{5}$ of the compression region are included (see Figure below).

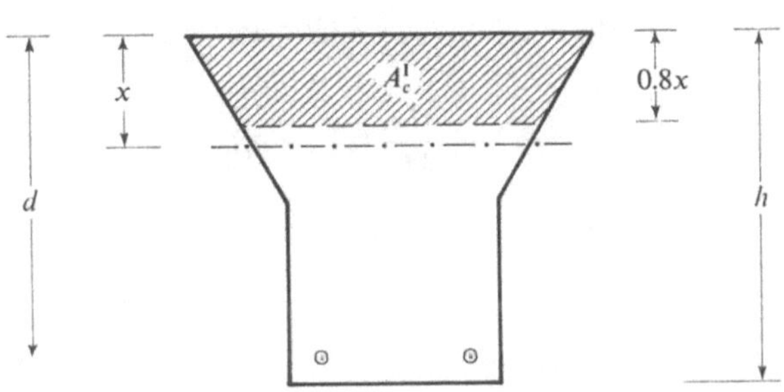

From this it follows that

$$b_m = \frac{A_c^I}{0.8x}.$$

14.2.2.5. *Calculation of the Most Probable Deformation of Partially Prestressed and Reinforced Concrete Elements*

Sections 14.2.2.3 and 14.2.2.4 contain only data for the determination of threshold values which are valid solely with the idealizing precondition that the elements under consideration will either remain in state I (uncracked) or change over completely to state II (fully cracked). Such threshold values are never reached in practical cases. The value to be expected is in between and may be calculated from the values determined in Eqs. (14.9) and (14.15) and/or (14.10) and (14.16), according to the following relationship:

At time t_0:

$$a_0 = a_0^{\mathrm{I}} + \varkappa_{\alpha 0}(a_0^{\mathrm{II}} - a_0^{\mathrm{I}}). \tag{14.17}$$

At time t:

$$a_t = a_t^{\mathrm{I}} + \varkappa_{\alpha t}(a_t^{\mathrm{II}} - a_t^{\mathrm{I}}). \tag{14.18}$$

The upper- and lower-bound values for deflections (a_0^{II}, a_t^{II}, and/or a_0^{I}, a_t^{I}) have to be calculated for the same load intensity and load position here.

The coefficients $\varkappa_{\alpha 0}$ and $\varkappa_{\alpha t}$, essentially depend on the magnitude of the cracking moment. They take into account uncracked regions (state I) as well as the additional effectiveness of concrete between cracks (state II). As far as the quality of bond of reinforcement is concerned, we have to assume well-deformed reinforcing bars. The dependence of coefficient $\varkappa_\alpha$ from the cracking moment M_R is shown in Fig. 14.6.

It shows that $\varkappa_\alpha$ can be described with sufficient precision through the expression $1 - M_R/M_F$. In simply supported beams as well as in normal continuous beams, M_F is the maximum moment at mid-span. For cantilevers we can use the fixed end moment instead of M_F as a good approximation. For continuous beams, which closely resemble the cantilever in behavior, it is suitable to use for M_F the mean value $(M_{\text{support}} + M_{\text{span}})/2$. This simplifies Eqs. (14.17) and (14.18) as follows:

$$a_0 = a_0^{\mathrm{II}} - \frac{M_{R0}}{M_F}(a_0^{\mathrm{II}} - a_0^{\mathrm{I}}), \tag{14.19}$$

$$a_t = a_t^{\mathrm{II}} - \frac{M_{Rt}}{M_F}(a_t^{\mathrm{II}} - a_t^{\mathrm{I}}). \tag{14.20}$$

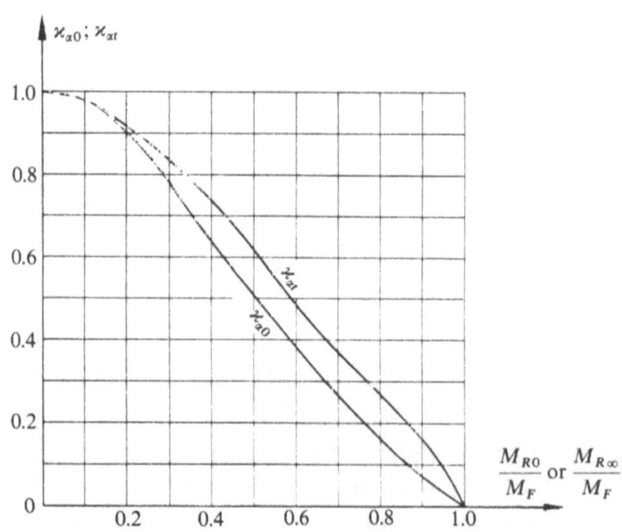

Figure 14.6. Effect of tensile stresses in concrete between cracks (M_F is the moment at mid-span).

The cracking moment at $t = 0$ can be calculated from Eq. (14.21):

$$M_{R0} = \frac{1}{6} \cdot b \cdot h^2 \cdot f_{tut} \cdot \varkappa_\mu - \frac{N \cdot I_{id}}{A_{id} \cdot y_{id}} \approx \frac{1}{6} bh^2 f_{tut} \varkappa_\mu - 0.2 \cdot N \cdot d, \quad (14.21)$$

$$M_{Rt} \approx 0.8 M_{Rti}. \quad (14.22)$$

M_{Rti} is obtained by using Eq. (14.21) for time t. The reduction factor 0.8 takes into account that over the course of time, a noticeable amount of tensile strength of the concrete is used up through internal stresses caused by temperature and shrinkage.

In the preceding equations we have: $f_{tut} = c\sqrt[3]{f'^2_{ct}}$ (N/mm^2) modulus of rupture of concrete according to [25] (see also Part B, Section 2.2); $c = 0.39$, lower bound value; $c = 0.57$, median value; $c = 0.62$, upper bound value; f'_{ct}, compressive strength at time t in N/mm^2; N, normal force (negative for compression); $\varkappa_\mu$, correction coefficient according to Fig. 14.7, dependent on the amount of reinforcement.

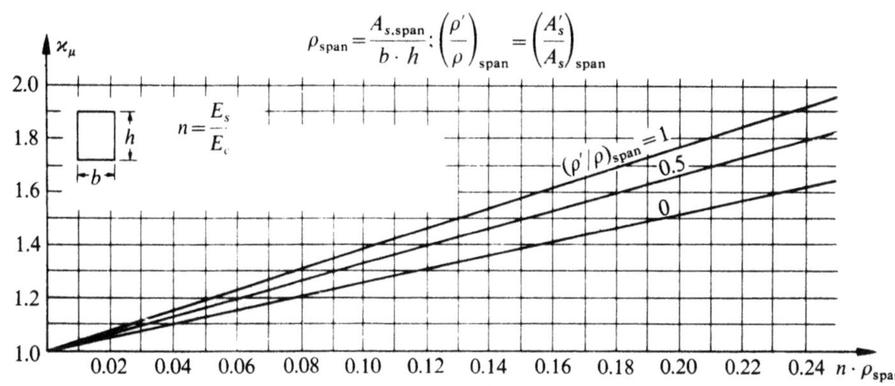

Figure 14.7. Coefficient $\varkappa_\mu$ to consider the effect of reinforcement on the cracking moment for beams with rectangular cross sections.

The reliability of value a, as determined from Eqs. (14.19) and/or (14.20), essentially depends on how accurately cracked or uncracked regions and the corresponding moments may be estimated. The following deliberations apply here:

We have to choose different input data, depending on whether the most likely value or an extreme value is to be determined for a.

The usual assumptions of the magnitude of live loads result in values larger than the average loads. To determine the most likely values of a, we can, therefore, usually reduce the live load moments. This is particularly true if M_w is derived from several load conditions which rarely reach maximum values at the same time. Very rarely occurring or short-term loads can often be disregarded totally.

In reality, the cracked region cannot be distinguished clearly from the uncracked region as far as deformability is concerned. The change is gradual because the distance between cracks increases towards the last crack and the crack length decreases.

The effects of temperature and humidity fluctuations, which generally are not included in the calculations, can promote cracking. This has to be taken into account by carefully selecting the value of c in Eq. (14.21) or the reduction factor in Eq. (14.22).

14.2.3. Stability Problems (Taking into Account Creep According to the Theory of Second Order)

14.2.3.1. General

For very slender compression members, the creep deformation caused by the sustained load frequently is very important. A closed solution for the calculation of creep deformation does not appear possible for two reasons: On the one hand, creep leads to a continued change in the effective cross section while the neutral axis moves away from the extreme compression fiber (see Section 14.3.1.3); on the other hand, the moment which causes the deflection simultaneously increases as a result of creep.

This is why we will always have to be satisfied with approximations. A particularly suitable way to solve these problems entails the use of Dischinger's stability assumptions which he developed as early as 1937 for the buckling safety of arched bridges [1]. For those very rare cases where further and more exact data are needed, Section 14.2.3.3 contains a step-by-step integration method.

14.2.3.2. Approximation Using Dischinger's Assumptions

The following explanations follow those equations, deducted by Dischinger [1] for creep on arched bridges, which strictly apply to homogeneous cross sections and elastic material behavior. These equations also take into account creep effects. They can be used as approximations for cracked reinforced concrete cross sections if we use realistic values for the flexural rigidity EI which satisfactorily describe average behavior.

The following deduction of Dischinger's equation is to demonstrate where and in which way the flexural rigidity becomes effective. We have to assume first, however, that the initial moment diagram is similar in shape to the buckling figure and that the flexural rigidity of the bar is constant.

Theory of the First Order

$$y_{10} = \frac{M_0}{N} + e_u = e_d + e_u \text{ (e_u is the unintended eccentricity).} \tag{14.23}$$

Theory of the Second Order

According to Vianello, the safety against buckling $\gamma = N_E/N = \pi^2 EI/l_{\text{eff}}^2 N$ in the elastic range is determined by the ratio of the ordinates of two deflection curves and/or the pertinent moments (Fig. 14.8a).

$$\gamma = \frac{y_i}{y_i + 1} = \frac{M_i}{M_{i+1}}. \tag{14.24}$$

From M_0, we obtain an additional moment $M_1 = M_0/\gamma$, in accordance with the theory of second order.

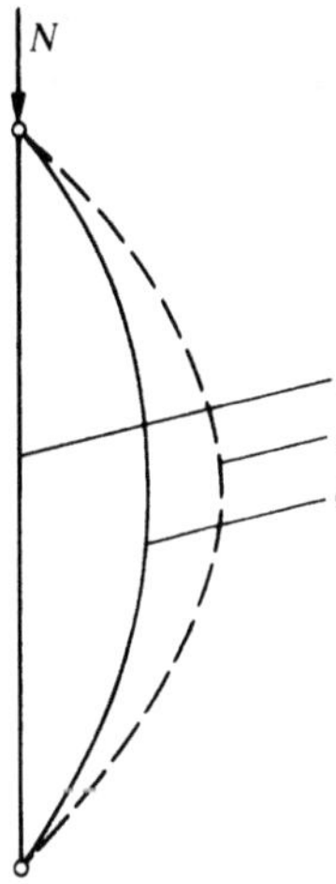

Figure 14.8a.

From this, we then obtain $M_2 = M_1/\gamma = M_0/\gamma^2$, etc. The total moment M_{II} is then:

$$M_{\text{II}} = M_0\left(1 + \frac{1}{\gamma} + \frac{1}{\gamma^2} + \frac{1}{\gamma^3} + \cdots\right) = M_0 \frac{\gamma}{\gamma - 1}, \tag{14.25}$$

and the additional moment

$$\Delta M = M_{\text{II}} - M_0 = \frac{M_0}{\gamma - 1} \tag{14.26}$$

and/or

$$y_{\text{II}0} = y_{\text{I}0} \frac{\gamma}{\gamma - 1}, \tag{14.27}$$

and the additional deflection is

$$\Delta y = \frac{y_{\text{I}0}}{\gamma - 1}. \tag{14.28}$$

Other Influences

Dischinger also proved [1] that, with the aid of these equations, we can solve even those problems where no affinity exists between the $y_{\text{I}0}$ line and the buckling figure. In that case, Eqs. (14.25)–(14.28) will look as follows:

$$M_{\text{II}} = M_0 \frac{\gamma + \nu}{\gamma - 1}, \tag{14.29}$$

$$y_{\text{II}0} = y_{\text{I}0} \frac{\gamma + \nu}{\gamma - 1}, \tag{14.30}$$

$$\Delta M = M_0 \frac{1 + \nu}{\gamma - 1}, \tag{14.31}$$

$$\Delta y = y_{\text{I}0} \frac{1 + \nu}{\gamma - 1}. \tag{14.32}$$

The influence of the ν values calculated by Dischinger increases $y_{\mathrm{II}0}$ by only about 10%, even under unfavorable circumstances (e.g., $\gamma = 2$ and horizontal concentrated load at half of the buckling length):

$$y_{\mathrm{II}0} \sim y_{\mathrm{I}0} \frac{\gamma}{\gamma - 1} 1.1, \tag{14.33}$$

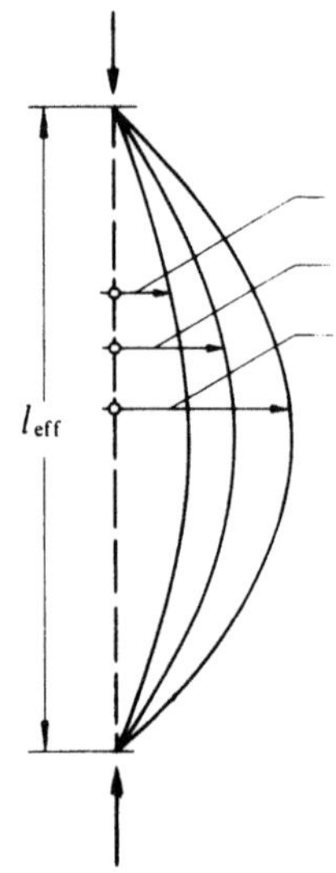

Figure 14.8b.

Taking Creep into Account

The deformation occurring in time interval dt is composed of a part due to creep and a part due to elastic deflection (see Fig. 14.8b).

$$dy = dy_c + dy_e, \tag{14.34}$$

whereby dy_e is caused by dy_c in the same way as Δy is caused by $y_{\mathrm{I}0}$, so that the following applies according to Eq. (14.28):

$$dy_e = dy_c \frac{1}{\gamma - 1} \text{ and/or } dy_c = dy_e(\gamma - 1). \tag{14.35}$$

Since the following is true for the creep deformation:

$$dy_c = y_e\, d\varphi, \tag{14.36}$$

we obtain the differential equation for the problem by setting Eq. (14.35) equal to Eq. (14.36):

$$dy_e = y_e\, d\varphi \frac{1}{\gamma - 1}. \tag{14.37}$$

Since for $t = 0$, $y_e = y_{\mathrm{II}0}$, the solution is

$$y_e = y_{\mathrm{II}t} = y_{\mathrm{II}0}\, e^{\varphi/\gamma - 1} = y_{\mathrm{I}0} \frac{\gamma}{\gamma - 1} e^{\varphi/\gamma - 1} \tag{14.38}$$

if the deflection curve $y_{\mathrm{I}0}$ is not affinitive to the buckling curve:

$$y_e = y_{\mathrm{I}0} \frac{\gamma + v}{\gamma - 1} e^{\varphi/\gamma - 1}. \tag{14.39}$$

Of course, we can use Eqs. (14.38) and (14.39) only for creep-inducing sustained loads.

Analogous to Eq. (14.30), permanent creep deformation can be deduced by setting Eq. (14.30) equal to Eq. (14.39).

$$y_{\mathrm{II}t} = (y_{\mathrm{I}0} + y_\varphi) \frac{\gamma + v}{\gamma - 1} = y_{\mathrm{I}0} \frac{\gamma + v}{\gamma - 1} e^{\varphi/(\gamma - 1)}. \tag{14.40}$$

Thus we obtain

$$y_\varphi = y_{\mathrm{I}0}(e^{\varphi/(\gamma - 1)} - 1) = (e_d + e_u)(e^{\varphi(\gamma - 1)} - 1). \tag{14.41}$$

Application to Reinforced Concrete
The equations listed above describe the effects of sustained load N_d, initial eccentricity e_d, and the unintended eccentricity e_u which may be assumed to be $e_u \approx l_{\mathrm{eff}}/300$.

In order to determine the permanent deflection due to creep under sustained load y_φ, we have to determine γ first. For this, we have to take into account the average flexural rigidity EI_w of the reinforced concrete member under working loads, which is subjected to a sustained load. Kordina offers the following in [21]:

$$EI_w = \left(0.6 + 20 \frac{A_s + A_s'}{bh}\right) E_c I_c, \tag{14.42}$$

$$\gamma = \frac{\pi^2 EI_w}{l_{\mathrm{eff}}^2 N_d}. \tag{14.43}$$

We can then proceed to demonstrate the safety against buckling considering an axial load which is increased by a factor of 1.75, a planned eccentricity e, an unintended eccentricity e_u, and by taking into consideration an additional deflection y_φ from creep under sustained loads [Eq. (14.41)].

This calculation method can be used to solve most buckling problems because creep under sustained working loads, occurs mostly in the uncracked state. Thus, the rigidity change due to cracking does not play a major role. However, it is suggested to insert the reduced value 0.8φ into Eq. (14.41) because of the inhibiting effect of the compression reinforcement and a partial changeover to the cracked stage.

14.2.3.3. More Accurate Solutions with Step-by-Step Integration
The following somewhat tedious method can be recommended only for cases where more accurate proof of buckling safety appears necessary, e.g., if the rigidity

of a cross section changes considerably or if the transition to the cracked state has to be followed more closely. With step-by-step integration, it is easy to consider the increase of action effects due to creep and also the shift of the neutral axis due to cracking (see the example in Section 14.3.1.3). In the example, at time $t = \infty$, the sustained load is multiplied by a safety factor and 1.75 times the life load is added. Usually, it is sufficient to iterate in four stages.

We can also take shrinkage into account. However, its influence is so insignificant that it is usually neglected.

Since the first steps of the analysis normally correspond to the range of working loads, initially linear stress distribution may be assumed. To determine the creep deformation, it is appropriate to use the method of mean creep-inducing action effects described in Section 3.2. As a rule, a separation into delayed elastic strain and flow is unnecessary.

The calculation for the individual time stages is carried out as follows:

Time Interval I

The sustained load acts on the system through a planned eccentricity e_d and the unintended eccentricity e_u. The bar deforms according to the theory of second order. A further increase in the bar deflection is caused by creep stage $\Delta\varphi_1$ (Fig. 14.9a).

Thus, we have to estimate a bar deflection at the beginning, $y_t = 0$, and at the end, $y_t = t$, of the time interval. In order to estimate these values, it is reasonable to use Eqs. (14.42), (14.43), and (14.30), deduced in Section 14.2.3.2 for estimated $y_{\text{II}0}$ and Eqs. (14.40) and (14.41) for $y_{\text{II}t}$ and $y_{\Delta\varphi_1}$, with the estimated initial values l_{eff} and EI_w. From that we obtain the distribution of the respective action effects along the length of the bar for the beginning and the end of the time interval. For the average value of action effects $(M_{t=0} + M_{t=1})/2$, we determine the average elastic curvature $1/r_{e\text{II}}$ and from that the increase $1/r_{\Delta\varphi}$ which develops in creep stage $\Delta\varphi_1$. Prior to this. however, the elastic curvatures for the action effects had to be determined for the individual cross sections at the beginning

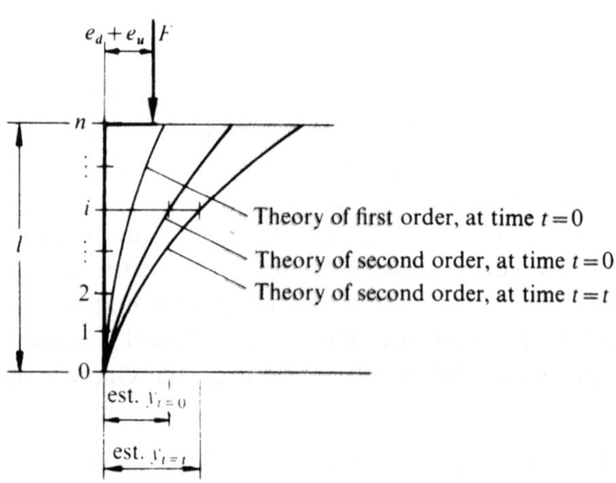

Figure 14.9a. Deformed state.

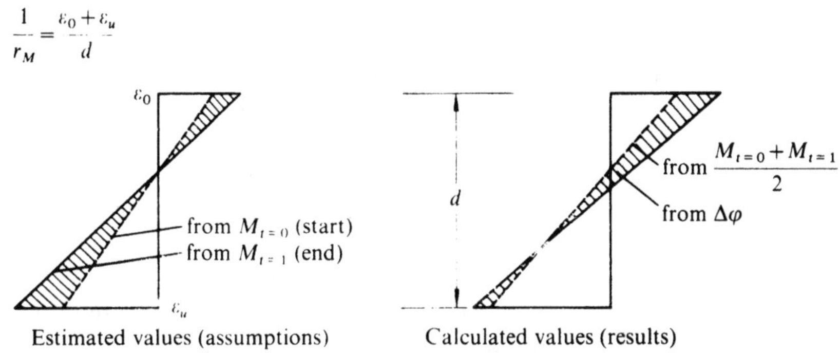

Figure 14.9b. Strains in section i, used to obtain the curvature and, through integration the deformation y of the bar.

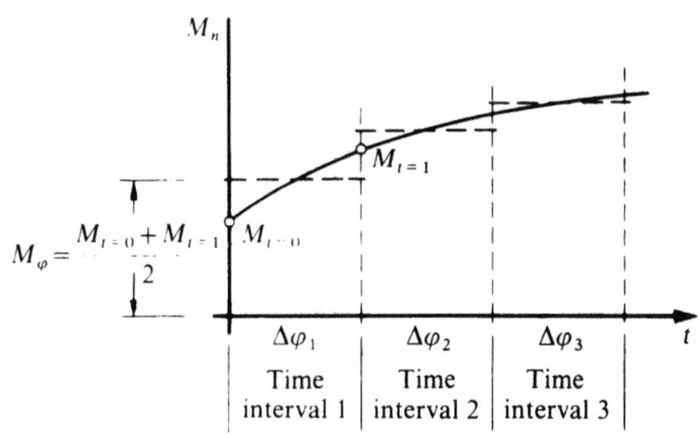

Figure 14.9c. Moments in section i.

$1/r_{M_{t=0}}$, and the end, $1/r_{M_{t=1}}$ of the time interval. From curvature, we can determine the bar deflections for calculated $y_{t=0}$ of the particular time interval and compare it with the previously estimated $y_{t=0}$. Likewise, the deformation at the end of the time interval, calculated $y_{t=1}$, can be determined and added to the increase from the creep deformation $y_{\Delta\varphi}$. The sum of calculated $y_{t=1}$ and $y_{\Delta\varphi}$ has to correspond approximately to the estimate of $y_{t=1}$. If not, the calculation has to be repeated (Fig. 14.9b and c).

Time Interval II

The deflections and/or action effects and thus also the curvatures and bar deformations at the beginning of the time interval are known from step I. We only have to estimate the deflection $y_{t=2}$ at the end of time interval II. In principle, though, we use the same calculation as in stage I. Interval II is concluded with a comparison of calculated and estimated values for the bar deformations.

Time Interval III

This creep interval proceeds as in time interval II. Usually, it is sufficient to divide φ into three intervals.

Time Interval IV, Last Interval
In addition, we now have to consider 0.75 times the sustained load and 1.75 times the short-term life load without creep. If we proceed from the deformation in interval III, the deflection increases through the load increase. This increase, however, has to be estimated. Subsequently, we have to determine the curvature caused by the additional action effects, whereby we also have to take into account the non-linear stress-strain diagram of concrete because of the high stresses. A comparison of estimated and calculated bar deflections concludes the calculation and demonstrates whether the structure is stable or not.

A considerable calculation effort is needed here, but prepared diagrams, approximation methods for an easier estimate of individual values, as well as various computer programs can substantially reduce the amount of work.

14.3. Examples

14.3.1. Deformation Calculations

14.3.1.1. Simply Supported Prestressed Concrete Beam Cast in Place
In accordance with the relations in Section 14.2, Eq. (14.1) is rewritten as follows:

$$a_{ct}^{\mathrm{I}} = a_{cod}^{\mathrm{I}}(\beta_\varepsilon + \varphi_\infty) + a_{cop}^{\mathrm{I}}(\beta_\varepsilon + \varphi_\infty) + a_{co\Delta p}^{\mathrm{I}})\left(1 + \frac{\varphi}{2}\right). \tag{14.44}$$

(For the sake of simplicity, β_ε was assumed to equal 1 in good approximation for load condition Δp.)

Assumptions: Concrete $f'_{ct} = 45\ \mathrm{N/mm^2}$, Type I cement; prestressing and form removal after 4 days; loss of prestress due to creep, 15%; $\varphi = 2.5$.

The mid-span deflection of the beam has to be determined for time $t = \infty$.

According to Eq. (14.4) and Table 14.2, the following applies to simply supported beams:

$$a_{cod}^{\mathrm{I}} = 13.0 \times 10^{-3}\frac{Ll^3}{E_cI_c} = 13.0 \times 10^{-3}\frac{dl^4}{E_cI_c}.$$

Prestressing is chosen such that it will just neutralize the dead weight at time $t = 0$.

$$a_{cop} = -a_{cod}^{\mathrm{I}} \text{ and } a_{co\Delta p} = 0.15a_{cop}^{\mathrm{I}}.$$

According to Fig. 14.2, $\beta_\varepsilon = 1.19$ and according to Table 1.1, $E_c = 3\,700\,000$ N/cm^2. Thus, we obtain for $l = 30$ m, $d = 15$ kN/m and $I_c = 0.040\ \mathrm{m^4}$.

$$a_{ct}^{\mathrm{I}} = \frac{dl^4}{E_cI_c}13.0 \times 10^{-3}\left[(1.19 + 2.5) - (1.19 + 2.5) + 0.15\left(1 + \frac{2.5}{2}\right)\right]$$

$$= \frac{15(30^4)10^{-1}}{3.7(10^6)0.040}13.0 \times 10^{-3}0.338 = 0.036\ \mathrm{m} = 36\ \mathrm{mm}\ (1.4\ \mathrm{in.}).$$

14.3.1.2. Prestressed Bridge Deck Slab with Fixed Ends

The prestressed deck slab of a bridge can be considered to be rigidly fixed between the main girders with a high torsional stiffness. For a span of $l = 8$ m, the moments are:

fixed end moment: $M_d = 40.2$ kNm; $M_p = 67.0$ kNm (1 kNm $= 7.291 \times 10^2$ ft lbs);
mid-span moment: $M_d = +20.1$ kNm; $M_p = 67.0$ kNm.

The 30-cm-thick slab is prestressed and the forms are removed after 10 days. At this time, the main girders are still propped up.

Additional data: concrete $f_c' = 35$ N/mm², Type I cement, sandstone gravel; expected values are $f_{c28}' = 41.0$ N/mm² and $w_c = 2.28$ t/m³.

From Eqs. (3.2) and (3.8), Part B, Section 3.1, we obtain:

$$E_{c28} = 19\,000(0.8)2.28\sqrt[3]{100(4100)} = 2\,552\,000 \text{ kN/m}^2;$$

$$E_{c10} = \beta_e E_{28} = 0.94E_{c28} = 2.4 \times 10^6 \text{ kN/m}^2;$$

$$E_{c10}\frac{I_c}{l^2} = 840 \text{ kN}.$$

The deflection at time $t = 0$, according to Table 14.2 and Eq. (14.2) is, from d,

$$a_{d0}^{\text{I}} E_{c10}\frac{I_c}{l^2} = +0.0625(20.1) = +1.26.$$

From p: since the distribution of the prestressing moments approximately corresponds to the load condition concentric load at mid-span (see Fig. 14.10), the following applies:

$$a_{p0}^{\text{I}} E_c \frac{I_c}{l^2} = 0.0417(67.0) = -2.80,$$

$$a_0 = \frac{+1.26 - 2.80}{840} = -0.00183 \text{ m} = 1.83 \text{ mm}.$$

Deflections at time $t = \infty$: From creep with $\varphi = 2$; $P_0 = 960$ kN, and $P_\infty = 870$ kN, we obtain the following from Eq. (14.6):

$$a_p \frac{E_c I_c}{l^2} = 1.26(2) - 2.80\frac{960 + 870}{2(960)}2 = 2.52 - 5.30 = 2.78,$$

$$a_\infty = a_0 - \frac{2.78}{840} = -0.00183 - 0.00331 = -0.00514 \text{ m} = -0.51 \text{ mm (0.02 in.)}.$$

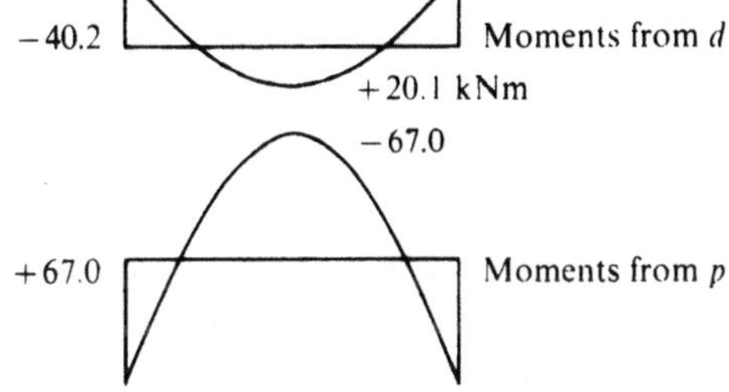

Figure 14.10.

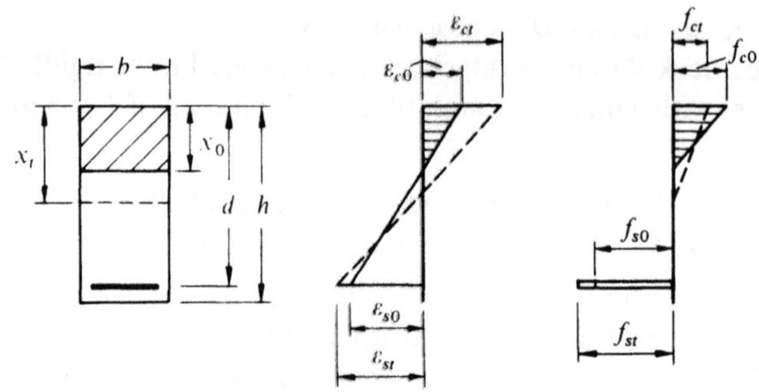

Figure 14.11.

14.3.1.3. Effects of Creep on the Curvature of Reinforced Concrete Rectangular Cross Sections

In a reinforced concrete cross section, the stresses which counteract a sustained flexural moment are redistributed noticeably through creep. Since the strain of the extreme compression fiber increases, the neutral axis moves downward (see Fig. 14.11). With that, the compression region increases and the extreme fiber stress decreases perceptibly. This slows down the strain increase at the extreme compression fiber. However, the lever arm of internal forces is reduced slightly during this process, which causes the steel stresses on the tension side to increase slightly.

These developments strongly influence the local curvature and thus the deflection. In the following, we demonstrate with a simple example of a rectangular cross section how they can be described mathematically. In order to simplify, we assume that at any point in time the stresses under working loads can be assumed to be linearly distributed. Deviations from this assumption occur primarily near the neutral axis and therefore have little effect.

Equilibrium Conditions

If we express the increased strain of the compression region due to creep by a reduction of the modulus of deformation E_D, we can express the position of the neutral axis as a function of the modular ratio, n:

$$\beta_x^2 - 2n\rho_0(1 - k_x) = 0, \tag{14.45}$$

where β_x is x/d, $\rho_0 = A_s/A_c$, and $n = E_s/E_D$.

However, n is a variable quantity because the modulus of deformation E_D is designed to include the sum of the elastic and time-dependent strains developing under a decreasing compressive stress.

Compatability Conditions

This condition can be satisfied with the method described in Section 2.3 for the mean creep-inducing stress. However, here we have to assume that the extreme fiber stress changes linearly with increasing φ. In reality, we have a slightly hyperbolic development, as shown in Fig. 14.12, which can be neglected here, along with the effects of delayed elasticity. One can, however, easily improve on the pre-

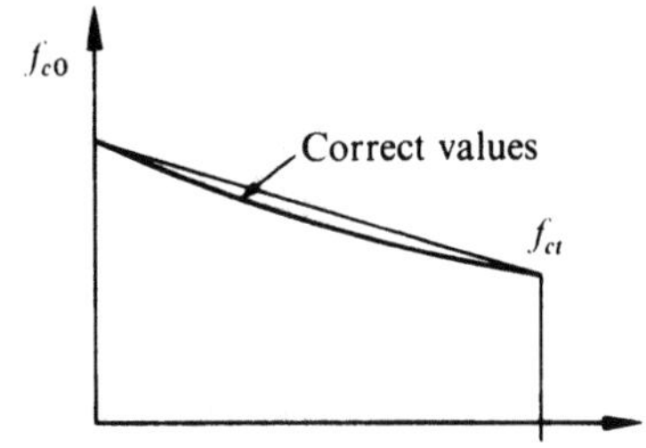

Figure 14.12.

cision by using the step-by-step integration method described in Section 2.3 and demonstrated with an example in Section 9.3.5.

For the extreme compression fiber strain, the following applies:

$$\varepsilon_{ct} = \frac{f_{c0}}{E_{c0}} + \frac{2(f_{ct} - f_{c0})}{E_{c0} + E_{ct}} + \frac{f_{c0} + f_{ct}}{2} \quad \frac{\varphi_t}{E_{c28}} \quad = \frac{f_{ct}}{E_{Dt}},$$

$$n_t = \frac{E_s}{E_{Dt}} = \frac{E_s}{E_{c28}}\left(\frac{f_{c0}}{f_{ct}\beta_{e0}} + \frac{2(f_{ct} - f_{c0})}{f_{ct}(\beta_{e0} + \beta_{et})} + \frac{f_{c0} + f_{ct}}{2f_{ct}}\varphi_t\right).$$

If we say $1/\beta_c = \beta_\varepsilon$, we then obtain:

$$n_t = \frac{E_s}{E_{Dt}} = \frac{E_s}{E_{c28}}\left[\frac{f_{c0}}{f_{ct}}\beta_{\varepsilon 0} + 2\left(1 - \frac{f_{c0}}{f_{ct}}\right)\frac{\beta_{\varepsilon 0}\beta_{\varepsilon t}}{\beta_{\varepsilon 0} + \beta_{\varepsilon t}} + \frac{\varphi_t}{2}\left(1 + \frac{f_{c0}}{f_{ct}}\right)\right]. \qquad (14.46)$$

Unless the beam is loaded at a very early age, we can neglect the age dependence of E_c, and we then have $\beta_{\varepsilon 0} \sim \beta_{\varepsilon t} = 1.0$.

$$n_t = \frac{E_s}{E_{c28}}\left[1 + \frac{\varphi_t}{2}\left(1 + \frac{f_{c0}}{f_{ct}}\right)\right]. \qquad (14.47)$$

Even if we disregard this simplification, we can take the values $\beta_{\varepsilon 0}$ and $\beta_{\varepsilon t}$, which correspond to the initial and the final points in time, from Fig. 14.2 and then obtain, with

$$k = 2\frac{\beta_{\varepsilon 0}\beta_{\varepsilon t}}{\beta_{\varepsilon 0} + \beta_{\varepsilon t}},$$

an equation in the form of

$$n_t = \frac{E_s}{E_{c28}}\left[k + (\beta_{\varepsilon 0} - k)\frac{f_{c0}}{f_{ct}} + \frac{\varphi_t}{2}\left(1 + \frac{f_{c0}}{f_{ct}}\right)\right].$$

Determination of the Location of the Neutral Axis

For simplification in the following we proceed from Eq. (14.47) for n_t. Furthermore, we use the following abbreviation: $v = f_{c0}/f_{ct}$:

$$n_t = n_0\left(1 + \frac{(1 + v)\varphi_t}{2}\right). \qquad (14.48)$$

The equilibrium between internal and external forces requires:

$$M_t = \tfrac{1}{2} b x_t f_{ct}\left(d - \frac{x_t}{3}\right) = \frac{1}{2} b d^2 f_{ct} \beta_{xt}\left(1 - \frac{\beta_{xt}}{3}\right),$$

$$x_t/d = k_x$$

$$M_0 = M_t \rightarrow f_{c0}\beta_{x0}\left(1 - \frac{\beta_{x0}}{3}\right) = f_{ct}\beta_{xt}\left(1 - \frac{\beta_{xt}}{3}\right),$$

$$v = \frac{f_{c0}}{f_{ct}} = \frac{\beta_{x0}(1 - \beta_{x0}/3)}{\beta_{xt}(1 - \beta_{xt}/3)}. \tag{14.49}$$

We can now eliminate the quantities n_t and v from Eqs. (14.45), (14.48), and (14.49) which results in an equation of the third order:

$$\beta_{xt}^3 - (4 + 3A)\beta_{xt}^2 + 3(1 - B)\beta_{xt} + 3B = 0. \tag{14.50}$$

Hereby is

$$A = \beta_{x0}\left(1 - \frac{\beta_{x0}}{3}\right)\frac{1}{\rho n_0 \varphi}, \tag{14.51}$$

$$B = \beta_{x0}\left(1 - \frac{\beta_{x0}}{3}\right)\frac{2 + \varphi}{\varphi}. \tag{14.52}$$

The following diagrams demonstrate an evaluation of these equations:

Figure 14.13 demonstrates the shift of the neutral axis with increasing φ.
Figure 14.14 demonstrates the decrease in the concrete compressive stress.
Figure 14.15 demonstrates the increase in steel stress.

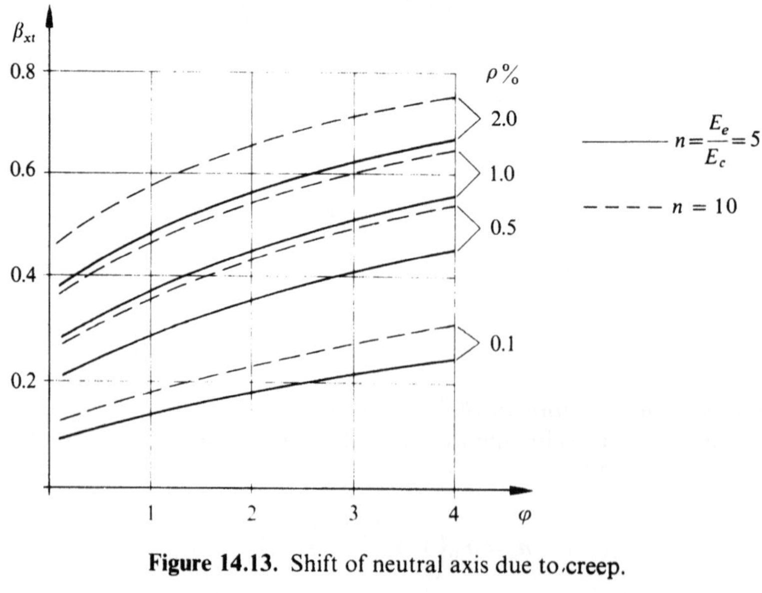

Figure 14.13. Shift of neutral axis due to creep.

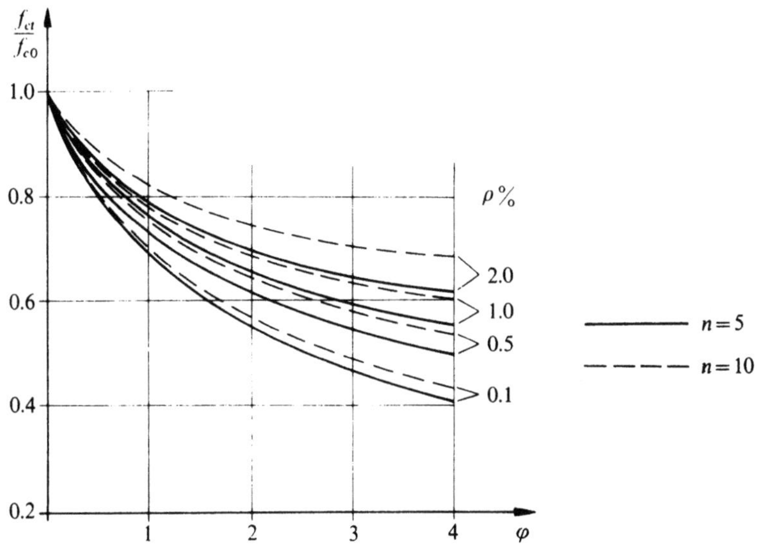

Figure 14.14. Decrease of concrete compressive stress due to creep.

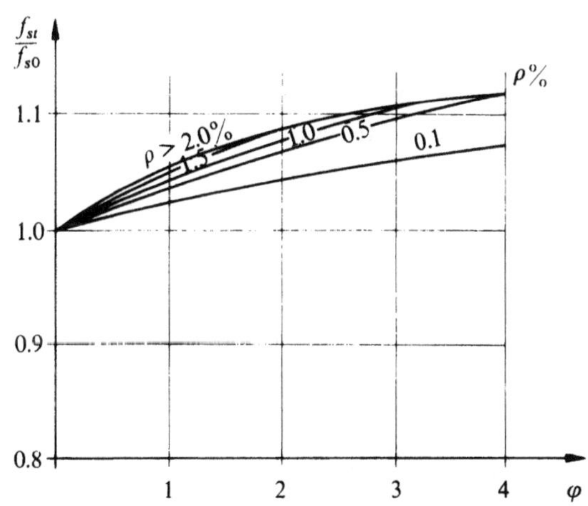

Figure 14.15. Increase of steel stress due to creep.

Figure 14.16 demonstrates the increase in the curvature $1/r_t$ with the aid of coefficient η and the equation

$$\frac{1}{r_t} = \varphi_t \eta \frac{1}{r_0}. \tag{14.53}$$

In order to simplify, the effectiveness of the concrete between cracks, which is particularly noticeable for $\rho < 1\%$, is neglected.

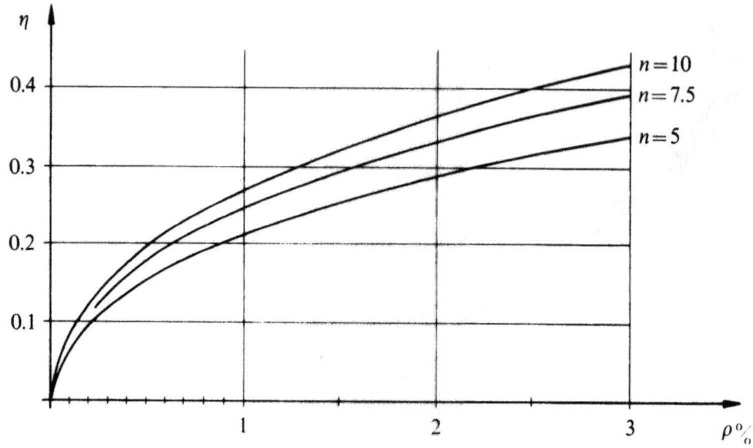

Figure 14.16. Dependence of coefficient η on the percentage of reinforcement and on the coefficient $n = E_s/E_c$ which depends on concrete compressive strength.

14.3.1.4. Simply Supported Prefabricated Reinforced Concrete Girder

For the roof girder made of reinforced concrete, and represented in Fig. 14.17, the following data apply: sustained load $d = 13.7$ kN/m; concrete, $f'_c = 45$ N/mm^2; $E_c = 37\,000$ N/mm^2; age at loading: 28 days; modulus of rupture (see Section 14.2.2.5): $f_{tut} = 0.57\sqrt[3]{45.0^2} = 5.8$ N/mm^2.

According to [8], this value applies only to beam heights of up to approximately $h = 20$ cm. For $h = 40$ cm, it has to be reduced by 25%. Thus, we obtain $f_{tut} \sim 4.5$ N/mm^2,

$$A_s = 28.8 \text{ cm}^2;\ A'_s = 4.1 \text{ cm}^2.$$

Under sustained load $x \sim 20$ cm. According to Section 14.2.2.4, reinforcement has to be related to the equivalent width of the compression zone that is:

$$b = \frac{A_{c0.8x}}{x} = \frac{0.8(20)12 + 32(11 + 5/2)}{20} = 31 \text{ cm},$$

$$\rho = \frac{28.8}{31(98)} = 0.95\%,\ \rho' = \frac{4.1}{31(9.8)} = 0.13\%,$$

$$\varphi = 3.0;\ \varepsilon_s = 25 \times 10^{-5}.$$

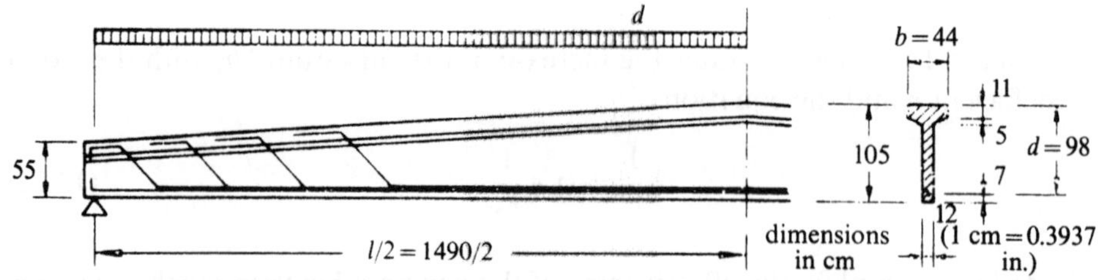

Figure 14.17.

For a roof girder with parallel flanches under uniform load, $\beta = 0.104$ in accordance with Table 14.2. For a girder shape as shown in Fig. 14.17, it pays to calculate this coefficient according to the formula given in Section 14.2.2.2. Integration is carried out according to Simpson and with the data contained in Fig. 14.18.

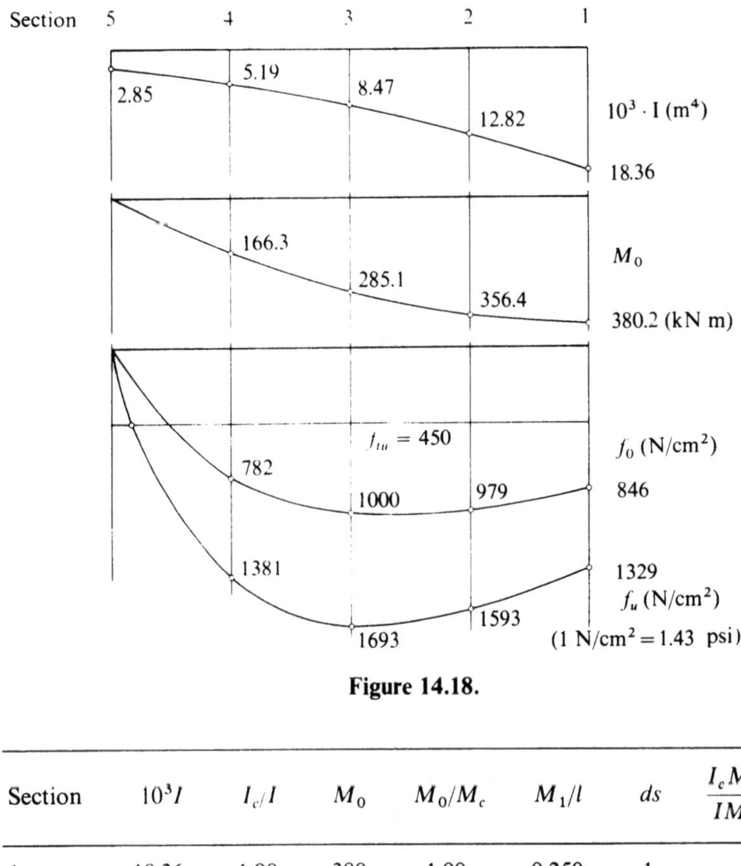

Figure 14.18.

Section	$10^3 I$	I_c/I	M_0	M_0/M_c	M_1/l	ds	$\frac{I_c M_0}{I M_c} \frac{M_1}{l} ds$
1	18.36	1.00	380	1.00	0.250	1	0.250
2	12.82	1.43	356	0.94	0.188		1.006
3	8.47	2.17	285	0.75	0.125		0.407
4	5.19	3.54	166	0.44	0.063		0.338
5	2.85	6.44	0	0	0		0.000
							$\sum = 2.051$

$$\beta = \frac{2.051}{3(4)} = 0.171.$$

Therefore, the coefficient β increases by (0.171–0.104)/0.104 ~ 65% for the selected girder shape. In Fig. 14.18 the tensile stresses caused by the sustained loads are given. From this it follows that for all practical purposes the girder is in a cracked state over its entire length.

In accordance with Section 14.2.2.2, we thus obtain for the moments from sustained load (380 kNm) and from the live load (278 kNm):

$$a_{c0}^{\mathrm{I}} = \beta \frac{Ml^2}{E_c I_c} = 0.171 \frac{(380 + 278)14.9^2 \times 10^{-4}}{3\,700\,000(0.01836)} = 0.0212 + 0.0155$$

$$= 0.0367 \text{ m},$$

for $(E_s/E_c)\rho_F = 5.7(0.95)/100 = 0.054$; $(E_s/E_c)\rho_F(d/h) = 0.05$; $d_c/h \sim 0.08$; $\beta_e = 1$; and for $\rho'/\rho = 0.14$, the following results are obtained in accordance with: Section 14.2.2.3, Equation (14.9):

$$a_0^{\mathrm{I}} = \varkappa_b^{\mathrm{I}} a_{c0}^{\mathrm{I}} = 0.91(0.0367) = 0.034 \text{ m};$$

Section 14.2.2.4, Eq. (14.12):

$$a_0^{\mathrm{II}} = \varkappa_b^{\mathrm{II}} \left(\frac{h}{d}\right)^3 a_{c0}^{\mathrm{I}} = 2.6\left(\frac{1.05}{0.98}\right)^3 0.0367 = 0.110 \text{ m};$$

Section 14.2.2.4, Eq. (14.13):

$$a_{d\varphi}^{\mathrm{II}} = a_{0d}^{\mathrm{II}} \varphi \varkappa_\varphi^{\mathrm{II}} = 2.6\left(\frac{1.05}{0.98}\right)^3 0.0212(3)0.18 = 0.037 \text{ m};$$

Section 14.2.2.4, Eq. (14.14):

$$a_s^{\mathrm{II}} \approx \frac{1.08(0.125)25 \times 10^{-5}(14.9^2)}{0.98}$$

$$= 0.008 \text{ m}; \; a_t = 0.110 + 0.037 + 0.008 = 0.155 \text{ m (6.1 in.)}.$$

Equations (14.17)–(14.20) do not have to be employed here because the beam is almost totally in a cracked state so that M_R/M_F equals zero.

14.3.2. Calculation of Required Camber

14.3.2.1. General

The camber u, which the scaffolding of a concrete structure should have prior to casting, has to offset all deformations occurring before or after casting the concrete, so that the structure will obtain the prescribed position and gradient. Among the deformations to be considered are settlement of the foundations, the deformation of the structure itself, taking into account the effects of creep and shrinkage, the deformation of the scaffolding, and settlement of the scaffolding foundations.

The examples dealt with in the following sections show how the effects of creep on the required camber can be calculated for more difficult problems.

14.3.2.2. Effects of a Change in the Static System

Frequently, the final static system deviates from the system which exists while the structure is under construction. Simply supported prefabricated beams, for example, can be changed into continuous ones. Then the deformation to be expected is influenced by both systems. It would be wrong to proceed on the assumption that in the second system only those loads are active which have been added later. Creep causes an ever-increasing share of the loads, applied prior to the system change, to also act in the final system (see Section 9.4.).

For such cases, we have to assume that the action effects which act prior to the system change ($t = 0$) developed due to superposition of two conditions:

(a) action effects S_{d_1} which would develop in the final system due to the loads during construction, e.g., dead weight, and
(b) action effects S_r which correspond to the difference between the action effects during the construction condition and those of the final condition.

In addition, we have action effects S_{d_2} which develop in the final system due to subsequently added loads. Under normal circumstances, creep does not noticeably change the action effects $S_{d_1} + S_{d_2}$, as mentioned in Section 2.1.2, Appendix II, and demonstrated in detail in Part A, Section 3.1.2.

Only the corresponding deformation increases to the value a_t. The imposed action effects S_r, on the other hand, are reduced considerably; that is, for uncracked cross sections, according to Section 2.1, by a factor γ_r according to Fig. 2.2. However, the corresponding deformation remains unchanged. To calculate the deformation, we therefore have, at time t, when the systems change occurs:

$$S_t = S_{td_1} + S_r, \qquad a_t = a_{td_1} + a_r, \tag{14.54}$$

and following the systems change, at time $t = \infty$,

$$S_\infty = S_{td_1} + S_{\infty r} + S_{\infty d_2}, \qquad a_\infty = a_{\infty d_1} + a_{\infty d_2} + a_r. \tag{14.55}$$

Since cracks usually develop at a later stage [16] and because most of the creep effects have occurred prior to that time in moderately thick concrete elements (50% of φ for $d_{\text{eff}} = 20$ cm after only 50 days), γ_r frequently provides a satisfactory solution for reinforced concrete.

Example. A floor system consisting of prefabricated girders which are connected later above their center support (Fig. 14.19). Load condition d_1 (dead weight) initially acts upon simply supported girders. The deformations $a^{(0)\text{II}}_{d_1 t=0}$ develop.

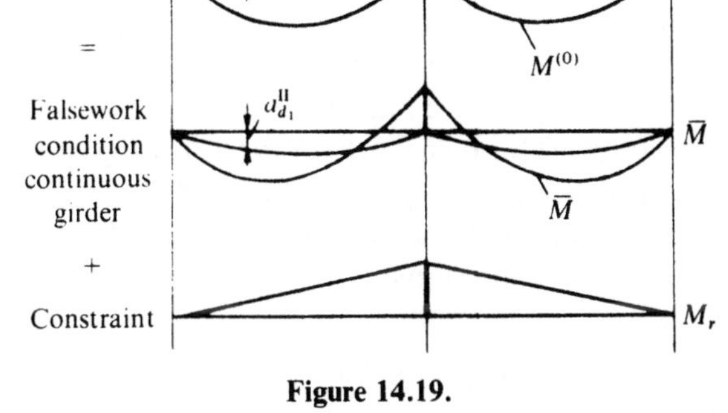

Figure 14.19.

After both spans are coupled above the center support after 28 days, a negative moment slowly develops there. By dividing the construction condition into an undisturbed load condition with full continuity and a constraint (see Fig. 14.19), we obtain moments $\overline{M}$ and $M_r = \overline{M} - M^{(0)}$. The new moment at the center support (see Section 9.4) $M_B = -M_{Br}(1 - \gamma_r)$.

We then obtain for the camber:

$$-u_{d_1} = a^{\mathrm{II}}_{d_1\infty} = a^{(0)\mathrm{II}}_{d_1 t=28} + a^{\mathrm{II}}_{d_1 t=28}\varkappa^{\mathrm{II}}_{\varphi}\varphi = (1 + \varphi\varkappa^{\mathrm{II}}_{\varphi})a^{\mathrm{II}}_{d_1 t=28} + a^{\mathrm{II}}_r. \quad (14.56)$$

Where:

$a^{(0)\mathrm{II}}_{d_1 t=28}$ the elastic deformation of the simply supported girder in a cracked condition under d_1;

$a^{\mathrm{II}}_{d_1 t=28}$ the elastic deformation of the continuous girder in a cracked condition under d_1;

$\varkappa^{\mathrm{I}}_{\varphi}$ see Section 14.2.2.4;

$a^{\mathrm{II}}_r = a^{(0)\mathrm{II}}_{d_1 t=28} - a^{\mathrm{II}}_{d_1 t=28}$, the deformation due to constraint.

Load condition d_2 (dead weight after completion) always acts upon the continuous system. The camber is

$$-u_{d_2} = a^{\mathrm{II}}_{d_2 t} = a^{\mathrm{II}}_{d_2 t=28}(1 + \varkappa^{\mathrm{II}}_{\varphi}\varphi). \quad (14.57)$$

A mathematical determination of the deformation of the cracked system can be carried out in accordance with the example in Section 14.3.1.4.

14.3.2.3. Free Cantilever Construction

In principle, for segmental cantilever construction, the deformations which determine the required camber are determined in the same way as if a scaffolding were used. We proceed from the concept that the deflection of a weightless cantilever has to be determined, which consists of concrete sections of different ages and which is subjected to new loads in different segments at different points in time. A similar method was chosen for Section 13.1.

We will explain this method in Fig. 14.20. When load group 1 has become effective, a curvature will develop in segment 1 only, while the remaining cantilever part follows the tangent at the end and is not deformed further. Load group 2 increases the curvature in section 1 and also deforms section 2. In the meantime, the first creep deformations also develop in section 1. Load group 3 has similar effects. However, here we have to take into account that differences in concrete age will lead to different creep of segments 1 and 2.

We therefore have to calculate the elastic and permanent deformations which can be expected for each structural condition ($a_{\mathrm{construction}}$) and have to compare it to the total deformation (a_{total}) of the final state which is to be the basis for determining the required camber. The form work for the following construction segment has to be cambered by $u = a_{\mathrm{total}} - a_{\mathrm{stressed}}$ (see Fig. 14.20).

If we denote the deformation of a section Δa, we obtain the following equations for the most important load conditions.

Load condition dead weight:

$$\Delta a_{dt} = \Delta a^{\mathrm{I}}_{d}(\beta_{\varepsilon} + \varphi_t).$$

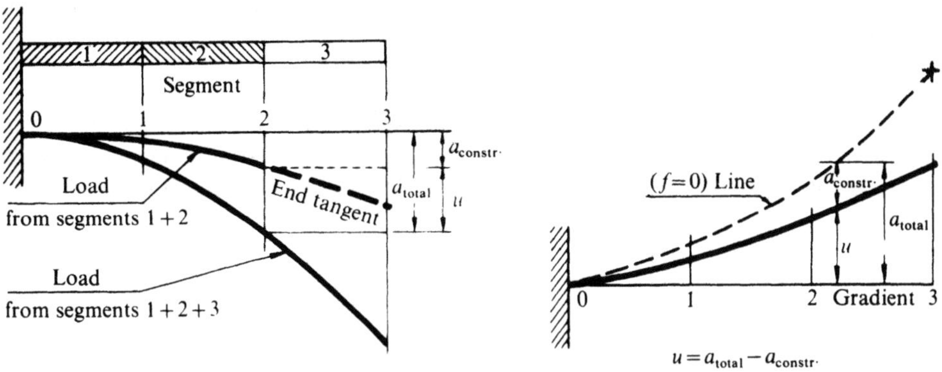

Figure 14.20. Relation between deflection and camber in free cantilever construction. The line $f = 0$ corresponds to the camber which would occur on a non-prestressed, weightless girder.

Load condition prestressing:

$$\Delta a_{pt} = \Delta a_p^1(\beta_\varepsilon + \varphi_t) + \Delta a_{\Delta pt}^1\left(\beta_{\varepsilon\Delta p} + \frac{\varphi_t}{2}\right), \qquad \beta_{\varepsilon\Delta p} \sim 1.$$

In contrast to the relatively exact analysis described above, the approximations described in the following are used in practice. Their effects on errors in curvature and camber calculations are described as well.

(a) Differences in Age at Loading Are Neglected

In the calculation, we proceed from a mean constant creep coefficient φ_m, a mean shrinkage strain ε_{sm}, and a constant modulus of elasticity E_{cm}. However, it is not easy to obtain a good approximation of the real deformations through suitable selection of φ_m and E_{cm}. Even if we select the mean creep coefficient φ_m such that the deflection curve corresponds as an average to the true curve, φ_m is too small for the initial load application of the sections and too big for loads applied later. The same is true for the choice of the mean modulus of elasticity. This means that the deformation of the individual concrete sections cannot be evaluated correctly. Thus, for a cantilevered system, we have a deviation from the exact deflection curve as shown in Fig. 14.21, even if the average of the approximated and exact

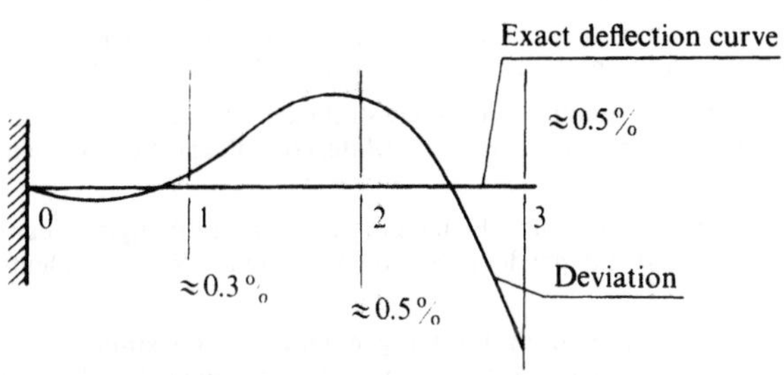

Figure 14.21. Deviation of u_{req} in percent.

curves coincide. As a rule, though, the deviations are within limits and are not bothersome to the naked eye.

(*b*) *The Effects of Creep Are Neglected During Construction*
The result of this approximation leads to a slight overestimate of the camber of the structure. Here, the effective thickness d_{eff} of the individual concrete sections plays a role. Usually, the error (see Fig. 14.22) is within permissible limits (e.g., $< 3\%$ of u_{req}).

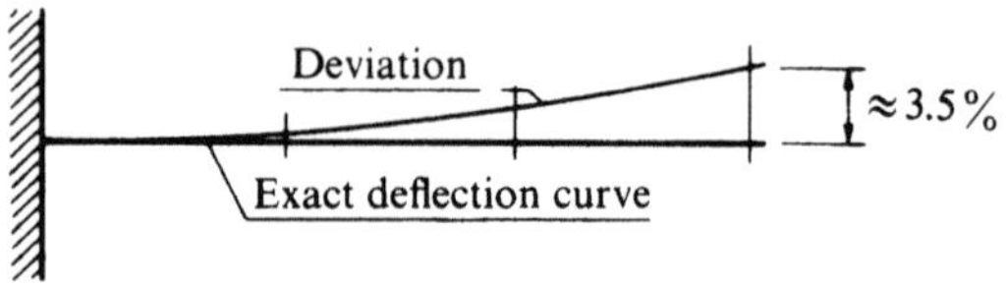

Figure 14.22. Deviation of u_{req} in percent.

References for Part C

[1] Rüsch, H.: "Einfluß von Kriechen und Schwinden auf die Schnittgrößen und Spannungen" (The influence of creep and shrinkage on action effects and stresses), Lecture Notes of the Institute for Concrete Construction, Technical University of Munich, 1967.

[2] Trost, H.: "Auswirkungen des Superpositionsprinzips auf Kriech- und Relaxationsprobleme bei Beton und Spannbeton" (Effects of the superposition principle on creep and relaxation problems for concrete and reinforced concrete), *Beton- und Stahlbetonbau*, **62**, No. 10, 1967.

[3] Rüsch, H. and Kupfer, H.: "Bemessung von Spannbetonbauteilen" (Dimensioning of reinforced concrete elements), Beton-Kalender, Verlag Wilhelm Ernst and Sohn, Berlin, 1980.

[4] Sattler: "Theorie der Verbundkonstruktionen" (Theory of composite construction), Verlag Wilhelm Ernst and Sohn, Berlin, 1953.

[5] Busemann, R.: "Kriechberechnung von Verbundträgern unter Benutzung von zwei Kriechfasern" (Calculation of creep of composite girders using two creep fibers), *Der Bauingenieur*, **25**, No. 11, 1950.

[6] Busemann, R.: "Spannungsumlagerungen infolge von Kriechen und Schwinden in Verbundkonstruktionen aus vorgespannten Fertigteilen und Ortbeton" (Stress redistribution as a result of creep and shrinkage in composite structures consisting of prefabricated parts and concrete cast in situ), *Beton und Stahlbetonbau*, **58**, No. 6, 1963.

[7] Trost, H. and Mainz, B.: "Zur Auswirkung von Zwängungen in Spannbetontragwerken" (The effects of constraint in prestressed concrete structures), *Beton- und Stahlbetonbau*, **65**, No. 8, 1970.

[8] Heilmann, H.: "Beziehung zwischen Zug- und Druckfestigkeit des Betons" (Relations between compressive and tensile strength of concrete), *Beton*, **19**, No. 2, 1969.

[9] Rüsch, H.: "Stahlbeton, Spannbeton (Werkstoffeigenschaften, Bemessungsverfahren)" (Reinforced Concrete, Prestressed Concrete (Materials, Properties, Dimensioning)), Werner-Verlag, Düsseldorf, 1972.

[10] Rüsch, H.: "Die wirklichkeitsnahe Bemessung für lastunabhängige Spannungen" (Realistic Determination of Load-Independent Stresses), Annual Report, Deutscher Betonverein, Wiesbaden, 1965.

[11] Falkner: "Zur Frage der Rißbildung durch Eigen- und Zwängsspannungen infolge Temperatur" (Concerning the problem of crack formation through internal and imposed action stresses caused by temperature), Proceedings Deutscher Ausschuß für Stahlbeton, No. 208, Berlin, 1969.

[12] Eibl, J.: "Zwängungen und Rißbildung von Stahlbetonstäben bei Behinderung der Längsverformung" (Imposed action and cracking of reinforced concrete beams with longitudinal constraint), *Die Bautechnik*, **46**, No. 11, 1969.

[13] Joas, H. and Petruschke, H.: "Tragfähigkeit von Bauteilen aus wasserundurchlässigem Beton" (Load capacity of structural elements made from water-impervious concrete), *Straße, Brücke, Tunnel*, **23**, No. 6, 1971.

[14] Lessing, A.: "Erfahrungen beim Bau eines vorgespannten Durchlaufträgers in abschnittweiser Herstellung" (Experiences with the construction of a prestressed continuous girder with step-by-step construction), Third Congress of the Fédération Internationale de la Précontrainte (FIP), Session II, Paper No. 18, Berlin, 1958.

[15] Rüsch, H.: "Die Ableitung der charakteristischen Werte der Betonzugfestigkeit" (Deduction of characteristic values for concrete tensile strength), *Beton*, **25**, No. 2, 1975.

[16] Rüsch, H. and Mayer, H.: "Die zeitliche Entwicklung der Durchbiegung von ausgeführten Stahlbeton-Traggliedern" (Time-dependent developments of deflections in existing reinforced concrete structures), *Beton- und Stahlbetonbau*, **59**, No. 16, 1964.

[17] Kunkel, K.: "Zur Berechnung von Schnittgrößen und Setzungen infolge zeitabhängiger Tragwerks- und Bodenverformungen" (Calculation of Action Effects and Settlement as a Result of time-dependent Deformations of the Structure and of the Soil), Dissertation, Technical University Darmstadt, Germany, 1978.

[18] Birkenmaier, M.: "Berechnung von Verbundkonstruktionen aus Beton und Stahl" (Analysis and Design of Composite Structures Made of Concrete and Steel), Verlag Leemann, Zürich, 1969.

[19] Haensel, J.: "Praktische Berechnungsverfahren für Stahlträgerverbundkonstruktionen unter Berücksichtigung neuerer Erkenntnisse zum Betonzeitverhalten" (Practical calculation methods for reinforced concrete composite structures by taking into consideration new findings of time-dependent properties of concrete), Scientific-Technical Information No. 15/2, The Institute for Civil Engineering, Ruhr University, Bochum, 1975.

[20] Mayer, H.: "Die Berechnung der Durchbiegung von Stahlbetonbauteilen" (Calculation of the Deflection of Reinforced Concrete Members), Proceedings Deutscher Ausschuß für Stahlbeton, No. 194, Berlin, 1967.

[21] Kordina, K. and Quast, U.: "Bemessung von schlanken Bauteilen-Knicksicherheitsnachweis" (Dimensioning of slender structural components—proof of buckling safety), Beton-Kalender, Verlag Wilhelm Ernst and Sohn, Berlin, 1972.

[22] Trost, H. and Mainz, B.: "Zweckmäßige Ermittlung der Durchbiegungen von Stahlbetonträgern" (Practical determination of the deflection of reinforced concrete beams), *Beton- und Stahlbetonbau*, **64**, No. 6, 1969.

[23] Ferry Borges: Non-linear behaviour of reinforced concrete structures, CEB Bulletin d'Information, No. 53, Paris, 1966.

[24] Comité Européen du Béton. Calculation and limitation of deflection, CEB Bulletin d'Information, No. 90, Paris, 1973.

[25] Heilmann, H., Hilsdorf, H., and Finsterwalder, K.: "Festigkeit und Verformung von Beton unter Zugspannungen" (Strength and deformation of concrete subjected to tensile stresses), Proceedings Deutscher Ausschuß für Stahlbeton, No. 203, Berlin, 1969.

[26] Dilger, W.: "Veränderlichkeit der Biege- und Schubsteifigkeit bei Stahlbetontragwerken und ihr Einfluß auf Schnittkraftverteilung und Traglast bei statisch unbestimmter Lagerung" (Variability of flexural and shear stiffness of reinforced concrete structures and their influence on distribution of action effects and on load capacity of statically indeterminate systems), Proceedings Deutscher Ausschuß für Stahlbeton, No. 179, Berlin, 1966.

[27] Guldan, R.: "Rahmentragwerke und Durchlaufträger" (Frames and Continuous Girders), Springer-Verlag, Heidelberg, Berlin 1959.

[28] Krämer, U., Thielen, G., and Grasser, E.: "Die Berechnung der Durchbiegung von biegebeanspruchten Stahlbetonbauteilen unter Gebrauchslast" (Calculation of deflection of reinforced concrete members subjected to bendung stress), *Beton und Stahlbetonbau*, **70**, No. 4, 1975; and "Hilfsmittel für die Berechnung und Bemessung von Stahlbetonbauteilen" (Aids for analysis and dimensioning of reinforced concrete members), Proceedings Deutscher Ausschuß für Stahlbeton, No. 240, Berlin, 1976.

Appendix I

Further Explanations to Part B

For readers in a hurry:

In Appendix I the information given in Part B will be explained in more detail. In particular the following topics will be dealt with:

Section 1. Examples regarding the modulus of elasticity.
Section 2. Justification of the prediction method of creep and shrinkage coefficients according to Rüsch–Jungwirth.
Section 3. Comparison of creep and shrinkage coefficients as predicted by various methods.

1. Modulus of Elasticity

In Part B, Section 3.1.8, relations to estimate the modulus of elasticity of concrete are summarized. In the following, the effect of type of aggregate and age at time of load application upon the modulus of elasticity, as well as on the modulus of deformation, will be illustrated.

How does the modulus of deformation of 7-day-old concrete change during a period of loading between 1 and 24 hr. The concrete is expected to have a nominal compressive strength of 30 N/mm² (4300 psi) at an age of 28 days. The following types of aggregates may be used: dense limestone, quartzite, or sandstone.

According to Eq. (3.8), Part B, the modulus of elasticity of the concrete at age t is

$$E_t = \beta_e E_{0.28}.$$

According to Eq. (3.2), Part B, we obtain

$$E_{0.28} = \underset{(1910)}{4.3} \quad \beta_a w \sqrt[3]{f'_c} \ \underset{(\text{psi})}{\text{MN/m}^2}.$$

The unit weight w is, according to Eqs. (3.12)–(3.14), Part B

for dense limestone: $$w = 2300 + \frac{f'_{c28}}{0.3}\ \mathrm{kg/m^3}$$

$$\left(143 + \frac{f'_{c28}}{700}\right)(\mathrm{lbs/ft^3});$$

for quartzite: $$w = 2200 + \frac{f'_{c28}}{0.3}\ \mathrm{kg/m^3}$$

$$\left(136 + \frac{f'_{c28}}{700}\right)(\mathrm{lbs/ft^3});$$

for sandstone: $$w = 2180 + \frac{f'_{c28}}{0.3}\ \mathrm{kg/m^3}$$

$$\left(135 + \frac{f'_{c28}}{700}\right)(\mathrm{lbs/ft^3}).$$

Finally, we have to take into account that in the equations to estimate the modulus of elasticity, f'_c describes the average compressive strength of the concrete. The average strength always has to be larger than the nominal strength of the concrete. If good quality control prevails we may assume, as an approximation, that

$$f'_{cm} = f'_c + 4\ \mathrm{N/mm^2},$$

i.e.

$$f'_{cm} = 30 + 4 = 34\ \mathrm{N/mm^2}\ (4800\ \mathrm{psi}).$$

Using the coefficients β_e from Fig. 3.3, Part B, and β_α from Section 3.1.8, Part B, we obtain the following values.

	w		β_a	β_e	$E_{0,28}$		$E_{0,7}$	
	kg/m³	lbs/ft³	–	–	MN/m²	ksi	MN/m²	ksi
Dense limestone	2 436	151	1.1	0.904	37 300	5410	33 800	4900
Quartzite	2 336	145	1.0	0.904	32 600	4730	29 500	4280
Sandstone	2 316	144	0.8	0.904	25 800	3740	23 400	3390

For the modulus of deformation, E_{Dt}, we obtain, from Eq. (3.10):

$$E_{Dt} = \frac{E_{0,t}}{1 + \beta_i(0.91 - 0.8\beta_t\beta_\infty)}.$$

Here, β_i has to be taken from Fig. 3.2, Part B. The coefficient β_t is given in Fig. 2.1, Part B, whereas β_∞ may be obtained from Section 3.1.8, Part B.

The development of the modulus of deformation with time is given in Fig. 1.

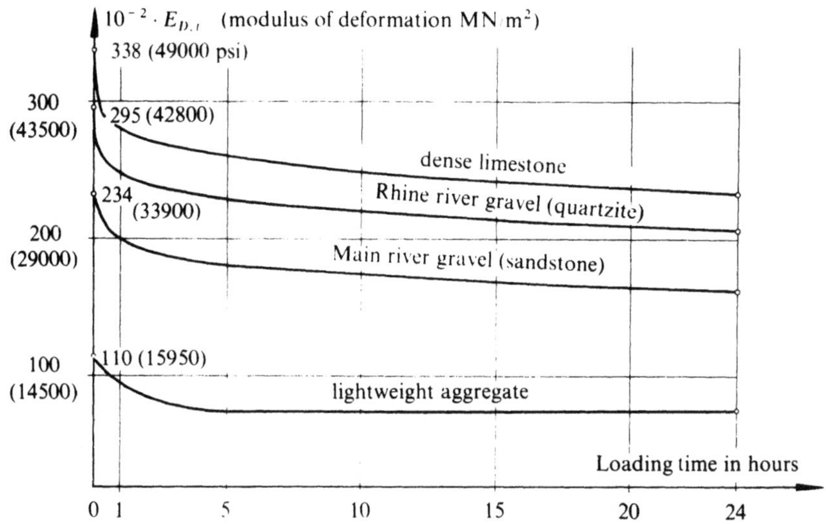

Figure 1. Development of modulus of deformation with time for concretes made with different types of aggregate.

The effect of the type of aggregate becomes even more apparent when lightweight concrete is made from lightweight aggregates. In Fig. 1, therefore, the modulus of deformation of a lightweight concrete with a unit weight of 1400 kg/m^3 (87 lbs/ft^3) is also plotted. Here, the modulus of elasticity $E_{0,28}$ has been estimated to be 11 000 MN/m^2 (1595 ksi) according to Part B, Section 4.3.4.

2. The Prediction Method of Rüsch–Jungwirth

This method has been described in detail in Part B, Section 4.3. In the following, the choice of parameters taken into account in this prediction method as well as the particular formulation will be explained. Furthermore, the method will be compared with other prediction methods.

2.1. Limits of Applicability

The method of Rüsch–Jungwirth may only be used, when taking into account the following limitations:

(a) The concrete stresses are at all times less than 50% of the concrete compressive strength.

(b) The temperature of the concrete is in the range between 0 °C (32 °F) to 30 °C (86 °F). Very short temperature variations should not exceed the limits −20 °C (−4 °F) to +50 °C (122 °F), respectively. If these limits are exceeded it is no longer sufficient to take into account the effect of temperature according to Eq. (2.2), Part B. To take into account deviations from these limitations, special rules have to be obeyed, as mentioned in Section 1.3, Part B.

(c) The relative humidity is in the range of 40–100%, or the concrete is stored in water.
(d) The concrete composition corresponds to that of a usual structural concrete. The method is not applicable to very rich concretes, very high slump mixes, or steam-cured concretes.

2.2. Choice of Parameters

In Part B, Sections 1 and 3, the most important parameters have been described, which significantly influence creep and shrinkage of concrete. These are:

Concrete composition;
Structure, porosity, and moisture content of the hydrated cement paste which in turn depends upon type of cement, water cement ratio, curing, and age of the concrete;
Properties of aggregates;
Dimensions of the structural member;
Environmental conditions.

Only some of these parameters are known to the designer at the time he analyzes the structure. Therefore, not all of them can be taken into account in a practical prediction method for creep and shrinkage coefficients to be used by a designing engineer.

In the Rüsch–Jungwirth method, therefore, only a limited number of parameters are used which may all be estimated reasonably well by the designer. These are:

Dimensions of the structural member;
Age of concrete at load application;
Temperature during hydration of the concrete;
Relative humidity while the structures are under load;
Type of cement;
Consistency of the fresh concrete as a measure of its composition.

2.3. Delayed Elastic Strain

The advantage of separating the delayed elastic strain from the total creep strain has already been discussed in Part B, Section 3.3. It is particularly apparent if the creep-inducing stress varies over the course of time. In the case of unloading of the concrete, we may then proceed according to Eq. (4.26), Part B.

The last term of Eq. (4.26), Part B needs further explanation. Upon unloading, creep recovery may not always be as large as $\varphi_{d0} = 0.4$, but only as large as the delayed elastic strain which had developed during the preceeding period of loading: $0.4\beta_{d(t_2-t_1)}$. Therefore, during the period of unloading $(t - t_2)$ creep recovery corresponds to $0.4\beta_{d(t_2-t_1)}\beta_{d(t-t_2)}$. (Refer to Example 4, Part C, Section 1.5). This simpler formulation does not correspond exactly to the principle of superposition as described in Section 3.7 and as used, for example, in the prediction method CEB/FIP 1970. However, the final values of delayed elastic strain are identical. There is a difference in the time development of delayed elastic strain which is significant only for short periods of loading or unloading respectively.

2.4. Flow

In the Rüsch–Jungwirth method, an effort has been made to formulate creep in such a way that the various processes become lucid to the designing engineer. This is particularly true for the application of the principle of superposition and for an estimate of creep under variable stresses.

Furthermore, the Rüsch–Jungwirth method is particularly suited for an application of the Dischinger method which is dealt with in Part C and Appendix II.

The estimated flow in the Rüsch–Jungwirth method is based upon the assumption that the flow of a particular type of concrete loaded at different ages may be described by a single function. Furthermore, the Rüsch–Jungwirth formulation assumes that for very high ages at loading, flow appraches a value of 0, indicating that creep is completely reversible. While the second assumption is justified, at least to some extent, by experiments, the first assumption, in general, is not entirely correct. This can be seen from the flow curves given as dashed lines in Fig. 3.5, Part B. Nevertheless, this assumption has been maintained because it significantly simplifies calculations of the effect of creep on structural behavior. Finally, it will be shown in Section 3 that despite these simplifying assumptions, the Rüsch–Jungwirth method is not any less accurate than other methods.

2.5. The Influence of Concrete Composition

Frequently, in prediction methods for creep and shrinkage of concrete, knowledge of the concrete composition, in particular, the water-cement ratio and the cement content, is required. Under normal conditions, such information is not available to the designing engineer at the time he has to do his design work. On the other hand, for structural concrete, the water-cement ratio does not deviate too much from approximately 0.5. Then the required cement content is primarily a function of the consistency of fresh concrete. Increasing the slump of the concrete requires an increase of the cement content and thus leads to increased creep and shrinkage strains. Therefore, creep may be given as a function of concrete slump, so that the effect of concrete composition can be taken into account by one rather than by two parameters. The slump of the concrete may be predicted by the designer, taking into account local conditions, structural dimensions, reinforcement, etc. Therefore, in Table 4.2, Part B, where the basic coefficients for creep and shrinkage are summarized, the slump is also introduced as a parameter.

2.6. The Influence of the Degree of Hydration of Concrete

It has been known for quite some time that the effect of age at loading can not be described only by the number of days which elapses between casting and loading the structural member. It is the degree of hydration of the concrete which is the more significant parameter. It depends on the temperature during hydration, on the curing conditions, and on the reactivity of a particular type of cement.

In Part B [38] the effect of temperature during hardening was taken into account using the well-known expression $\Delta t(T + 10)$, with T in °C, which is a measure of the maturity of the concrete. In order to take into account the effect of the type of cement, different creep relationships are given. Since only the parameters of temperature and type of cement affect the degree of hydration, they have been combined in the Rüsch–Jungwirth method and are taken into account by the parameter "effective concrete age, t."

This approach is facilitated by the observation that the reactivity of a particular type of cement may be taken into account by a time-independent factor: Two types of cement C_a and C_b may hydrate at different rates. Under the assumption that strength and degree of hydration are, as a first approximation, proportional, the degree of hydration may be expressed by $\bar{\alpha} = f'_{ct}/f'_{ct=\infty}$. Here f'_c is the compressive strength at time t, whereas $f'_{ct=\infty}$ is the final strength of the concrete after complete hydration. Cement C_a may reach a particular value of $\bar{\alpha}$ after t_a, whereas cement C_b may reach the same value of $\bar{\alpha}$ after a time t_b. An evaluation of experiments showed that within the range of required accuracy the ratio $\beta_{ce} = t_a/t_b$ is almost independent of the particular value of $\bar{\alpha}$. If we take for $\bar{\alpha}$ values which are valid for a Type I cement, then the values β_{ce} for Type III cements are in the range 2–3.

Thus the effective age of concrete t made of a Portland cement other than Type I and hydrating at a temperature differing from 20 °C (68 °F) may be estimated from Eq. (2.2), Part B.

For T in °C,

$$t = \beta_{ce} \sum \frac{T(°\mathrm{C}) + 10}{20} \Delta t.$$

For T in °F

$$\beta_{ce} \sum \frac{T(°\mathrm{F}) + 14}{54} \Delta t.$$

Example 2 in Part C, Section 1.5 shows that the second part of this formula is needed only when the concrete is cast in the winter. Then care has to be taken that the concrete structure is exposed to sustained loads only after the maturity value (numerator of the above equation) has reached the value assumed in the design of the structure.

2.7. The Influence of Dimensions of the Structural Member

The effect of member thickness on the delayed elastic strain is so small that it normally can be neglected. Also, basic flow is unaffected by structural dimensions. However, for drying flow and shrinkage, we may assume that for a given moisture differential, thin members dry faster than thick ones. The effect of members thickness decreases as the moisture differential decreases; it disappears entirely when the member is stored in water. These observations may be taken into account by introducing the new parameter effective thickness, d_{eff}. The relative thickness $\bar{l}$

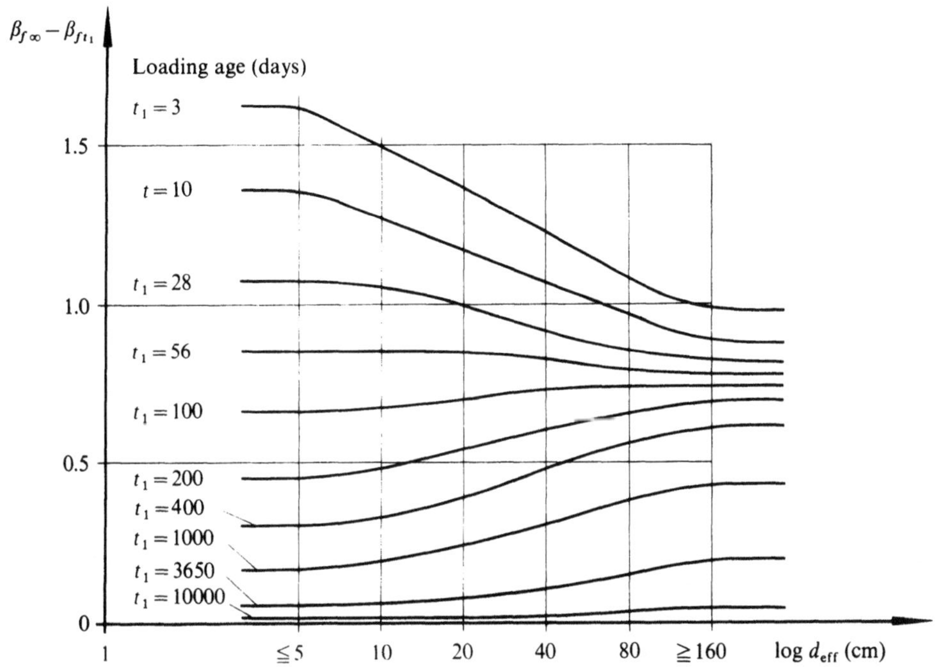

Figure 2. The influence of effective thickness on flow of concrete loaded at different ages.

which has been used in Part B, Section 3.4 depends only on structural dimensions. The effective thickness d_{eff} is defined in Eq. (4.27), Part B. It is a function of cross-sectional area, circumference, and moisture differential.

The influence of effective thickness on flow may be seen from Fig. 2. Here, the flow coefficients are given as a function of effective thickness and age at loading. Figure 2 shows that only for young concretes, flow decreases with increasing dimensions. The more the concrete has been dried prior to loading, the smaller the effect of structural dimensions. Figure 2 is quite useful for practical applications because it facilitates interpolation.

3. Comparison of Prediction Methods for Creep

Various methods to predict creep coefficients of concrete have been compared with each other and with experimental data. Some of the results of these studies are given in [23] Part (B). They are summarized in the following. The following methods have been analyzed:

(1) The method proposed by Bažant–Panula (BP), Part B [34];
(2) The method proposed by CEB-1970 (CEB-70), Part B [44];
(3) The method given in the CEB/FIP model code 1978 (CEB-78);
(4) The Rüsch–Jungwirth method (RJ), Part B [46];
(5) The method proposed by the British Concrete Society (BCS), Part B [47];
(6) The ACI 209 method (ACI), Part B [50].

3.1. Comparison of the Prediction Methods with Each Other

For the comparison, a structural concrete with the following composition was chosen:

Normal-weight aggregate; Type I Portland cement.
Cement content: 337 kg/m^3 (6 lbs/yd^3).
Aggregate content: 1818 kg/m^3 (3064 lbs/yd^3).
Water-cement ratio: 0.55.
Duration of moist curing: 7 days.

Further details may be taken from [23] Part B.

Figure 3 shows the effect of concrete age at loading on the creep coefficient for cylindrical specimens, $d = 200$ mm (7.87 in.) at a relative humidity of 60%. In order to exclude the effect of other parameters, the creep coefficient $\varphi_{28,\infty}(t_0)$ has been given as a fraction of the creep coefficient of a concrete loaded at an age of 360 days, $\varphi_{28,\infty}(t_0 = 360)$. All methods demonstrate the expected decrease of creep with increasing age at load application. However, there are substantial differences between the various prediction methods, particularly for young concretes.

In Fig. 4, the effect of relative humidity on creep of concrete is shown. There, as predicted by various methods, the final creep at a humidity H, $\varphi_\infty(H)$ is given as a fraction of the final creep coefficient for a relative humidity of 100%, $\varphi_\infty(H = 100)$. Again, substantial differences may be observed for different methods.

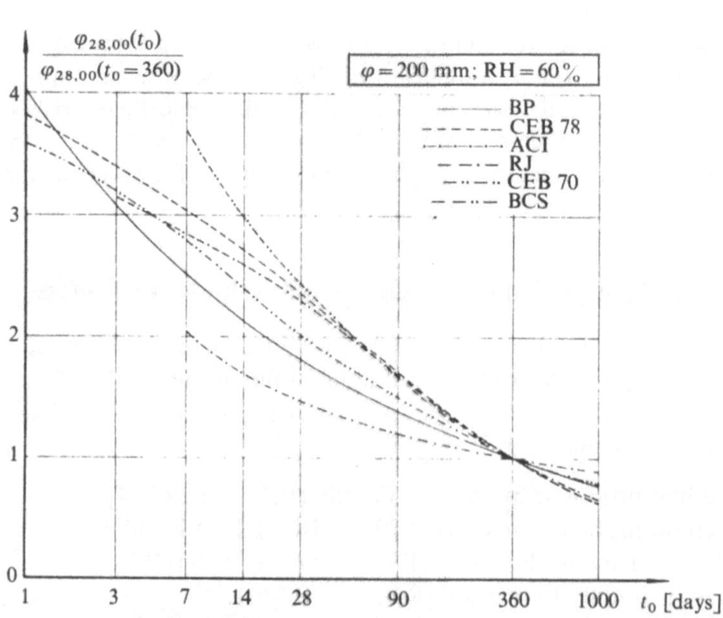

Figure 3. Effect of concrete age at loading on creep as predicted by various methods.

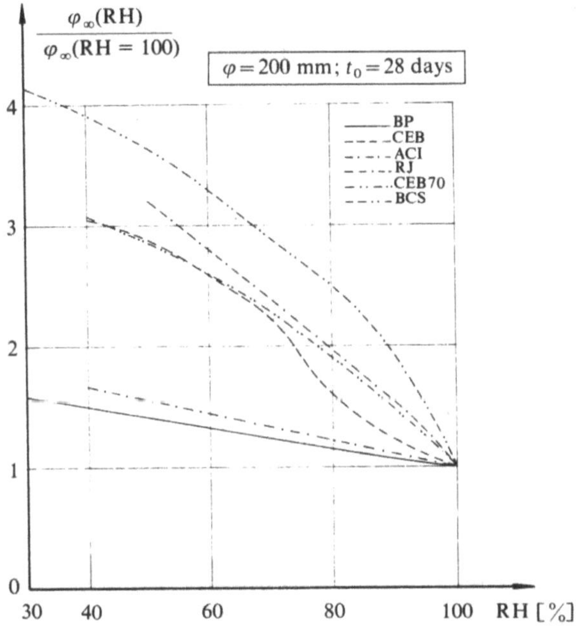

Figure 4. Effect of relative humidity on concrete creep as predicted by various methods.

Finally, Fig. 5 shows the effect of member thickness on creep. Only the methods BP, CEB 78, and RJ take into account that the effect of member thickness also depends on the age at loading and on the relative humidity. Again, significant differences may be observed, particularly for small members.

In conclusion we may state that methods 1 and 6, in most cases, predict creep values which are smaller than those predicted by methods 2, 3, 4, and 5. The values predicted by the latter methods are in close agreement.

3.2. Comparison with Experimental Results

In order to evaluate the validity of creep strains as predicted by different methods, the creep functions as determined by some prediction methods have been compared with 102 creep experiments.

In the choice of creep experiments, a wide range of parameters affecting creep has been chosen.

The methods have been statistically evaluated by calculating "standard deviations" as well as "coefficients of variation" for the differences between experimental and predicted values. For a particular experiment, the difference between measured and predicted strain has been determined for certain predetermined intervals of duration of loading. From this, an average standard deviation s_i and a coefficient of variation V_i have been determined for each method and for each experiment. The total coefficient of variation of a particular method $\bar{V}$ has been determined as the arithmetic mean of individual values of V_i. In

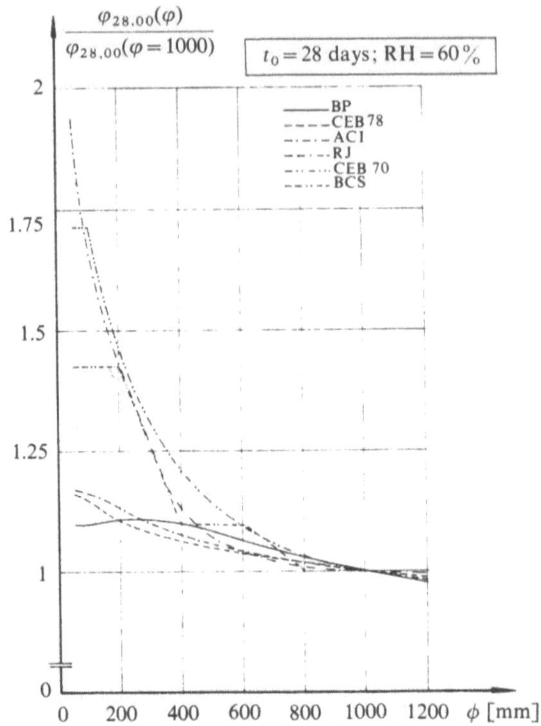

Figure 5. Effect of dimensions of member on concrete creep as predicted by various methods.

Table 1, the average values $\bar{V}$ are summarized. For the BCS method, no value of $\bar{V}$ is included, since this simple method allows only an estimate of final creep and not the development of creep with time.

Table 1

Method	BP	CEB 70	CEB 78	RJ	BCS	ACI[a]
$\bar{V}$ (%)	32.0	23.1	24.1	25.1	-	24.8

[a] Based on 72 experiments.

3.3. Conclusions

At first sight, it is surprising that despite the large and systematic differences in the creep strains as predicted by various methods and as demonstrated in Section 3.1, the methods differ relatively little with regard to their accuracy in predicting experimental data. From this, we may at least conclude that none of the methods investigated is already sufficiently optimized. They may be improved further by adapting them better to experimental data. Both ACI and CEB are conducting a joint effort in this respect. Furthermore, none of the prediction methods are at this stage very accurate. Therefore, in extreme cases where creep and shrinkage may endanger the integrity of a structure, variations in creep and shrinkage coefficients of at least $\pm 20\%$ should be taken into account in the analysis.

Considering these large deviations in the laboratory studies, the differences between predicted values and concrete strains as observed on actual structures and as described in Part B, Section 6 are surprisingly small. Apparently, because of their larger dimensions, actual structural members are more stable in their behavior, so that their creep and shrinkage properties are easier to predict than those of the small and more sensitive laboratory specimens.

Appendix II

Realistic Calculation of Creep and Shrinkage Effects

For readers in a hurry:

As an introduction and survey, Appendix II describes the methods developed for the mathematical treatment of creep and shrinkage problems. Part C is designed to deal with their application in more detail. Described are:

Section 1. the special terms.
Section 2. Dischinger's solution of the differential equation and its adaptation to seperation of creep into delayed elastic strain and flow.
Section 3. The finite-difference method.
Section 4. The Trost method using the relaxation coefficient.
Section 5. The Bažant method using the effective modulus.
Section 6. The graphic-analytical method by Busemann.
Section 7. Application of these methods to frequently occurring special cases (statically indeterminate structures, varying creep properties of materials within a structural member, slowly developing imposed action effect, influence of creep of prestressing tendon).
Section 8. Comparison of different methods and of experimental results.
Section 9. Outlook.
Section 10. References.

1. Introduction

1.1. Survey

A strictly mathematical formulation of the consequences (see Part A) of true time-dependent strains (see Part B) of concrete is difficult. The mathematical effort can grow to such an extent that it becomes unworkable. Just consider the inhomogeneous behavior of reinforced concrete cross sections in a multiple statically indeterminate structure which is erected in segments. Therefore, the analytical methods developed for use in engineering practice always contain more or less rough approximation, either of the mathematical treatment or of the material properties. One of the least accurate approximations, though frequently used in the past is the consideration of creep by reducing the modulus of elasticity with the formula $E_{c\,\mathrm{id}} = E_c/(1 + \varphi)$.

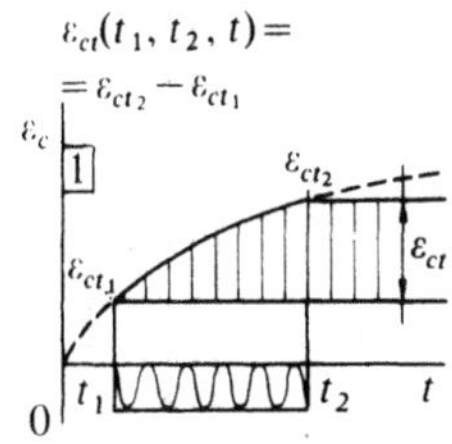

Figure 1.1 Rate of creep.

Dischinger [1] was probably the first to consider the influence of creep on action effects with the aid of a differential equation. He takes into account the age dependence of the modulus of elasticity, but not, in accordance with the state of knowledge at that time, the effects of delayed elasticity [2]. This borderline case in which creep is considered a totally permanent strain is also called the "Theory of Aging" or rate of creep (see Fig. 1.1). Here creep depends only on concrete age at the time of load application and on the duration of loading. The solution provided by Dischinger is easy to handle. However, for extremely variable stress and, particularly, for relaxation problems, his results are inaccurate because they do not consider the delayed elastic strain ε_d. For multiple statically indeterminate problems, coupled differential equations result which are less suited to practical applications.

The delayed elasticity approach proposed by Boltzmann [3], which applies to old concrete, can be considered an extreme case considering creep to consist only of delayed elastic strains (Fig. 1.2).

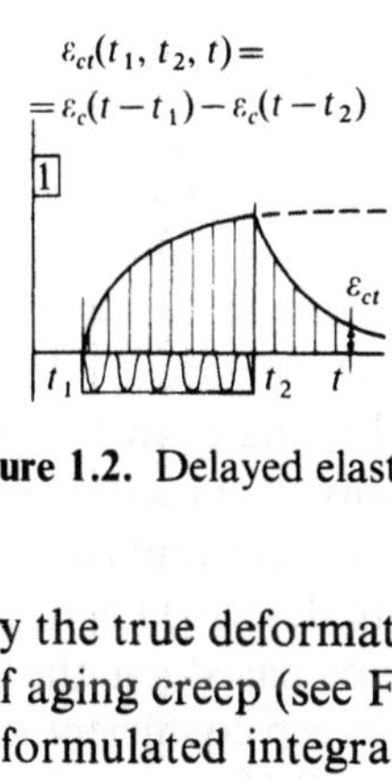

Figure 1.2. Delayed elasticity.

Attempts to cover accurately the true deformation behavior of concrete can be combined under the heading of aging creep (see Fig. 1.3). Arutjunjan [4], Alexandrowski [5], and others have formulated integral equations which are based on creep functions.

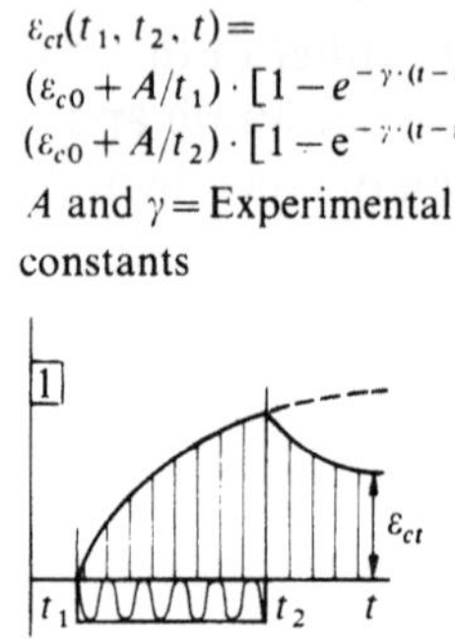

Figure 1.3. Aging creep.

In Germany, Trost [6] and Zerna [7], and in the United States, Bažant, have made successful efforts to describe creep and relaxation problems on the basis of rheological models. In his work along the same lines, Schade [8] describes the relation between creep and relaxation and explains why it is preferable to use integral and differential equations to describe the problems. He also points out contradictions in generally accepted assumptions. A comprehensive summary of analytical methods is contained in a report by the CEB [9].

It turns out that the most basic form of an integral equation which contains concrete creep behavior, dependent on two time functions (Part B, Section 4) can not be solved in closed form. With the aid of an easily used algebraic equation and/or by studying the basic form of the creep function, Trost [10] and Zerna [11] were able to develop closed approximate solutions. Bažant's proposal to base creep calculations on effective moduli which have been deduced theoretically [12], rather than to use effective creep coefficients as proposed in [10], deserves special attention.

The differential equation which is connected with the integral equation can also be solved through finite-difference methods. The great numerical effort can be handled with the aid of computers. With the aid of suitable methods, e.g., using the creep functions according to Arutjunjan, best use may be made of computers [8]. We have to make sure, though, that the material properties of the concrete are not subjected to redundancy, either through the analysis itself, or when taking the material characteristics of creep from specifications.

One could easily get the impression that creep problems can only be solved satisfactorily with a tremendous calculation effort. This is particularly true since the methods discussed so far mainly take into account only the deformation behavior of a homogeneous concrete section, while, in reality, composite effects and static indeterminacy problems dominate. In the following, we can demonstrate simple and lucid methods for solving practical creep and shrinkage problems, which correspond to the rheological concepts developed in Part B. The sequence in which the different methods are described is irrelevant here.

The following explanations are based on the information given in Part B, Section 4.2 for an estimate of the creep coefficient φ and the shrinkage strain ε_s, in accordance with the formulas $\varphi_t = 0.4\beta_d + \varphi_{f_0}\Delta\beta_f$; $\varepsilon_{st} = \varepsilon_{s0}\Delta\beta_s$.

1.2. Terminology

Subscripts:

- c concerns that part of the structural member which creeps and which consists of concrete. It is thus a substitute for all material which creep;
- sus a permanently acting load (sustained load);
- e concerns the elastic part of the structural member or, a consequence of the elastic deformation;
- id idealized characteristics of cross-section properties, e.g., $A_{id} = A_c + (n - 1)A_s$; $n = E_s/E_c$; A is the cross section
- r indication of a suddenly appearing imposed action (Part A, Section 1), e.g., foundation settlement;
- s unstressed reinforcement or indication of an imposed action which develops nearly affinitive to creep, e.g., shrinkage;
- w indication of loads of short duration;
- p indication of prestressed steel, it replaces in this special case e and/or s;
- cs indication of creep and shrinkage.

Deformation characteristics:

- E_e, E_s, E_c, E_p the modulus of elasticity of elastic structural elements and of those experiencing creep;
- ε_e elastic strain of elastic structural elements and of those experiencing creep;
- ε_c creep strain;
- φ_t creep coefficient $\varphi_t = \varepsilon_{ct}/\varepsilon_{e28}$, valid at time t;
- ε_s shrinkage.

Displacement values:

- $\delta_{e_11}, \delta_{s_11}, \delta_{p_11}$ elastic displacement of the elastic element (e.g., steel) at point 1, caused by force $X = 1$. It is determined either on the statically determinate basic system or on an element dx (see Fig. 2.1);

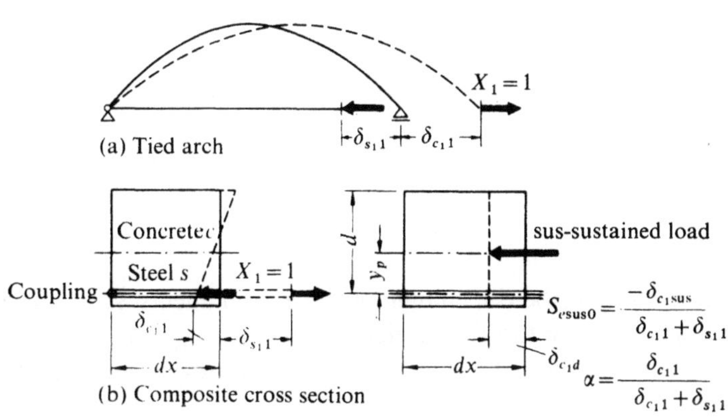

Figure 2.1. Elastic displacements for local coupling, e.g., a tied arch and for continuous coupling, e.g., a composite cross section.

$\delta_{c_1 1}$ as above, but displacement caused by $X_1 = 1$ of that element which creeps (e.g., concrete arch) calculated with E_c;

$\delta_{e_1 w}, \delta_{p_1 w}$, $\delta_{c_1 w}, \delta_{s_1 w}$ as above, but displacements caused by the effect of short-term loads;

$\delta_{e_1\,sus}, \delta_{p_1\,sus}$, $\delta_{c_1\,sus}, \delta_{s_1\,sus}$ as above, but displacements caused by the effect of sustained loads;

$\delta_{e_1 r}, \delta_{p_1 r}$, $\delta_{c_1 r}, \delta_{s_1 r}$ as above, but displacements caused by suddenly occurring imposed actions;

$\delta_{e_1 s}, \delta_{p_1 s}$, $\delta_{c_1 s}, \delta_{s_1 s}$ as above, but displacements caused by an imposed action which develops affinitive to creep.

Ratios:

α stiffness value or stiffness ratio, $\alpha = \delta_{c_1 1}/(\delta_{c_1 1} + \delta_{e_1 1})$;

Action effects:

S_{et} the total value of the action effects which act on the elastic element during interaction of both elements at time t, taking into consideration creep and shrinkage and/or the coupling force which appears at time t;

S_{e0} as above, but at time $t = 0$;

S_{ew} as above, but only as the result of short-term loads;

$S_{e\,sus\,t}, S_{e\,sus\,0}$ as above, but only as the result of sustained loads;

S_{ert}, S_{er0} as above, but only as the result of a suddenly developing imposed action;

S_{est} as above, but only as the result of an imposed action which develops affinitive to creep;

S_{ecst} creep- and shrinkage-induced change in action effects $S_{ecst} = S_{et} - S_{e0}$;

$S_{es}^{(e)}$ as above, auxiliary action effect to determine the effect of shrinkage while neglecting creep: $S_{es}^{(e)} = -\delta_{c_1 s}/(\delta_{c_1 1} + \delta_{e_1 1})$.

Stresses:

the subscripts to stresses f correspond to those of action effects S;

Creep factors:

$\gamma_{sus}, \bar{\gamma}_{sus}$ factors to describe the change of the action effects $S_{e\,sus}$ caused by creep and shrinkage in the elastic element: $\gamma_{sus} = S_{e\,sus\,t}/S_{e\,sus\,0}$;

$\gamma_r, \bar{\gamma}_r$ as above, but for the effect of suddenly developing imposed actions $\gamma_r = S_{ert}/S_{er0}$;

$\gamma_s, \bar{\gamma}_s$ as above, but for effect of imposed actions developing nearly affinitive to the development of φ: $\gamma_s = S_{est}/S_{es}^{(e)}$.

Dimensions:

$A_e, A_c, A_{id}, A_p, A_s$ cross-sectional area;
I_e, I_c, I_{id} moment of inertia;
y_p distance between centers of gravity of cross sections of elements which do or do not creep.

2. The Differential Equation According to Dischinger

2.1. Original Solution

2.1.1. Introduction

The differential equation solved by Dischinger in 1937 [1] describes the redistribution of action effects and/or stresses developing in a structure as the result of creep and shrinkage, if purely elastic structural members interact with others subject to creep. The coupling of these members can be local (e.g., a concrete arch with a steel tie or unbonded prestressing) or continuous (e.g., reinforced concrete, prestressing with bond, composite girders). Since the use of the equation is not limited to concrete construction, we use the subscripts e for the elastic element (e.g., reinforcement or prestressing steel); and c for the member which is subjected to creep.

2.1.2. Local Coupling of a Purely Elastic Element with a Creeping Element

The differential equation is deduced for the simple example of an unbonded prestressed girder (see Fig. 2.2). Internally, it is a single statically indeterminate system. The statically determinate basic system is developed by cutting the tension member at point 1. The pertinent displacements are given the indices c for the creeping element and e for the elastic element:

$\delta_{c_1 w}$ for short-term load;
$\delta_{c_1 \mathrm{sus}}$ for sustained loads causing creep;
$\delta_{c_1 r}, \delta_{e_1 1}$ for suddenly developing imposed action;
$\delta_{c_1 1}, \delta_{e_1 1}$ for statically indeterminate action effect $S_c = 1$;
$\delta_{11} = \delta_{c_1 1} + \delta_{e_1 1} = \delta_{c_1 1} + \delta_{p_1 1}$.

At time $t = 0$, before the effects of creep and shrinkage commence, the following applies to the force effective in the elastic member:

$$S_{e0} = \frac{\delta_{c_1 w} + \delta_{c_1 \mathrm{sus}} + \delta_{c_1 r}}{\delta_{11}}.$$

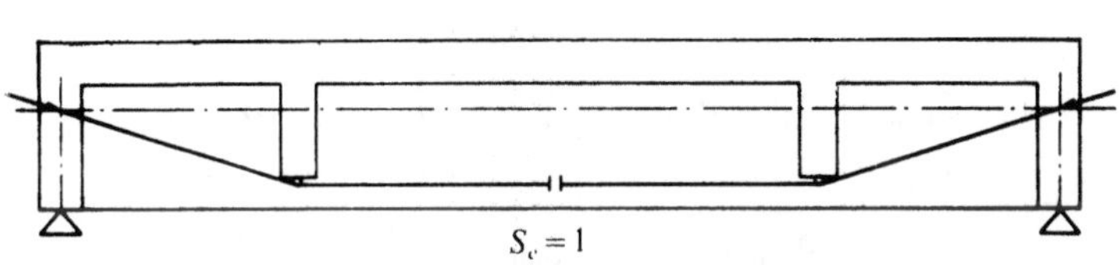

Figure 2.2.

At an arbitrary later point in time the following additional displacements occur in time interval dt, caused by a change of $d\varphi$ of the creep coefficient and of $d\varepsilon_s$ of shrinkage. As a simplification, we assume E_c is approximately constant (for the exact solution, see Part C, Sections 10.3.5, 10.4, and 10.5):

caused by creep under sustained load: $\delta_{c_1\,\mathrm{sus}}\,d\varphi$;
caused by creep under force S_e: $S_{et}\delta_{c_1 1}\,d\varphi$;
caused by shrinkage, under the simplifying assumption that it develops affinitive to creep: $(\delta_{c_1 st}/\varphi_t)\,d\varphi$;
caused by the elastic deformation due to a change of S_e: $dS_{et}\delta_{11}$.

We have to point out here that the above list of displacements does not take into account the fact that the portion φ_d of total creep $\varphi = \varphi_d + \varphi_f$ is of a reversible nature (see Part B, Section 4.2). This peculiarity of concrete creep was not known until 20 years after Dischinger's publication [1]. Section 2.2 demonstrates how the resulting deficiencies can be removed. The above list also neglects the fact that over the course of time, the elastic properties of concrete change as well, as a result of aging and drying. In [1], Dischinger demonstrated how this effect could be considered. However, in most practical cases, it has been neglected and rightly so. In Part B, Section 7, we provided an explanation for this and list a number of cases where it cannot be neglected.

The sum of all displacement values which develop in time interval Δt must equal 0 if no gap is to appear at the point of coupling. This requirement leads to the following differential equation:

$$\frac{dS_{et}}{d\varphi}\delta_{11} + S_{et}\delta_{c_1 1} + \delta_{c_1\,\mathrm{sus}} + \frac{\delta_{c_1 st}}{\varphi_t} = 0.$$

The solution for an equation $y' + ay + b = 0$ reads as follows:

$$y = y_0 - (1 - e^{-ax})\,(y_0 + b/a).$$

With the boundary condition that for $\varphi = 0$, the coupling force will be $S_{et} = S_{e0}$, we then obtain:

$$S_{et} = S_{e0} - \left(1 - e^{-(\delta_{c_1 1}/\delta_{11})\varphi_t}\right)\left(S_{e0} + \frac{\delta_{c_1\,\mathrm{sus}}}{\delta_{c_1 1}} + \frac{\delta_{c_1 st}}{\varphi_t\delta_{c_1 1}}\right).$$

The equation becomes more lucid if we substitute as follows:

$$S_{e0} = S_{e\,\mathrm{sus}\,0} + S_{er0},\ S_{ecst} = S_{et} - S_{e0},$$

$$\alpha = \frac{\delta_{c_1 1}}{\delta_{11}},$$

$$\frac{\delta_{1\,\mathrm{sus}}}{\delta_{c_1 1}} = \frac{\delta_{1\,\mathrm{sus}}}{\delta_{11}}\frac{\delta_{11}}{\delta_{c_1 1}} = -S_{e\,\mathrm{sus}\,0}\frac{\delta_{11}}{\delta_{c_1 1}} = -\frac{S_{e\,\mathrm{sus}\,0}}{\alpha},$$

$$\frac{\delta_{c_1 st}}{\delta_{c_1 1}} = \frac{\delta_{c_1 st}}{\delta_{11}}\frac{\delta_{11}}{\delta_{c_1 1}} = -S_{cst}^{(e)} = -\frac{S_{est}^{(e)}}{\alpha}.$$

In this case, $S_{est}^{(e)}$ is an auxiliary value which determines the effect of shrinkage developing at time t in the member which creeps, assuming purely elastic behavior, i.e., neglecting creep.

Thus, we obtain the following for the change of action effects caused by shrinkage and creep in the elastic member:

$$S_{ecst} = (1 - e^{-\alpha\varphi t})\left(\frac{1-\alpha}{\alpha} S_{e\,\mathrm{sus}\,0} - S_{er0} + \frac{1}{\alpha\varphi_t} S_{est}^{(e)}\right), \tag{2.1}$$

and for the sum of all action effects present in the elastic member, which interacts with the creeping member, at time t:

$$S_{et} = S_{ewt} + \frac{1-(1-\alpha)e^{-\alpha\varphi_t}}{\alpha} S_{e\,\mathrm{sus}\,0} + e^{-\alpha\varphi_t} S_{er0} + \frac{1-e^{-\alpha\varphi_t}}{\alpha\varphi_t} S_{est}^{(e)}. \tag{2.2}$$

The dimensionless values α and φ which appear in Eqs. (2.1) and (2.2) can now be combined into coefficients $\bar{\gamma}$ as follows:

$$S_{et} = S_{ewt} + \bar{\gamma}_{\mathrm{sus}} S_{e\,\mathrm{sus}\,0} + \bar{\gamma}_r S_{er0} + \bar{\gamma}_s S_{est}^{(e)} = A + B + C + D, \tag{2.3}$$

$$S_{ecst} = (1 - \bar{\gamma}_{\mathrm{sus}}) S_{e\,\mathrm{sus}\,0} + (1 - \bar{\gamma}_r) S_{er0} + \bar{\gamma}_s S_{est}^{(e)}, \tag{2.3a}$$

where,

$$\bar{\gamma}_{\mathrm{sus}} = \frac{1-(1-\alpha)e^{-\alpha\varphi}}{\alpha}, \tag{2.4}$$

$$\bar{\gamma}_r = e^{-\alpha\varphi}, \tag{2.5}$$

$$\bar{\gamma}_s = \frac{1-e^{-\alpha\varphi}}{\alpha\varphi}. \tag{2.6}$$

As will be shown in the following sections, the γ coefficients[1] can be calculated in accordance with various methods. The results are plotted in Figs. 8.2 through 8.4, Section 8.2.

The action effects occurring at time t thus consist of the independent parts A through D which are influenced differently by creep [Eq. (2.3)]. This important observation applies to all analytical methods described in the following sections. We will discuss the importance of parts A–D later on.

In Eqs. (2.1)–(2.6), α is a dimensionless coefficient (stiffness ratio) which was introduced by Dischinger to characterize the deformation behavior of the interacting members. If we use the subscript c for the member which creeps and s for the elastic member, we then obtain the following for local coupling (e.g., a concrete arch with steel tie rod in accordance with Fig. 2.1a) and with the aid of the displacement values:

$$\alpha = \frac{\delta_{c_1 1}}{\delta_{c_1 1} + \delta_{s_1 1}}. \tag{2.7}$$

[1] The term $\bar{\gamma}$ is used to prevent confusion with the more exact γ coefficients deduced in Section 2.2.2.

In this case, $\delta_{c_1 1}$ and $\delta_{s_1 1}$ represent the elastic displacements developing in both members c and s as a result of the coupling force $X_1 = 1$ at the point of coupling. Thus the value α describes the contribution of the steel to the total behavior of the structural element.

The great importance of the stiffness ratio α which ranges between 0 and 1 becomes particularly obvious if we discuss its effect on parts A–D [occurring in Eq. (2.3)] on the basis of the γ coefficients which are shown in Figs. 8.2–8.4 and/or Figs. 2.5–2.7.

Part A of Eq. (2.3) combines the effects of loads which are not influenced by creep because of their short duration (e.g., wind, peak live load values, short-term imposed action).

Part B describes the effect of sustained loads. How these action effects can be changed by creep depends on the value α. With $\alpha = 1$, $\bar{\gamma}_{\text{sus}} = 1$ (see Fig. 8.2), i.e., no change takes place. If $\alpha = 1$, $\delta_{s_1 1} = 0$, i.e., coupling occurs with an infinitely rigid member, such as a rigid abutment. For the other extreme case with $\alpha = 0$, $\bar{\gamma}_{\text{sus}} = \varphi$. Thus, the action effects in the elastic member for which it is assumed that $E = 0$ ($\delta_{s_1 1} = \infty$) approach $S_{\text{sus}\,t} = \varphi S_{\text{sus}\,0}$, if $\delta_{s_1 1}$ approaches ∞. However, S_{s0} then equals 0 and S_{st} also equals 0.

Part C describes the effect of a suddenly appearing imposed action (see Part A, Section 1). The reduction in the pertinent action effects is particularly large if $\alpha = 1$. This is the limiting case of relaxation which occurs only if a deformation is imposed through infinitely rigid abutments on an element which creeps. If elastic and creeping materials interact, then $\alpha < 1$ and the reduction of action effects is also less (Fig. 8.3).

Part D, the final expression in Eq. (2.3), concerns a gradually developing imposed action, the development over time of which can be assumed to be nearly similar to creep. Shrinkage or gradual settlement of foundations are typical examples. The difference as compared to part C is only quantitative. Of course, it is understood that those parts of the imposed action which develop late are much less reduced by creep (Fig. 8.4) than those which are effective from the start. The imposed action which does not develop similar to creep is dealt with in Section 7.3.

2.1.3. Continuous Coupling of a Purely Elastic Member with a Creeping Member

This problem is of relevance for continuously bonded members. It can be solved simply by analogies. We want to demonstrate this first using the stiffness ratio $\alpha = \delta_{c_1 1}/(\delta_{c_1 1} + \delta_{e_1 1})$.

The effects of α can be illustrated as follows. If an external force N acts on a structure at the point of coupling, the elastic member takes up the force αN and the creeping member takes up $(1 - \alpha)N$. This statement can be checked easily from compatibility requirements:

$$\alpha N \delta_{e_1 1} = (1 - \alpha) N \delta_{c_1 1},$$

$$\alpha = \frac{\delta_{c_1 1}}{\delta_{c_1 1} + \delta_{e_1 1}}$$

as in Eq. (2.7).

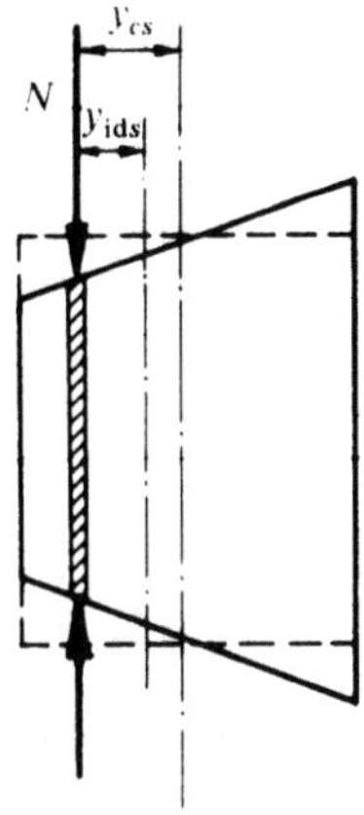

Figure 2.3.

If we apply the initial conceptions to composite cross sections, we obtain (Fig. 2.3) $\alpha N = A_s f_s$, and from that,

$$\alpha = A_s \frac{f_s}{N} = \frac{A_s}{N} n\left(\frac{N}{A_{\text{id}}} + \frac{M}{I_{\text{id}}} y_{\text{id}\,s}\right),$$

$$\alpha = n \frac{A_s}{A_{\text{id}}}\left(1 + \frac{A_{\text{id}}}{I_i} y_{\text{id}\,s}^2\right), \tag{2.8}$$

$$n = \frac{E_s}{E_c}.$$

Where,

A_s is the cross section of the elastic part, e.g., reinforcement. For prestressed concrete, we replace A_s with A_p;

$A_{\text{id}} = A_c + (n - 1)A_e$, the idealized cross section, in general;

$A_{\text{id}} = A_c + (n - 1)A_s$ (for reinforced concrete);

$A_{\text{id}} = A_c + (n - 1)A_p$ (for prestressed concrete);

$I_{\text{id}} = I_c + (n - 1)A_e y_{ce} y_{\text{id}\,e}$, the idealized moment of inertia, in general;

$I_{\text{id}} = I_c + (n - 1)A_s y_{cs} y_{\text{id}\,s}$ (for reinforced concrete);

$I_{\text{id}} = I_c + (n - 1)A_p y_{cp} y_{\text{id}\,p}$ (for prestressed concrete);

y_{ce}, y_{cs}, y_{cp} are the distances of the center of gravity of the cross section of the elastic element from the axis of gravity of the cross section of the creeping element;

$y_{\text{id}\,e}, y_{\text{id}\,s}, y_{\text{id}\,p}$ are the distances of the center of gravity of the cross section of the elastic element from the axis of gravity of the idealized cross section.

With a similar analogous conclusion, we can further develop Eqs. (2.2) and (2.3) for the analysis of bonded prestressed members. Here, it is senseless to work with action effects. It is better to replace them with the stresses they cause using the following equation:

$$f_e = nS_e\left(\frac{1}{A_{\text{id}}} + \frac{y_{\text{id}\,e}^2}{I_{\text{id}}}\right), \quad \text{or} \quad S_e = \frac{f_e}{n(1/A_{\text{id}} + y_{\text{id}\,e}^2/I_{\text{id}})}.$$

In this case, $n = E_e/E_c$ and $y_{\text{id}\,e}$ is the distance of the elastic member from the axis of gravity of the idealized cross section, consisting of the member which creeps and n-times the cross section of the elastic member. If we replace the S terms in Eq. (2.3) with the above expression, we then obtain, after some reductions, the following equation:

$$f_{et} = f_{ewt} + \bar{\gamma}_{\text{sus}} f_{e\,\text{sus}\,0} + \bar{\gamma}_r f_{er0} + \bar{\gamma}_s f_{est}^{\prime(e)}.$$

The auxiliary value $f_{est}^{\prime(e)}$, which has to be determined for the elastic region by neglecting creep, gives us the concept that the force $\varepsilon_s E_e A_e$ can be considered an external force acting on the cross section. Using this concept the force $\alpha\varepsilon_s E_e A_e$ disappears because of the resilience of the steel. We then obtain the following equation:

$$f_{est}^{\prime(e)} = (1 - \alpha)\varepsilon_{st} E_e.$$

With that, the creep equation becomes

$$f_{et} = f_{ewt} + \bar{\gamma}_{\text{sus}} f_{e\,\text{sus}\,0} + \bar{\gamma}_r f_{er0} + \bar{\gamma}_s \varepsilon_{st} E_e(1 - \alpha). \tag{2.9}$$

The stress change caused by creep and shrinkage is

$$f_{ecs} = (\bar{\gamma}_{\text{sus}} - 1) f_{e\,\text{sus}\,0} + (\bar{\gamma}_r - 1) f_{er0} - \bar{\gamma}_s \varepsilon_{st} E_e(1 - \alpha). \tag{2.10}$$

For coefficients $\bar{\gamma}_{\text{sus}}$, $\bar{\gamma}_r$, and $\bar{\gamma}_s$, Eqs. (2.4)–(2.6) still apply.

2.1.4. Local Coupling of Two Structural Elements with Different Creep and Shrinkage Behavior (Dischinger–Kupfer Method)

The following deductions are a further development of the Dischinger equations. They are based on an unpublished work by H. Kupfer, in 1950, which deals with tunnel linings made of concrete which are prestressed against rock which also creeps.

As a conceptual model, the case of two cantilever beams is used here (see Fig. 2.4), a case which occurs in bridge construction. Without coupling at their free ends, they experience different deflections as a result of creep if they differ in age, thickness, or concrete composition. The two elements are called U and W. The following analysis only deals with the phenomena caused by subsequent

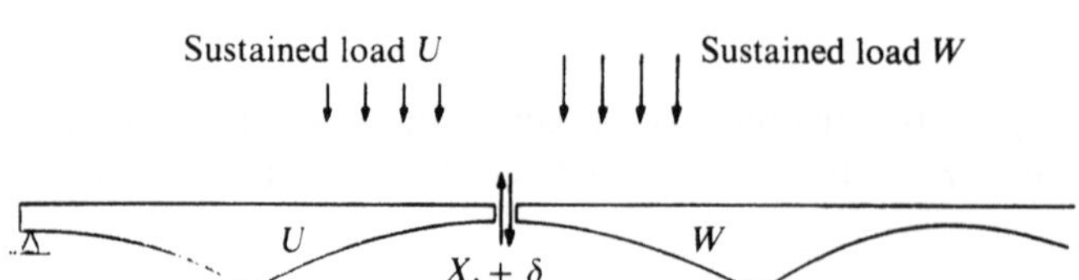

Figure 2.4.

coupling. The creep effects which appear prior to that have an effect only on the initial condition, that is, on action effects and displacements existing at the time of coupling.

Displacements in the uncoupled system (basic system), with relative elastic displacement under coupling force $X = 1$, are given by

$$\delta_{11} = \delta_{U_1 1} + \delta_{W_1 1}.$$

The relative elastic displacement under sustained external load sus, including the mean value of prestressing, if needed, is

$$\delta_{1\,\mathrm{sus}} = \delta_{U_1\,\mathrm{sus}} + \delta_{W_1\,\mathrm{sus}}.$$

The relative displacement as a result of an imposed action at the point of coupling (e.g., differential settlement of supports) is

$$\delta_{1r} = \delta_{U_1 r} + \delta_{W_1 r}.$$

The relative displacement as a result of shrinkage effects which can still be expected after coupling is

$$\delta_{1s} = \delta_{U_1 s} + \delta_{W_1 s}.$$

With these displacement values, the statically indeterminate coupling forces X can be determined for the individual load conditions in accordance with the theory of elasticity. They do not yet include the influence of creep.

The coupling force as a result of external load U is

$$X_{U0} = -\frac{\delta_{U_1\,\mathrm{sus}}}{\delta_{11}}.$$

The coupling force as a result of external load W is

$$X_{W0} = -\frac{\delta_{W_1\,\mathrm{sus}}}{\delta_{11}}.$$

The coupling force as a result of imposed action is

$$X_{r0} = -\frac{\delta_{1r}}{\delta_{11}}.$$

The fictitious coupling force as a result of shrinkage is

$$X_{st}^{(e)} = -\frac{\delta_{1st}}{\delta_{11}}.$$

If we denote the creep coefficients (determined for the time period following coupling) φ_U and φ_W, the following displacements occur in the direction of the coupling force X in the basic system as a result of creep during the time interval dt:

Creep deformation under external load sus:

$$\delta_{U_1\,\mathrm{sus}}\, d\varphi_U + \delta_{W_1\,\mathrm{sus}}\, d\varphi_W.$$

Creep deformation under coupling force X_t:

$$X_t \delta_{U_1 1}\, d\varphi_U + X_t \delta_{W_1 1}\, d\varphi_W.$$

Shrinkage:

$$\frac{\delta_{U_1 st}}{\varphi_{Ut}} d\varphi_U + \frac{\delta_{W_1 st}}{\varphi_{Wt}} d\varphi_W.$$

Elastic recovery as a result of a change of X_t:

$$dX_t \delta_{U_1 1} + dX_t \delta_{W_1 1} = dX_t \delta_{11}.$$

Compatibility requires that the sum of these displacements will equal zero:

$$dX_t(\delta_{U_1 1} + \delta_{W_1 1}) + X_t(\delta_{U_1 1}\, d\varphi_U + \delta_{W_1 1}\, d\varphi_W) + \delta_{U_1 \mathrm{sus}}\, d\varphi_U + \delta_{W_1 \mathrm{sus}}\, d\varphi_W$$
$$+ \frac{\delta_{U_1 st}}{\varphi_{U_1}} d\varphi_U + \frac{\delta_{W_1 st}}{\varphi_{Wt}} \delta\varphi_W = 0. \tag{2.11}$$

To simplify, we now assume that creep in the structural elements U and W will be similar in time despite the different creep factors φ_t. This applies to shrinkage as well. A more exact solution, however, would be possible at any time with the methods described in Section 3.

$$\frac{d\varphi_U}{d\varphi_W} = \frac{\varphi_U}{\varphi_W}.$$

In order to simplify the formulas, we introduce the system creep coefficients.

$$\varphi_{UW} = \frac{\delta_{U_1 1}\varphi_U + \delta_{W_1 1}\varphi_W}{\delta_{U_1 1} + \delta_{W_1 1}}. \tag{2.12}$$

The following also applies:

$$\frac{d\varphi_U}{d\varphi_{UW}} = \frac{\varphi_U}{\varphi_{UW}}; \qquad \frac{d\varphi_W}{d\varphi_{UW}} = \frac{\varphi_W}{\varphi_{UW}} \quad \text{or} \quad \frac{d\varphi_U}{\varphi_{Ut}} = \frac{d\varphi_W}{\varphi_{Wt}} = \frac{d\varphi_{UW}}{\varphi_{UWt}}.$$

Thus, the expressions in differential Eq. (2.11) can be transformed as follows

$$\delta_{U_1 1} + \delta_{W_1 1} = \delta_{11},$$
$$\delta_{U_1 1}\, d\varphi_U + \delta_{W_1 1}\, d\varphi_W = \delta_{11}\, d\varphi_{UW},$$
$$\delta_{U_1 \mathrm{sus}}\, d\varphi_U = -X_{U0}\,\delta_{11} \frac{\varphi_U}{\varphi_{UW}} d\varphi_{UW},$$
$$\delta_{W_1 \mathrm{sus}}\, d\varphi_W = -X_{W0}\,\delta_{11} \frac{\varphi_W}{\varphi_{UW}} d\varphi_{UW},$$
$$\delta_{U_1 st} \frac{d\varphi_U}{\varphi_{Ut}} + \delta_{W_1 st} \frac{d\varphi_W}{\varphi_{Wt}} = \frac{d\varphi_{UW}}{\varphi_{UWt}} (\delta_{U_1 st} + \delta_{W_1 st}) = \frac{d\varphi_{UW}}{\varphi_{UWt}} \delta_{1st} = -\frac{d\varphi_{UW}}{\varphi_{UWt}} X_{st}^{(e)} \delta_{11}.$$

If we divide the differential equation by $\delta_{11}\, d\varphi_{UW}$, we obtain

$$\frac{dX_t}{d\varphi_{UW}} + X_t - X_{U0} \frac{\varphi_U}{\varphi_{UW}} - X_{W0} \frac{\varphi_W}{\varphi_{UW}} - X_{st}^{(e)} \frac{1}{\varphi_{UWt}} = 0.$$

This linear differential equation has the same form, $y' + ay + b = 0$, that appears in Dischinger's solution described in Section 2.1.2. If we use the boundary condition that the imposed action $X_t = X_0$ for $\varphi_{UW} = 0$, we obtain the following solution:

$$X_t = X_0 - (1 - e^{-\varphi_{UW}})\left(X_0 - X_{U0}\frac{\varphi_U}{\varphi_{UW}} - X_{W0}\frac{\varphi_U}{\varphi_{UW}} - \frac{X_{st}^{(e)}}{\varphi_{UWt}}\right).$$

With the aid of substitutions $X_0 = X_{r0} + X_{\text{sus}\,0}, X_{cst} = X_t - X_0, \alpha_U = \varphi_{UW}/\varphi_U$ and $\alpha_W = \varphi_{UW}/\varphi_W$, we obtain a clear expression for the coupling forces caused by creep and shrinkage:

$$X_{cst} = (1 - e^{-\varphi_{UW}})\left[\frac{1 - \alpha_U}{\alpha_U}X_{U0} + \frac{1 - \alpha_W}{\alpha_W}X_{W0} - X_{r0} + \frac{1}{\varphi_{UWt}}X_{st}^{(e)}\right]. \quad (2.13)$$

The total force at any arbitrary selected point in time is

$$X_t = X_{Wt} + X_{\text{sus}\,0} + X_{r0} + X_{cst},$$

$$X_t = X_{Wt} + \frac{1 - (1 - \alpha_U)e^{-\varphi_{UW}}}{\alpha_U}X_{U0} + \frac{1 - (1 - \alpha_W)e^{-\varphi_{UW}}}{\alpha_W}X_{W0}$$

$$+ e^{-\varphi_{UW}}X_{r0} + \frac{(1 - e^{-\varphi_{UW}})}{\varphi_{UWt}}X_{st}^{(e)}. \quad (2.14)$$

Both equations are of the same type as Eqs. (2.1) and (2.2) in Section 2.1.2. The essential difference is that α has been replaced by the values α_U and α_W and $\alpha\varphi$ by φ_{UW}. Thus, for the analysis, we can make use of the graphical representation of the $\bar{\gamma}$ values (Figs. 8.2 through 8.4) if we use the following parameters.

Coefficient	Parameters α	Parameters φ
$\bar{\gamma}_{\text{sus}\,U}$	$\alpha = \varphi_{UW}/\varphi_U$	$\varphi = \varphi_U$
$\bar{\gamma}_{\text{sus}\,W}$	$\alpha = \varphi_{UW}/\varphi_W$	$\varphi = \varphi_W$
$\bar{\gamma}_r$	$\alpha = 1$	$\varphi = \varphi_{UW}$
$\bar{\gamma}_s$	$\alpha = 1$	$\varphi = \varphi_{UW}$

2.2. Expanded Dischinger Equations Taking into Account the Effect of Delayed Elasticity

2.2.1. *Basis*

At the time when Dischinger developed his creep equation, it was not known that only part of the time-dependent deformation in concrete, caused by creep, is irreversible. Therefore, for large changes in stress, this equation predicts losses of stress due to creep that are too high. This is true, however, only if $\alpha > 0.3$ and if imposed action stresses exist. In the following, we will demonstrate how this deficiency can be removed easily.

In Section 2 of Part B, we stated that the creep strain ε_c consists of delayed elasticity ε_d and flow ε_f, whereby ε_d develops much faster than ε_f. The half-life of ε_d is approximately 30 days and that of ε_f approximately 300 days. In general, it is therefore no great mistake to combine ε_d with the elastic strain ε_e; that is, if we assume as a simplification that ε_d occurs immediately after load application. Similar proposals were made earlier [13] and [14]. This idea is particularly fruitful in connection with the Dischinger equation because we obtain very satisfying results without basic changes in the formulas. Even in the most unfavorable cases, the results of the approximation remain well below the unavoidable scatter of the assumptions for φ_t and ε_{st}. This is why this simplification was adopted by the new German Standards for Prestressed Concrete, DIN 4227, and the CEB/FIP Model Code, 1978.

2.2.2. *Local Coupling of a Purely Elastic Member and a Member Which Creeps*

The following comparison shows the transformation of the original Dischinger equation to the expanded version with the aid of vivid analogies. Of course, we can obtain the same results by again solving the differential equation with the expanded assumptions. A corresponding example follows at the end of this section.

Even though the following statements mainly refer to concrete construction, we use subscripts c and e for the formulas. Depending on whether a reinforced concrete or a prestressed concrete problem has to be solved, e can be replaced by subscripts s or p.

Strains

Old version:

$$\varepsilon_{ct} = \varepsilon_e(1 + \varphi_t).$$

Expanded version:

$$\begin{aligned} \varepsilon_{ct} &= (\varepsilon_e + \varepsilon_d) + \varepsilon_f = \varepsilon_t'(1 + \varphi_t'); \\ \varepsilon_d &= 0.4\varepsilon_e; \\ \varepsilon_e' &= \varepsilon_e + 0.4\varepsilon_e = 1.4\varepsilon_e; \\ \varepsilon_{ct} &= 1.4\varepsilon_e(1 + \varphi_t'). \end{aligned} \tag{2.15}$$

Old version:

$$E_c.$$

Expanded version:

$$E_c' = \frac{E_c}{1.4}. \tag{2.16}$$

Old version:

$$\varphi_t = \frac{\varepsilon_{ct} - \varepsilon_e}{\varepsilon_e}.$$

Expanded version:

$$\varphi_t' = \frac{\varepsilon_{ct} - 1.4\varepsilon_e}{1.4\varepsilon_e} = \frac{\varphi_t - 0.4}{1.4}. \tag{2.17}$$

Stiffness Ratio

Old version:

$$\alpha = \frac{\delta_{c11}}{\delta_{c11} + \delta_{e11}}.$$

Expanded version: $$\alpha' = \frac{1.4\delta_{c_1 1}}{1.4\delta_{c_1 1} + \delta_{c_1 1}(1-\alpha)/\alpha} = \frac{1.4\alpha}{1 + 0.4\alpha} \tag{2.18}$$

Action Effects

Old version: $$S_{e\,\mathrm{sus}\,0} = -\frac{\delta_{c_1\,\mathrm{sus}}}{\delta_{c_1 1} + \delta_{e_1 1}} = -\frac{\delta_{c_1\,\mathrm{sus}}}{\delta_{c_1 1}}\alpha.$$

Expanded version: $$S'_{e\,\mathrm{sus}\,0} = S_{e\,\mathrm{sus}\,0}\frac{1.4}{1 + 0.4\alpha}. \tag{2.19}$$

Action effects according to the expanded version are expressed by the same terms used for the old version.

Old version: $$S_{er0} = -\frac{\delta_{c_1 r}}{\delta_{c_1 1} + \delta_{e_1 1}}.$$

Expanded version: $$S'_{er0} = S_{er0}\frac{1}{1 + 0.4\alpha}. \tag{2.20}$$

Old version: $$S^{(e)}_{est} = -\frac{\delta_{c_1 s}}{\delta_{c_1 1} + \delta_{e_1 1}}.$$

Expanded version: $$S^{(e)\prime}_{est} = S^{(e)\prime}_{es\infty}\frac{\varphi_t}{\varphi_\infty} = S^{(e)}_{est}\frac{1}{1 + 0.4\alpha}$$
$$= S^{(e)}_{es\infty}\frac{\varphi_t}{\varphi_\infty}\frac{1}{1 + 0.4\alpha}. \tag{2.21}$$

γ Values

A corresponding change can be performed for $\bar{\gamma}$ values as well.

Old version: $$\bar{\gamma}_{\mathrm{sus}} = \frac{1 - (1-\alpha)e^{-\alpha\varphi_t}}{\alpha}.$$

Expanded version: $$\gamma'_{\mathrm{sus}} = \frac{1 + 0.4\alpha - (1-\alpha)e^{-\alpha[(\varphi_t - 0.4)/(1 + 0.4\alpha)]}}{1.4\alpha}.$$

Old version: $$\bar{\gamma}_r = e^{-\alpha\varphi_t}.$$

Expanded version: $$\gamma'_r = e^{-\alpha[(\varphi_t - 0.4)/(1 + 0.4\alpha)]}.$$

Old version: $$\bar{\gamma}_s = \frac{1 - e^{-\alpha\varphi_t}}{\alpha\varphi_t}.$$

Expanded version: $$\gamma'_s = \frac{(1 + 0.4\alpha)(1 - e^{-\alpha[(\varphi_t - 0.4)/(1 + 0.4\alpha)]})}{\alpha(\varphi_t - 0.4)}.$$

When considering the new formulation of action effects as a result of the expanded version, and the replacement of γ' through γ, we obtain the following new equations to replace Eqs. (2.3) as well as (2.4)–(2.6).

$$S_t = S_{ewt} + \gamma_{\mathrm{sus}} S_{e\,\mathrm{sus}\,0} + \gamma_r S_{er0} + \gamma_s S^{(e)}_{est}, \tag{2.22}$$

$$\gamma_{sus} = \frac{1}{\alpha} - \frac{1-\alpha}{\alpha(1+0.4\alpha)} e^{-\alpha[(\varphi_t - 0.4)/(1+0.4\alpha)]}, \tag{2.23}$$

$$\gamma_r = \frac{1}{1+0.4\alpha} e^{-\alpha[(\varphi_t - 0.4)/(1+0.4\alpha)]}, \tag{2.24}$$

$$\gamma_s = \frac{1 - e^{-\alpha[(\varphi_t - 0.4)/(1+0.4\alpha)]}}{\alpha(\varphi_t - 0.4)}. \tag{2.25}$$

With the example of pure relaxation ($\alpha = 1$), we want to demonstrate briefly the possibility, indicated at the beginning of this section, of deducing the expanded version directly from the differential equation. For simplification, the more general form $\varepsilon_d = 0.4\beta_d \varepsilon_e$ is used here for the effects of ε_d in place of the equation $\varepsilon_d = 0.4\varepsilon_e$, used above.

According to Section 2.1.2, it follows from Dischinger's equation:

$$\frac{dS}{d\varphi} + S = 0.$$

In the expanded version, it is

$$\frac{dS}{d\varphi'} + S = 0.$$

The solution is

$$S_t = Ce^{-\varphi_t'} = Ce^{-[(\varphi_t - 0.4\beta_d)/(1+0.4\beta_d)]}.$$

The boundary condition can be introduced in either of two ways which, however, have little effect on the result:

Case a. $t = 0$;

$$S_{t=0} = S_0;\ \varphi_t = 0$$

leads to

$$C = S_0 e^{-[0.4\beta_d/(1+0.4\beta_d)]}$$

and

$$S_r = S_0 e^{-[\varphi/(1+0.4\beta_d)]}. \tag{2.26}$$

Case b, $t = 0$:

$$S_{t=0} = \frac{S_0}{1+0.4\beta_d};\ \varphi_t' = 0$$

results in

$$C = S_0 \frac{1}{1+0.4\beta_d}$$

and

$$S_r = S_0 \frac{1}{1+0.4\beta_d} e^{-[(\varphi_t - 0.4\beta_d)/(1+0.4\beta_d)]}. \tag{2.27}$$

Equation (2.27) coincides fully with Eqs. (2.22) and (2.24) for $\alpha = 1$, and $\beta_d = 1$. How to proceed when this is not the case will be demonstrated in the following.

In the case of short-term loads (Δt, for example, less than half a year), we replace the value 0.4, appearing in Eq. (2.24), with the expression $0.4\beta_d$, in accordance with Eq. (2.27), The same is true for Eqs. (2.23) and (2.25).

How little this matters in reality will be demonstrated with an extreme example. If we want to calculate the effects of creep on imposed action, after a 25-day load application, on thin structural elements with $\alpha = 1$ (relaxation) and $\varphi_t = 0.4\beta_d + \varphi_{f0}\Delta\beta_f = 0.4(0.5) + 2(0.4) = 1$, then the more accurate calculation improves the γ coefficient for the stress reduction only from $\gamma_r = 0.46$ to $\gamma_r = 0.43$.

However, we can imagine exceptions where values of $\varphi_t < 0.4$ occur. It can happen with very thick structural elements or with very old concrete after a

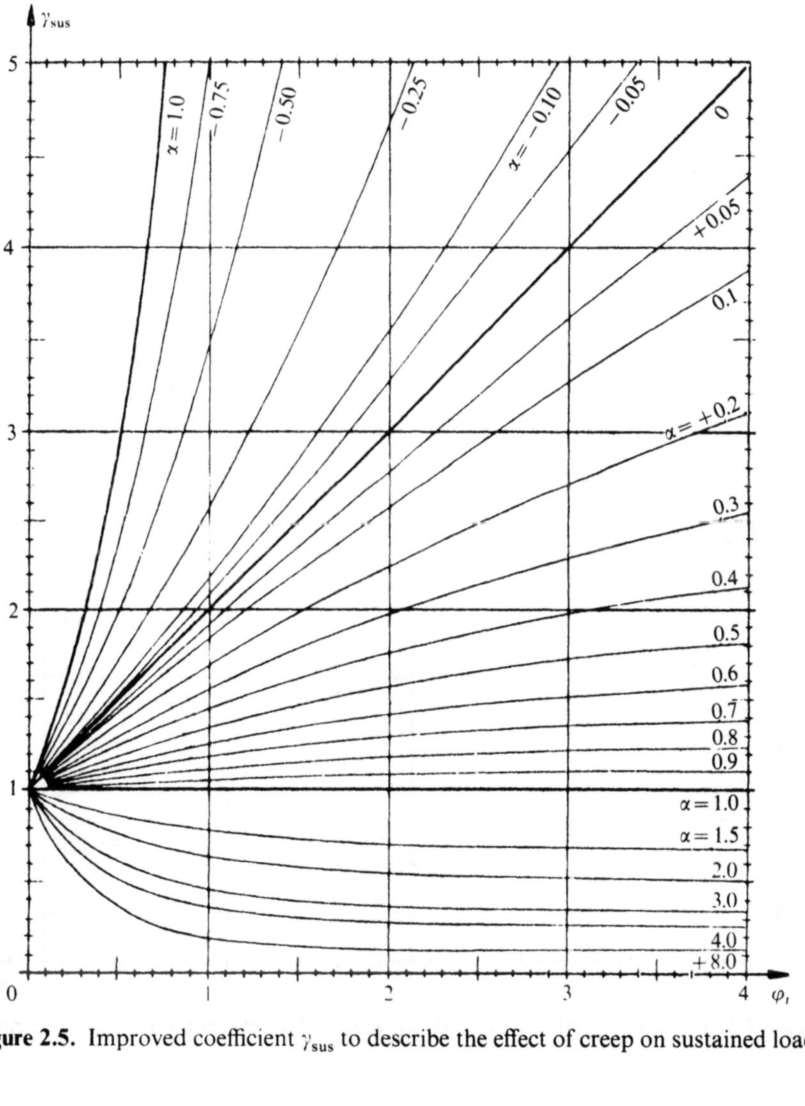

Figure 2.5. Improved coefficient γ_{sus} to describe the effect of creep on sustained load.

short duration of loading. After about 10 days of load application, for example, φ_t can equal $0.4(0.4) + 2.0(0.02) = 0.20$. For such exceptional cases, only the more accurate method can provide satisfactory results. For $0.4\beta_d \to 0$, the γ values change to $\bar{\gamma}$ values.

The γ values calculated from Eqs. (2.23)–(2.25), with consideration of the above suggested improvements, are plotted in Figs. 2.5–2.7. There, various values

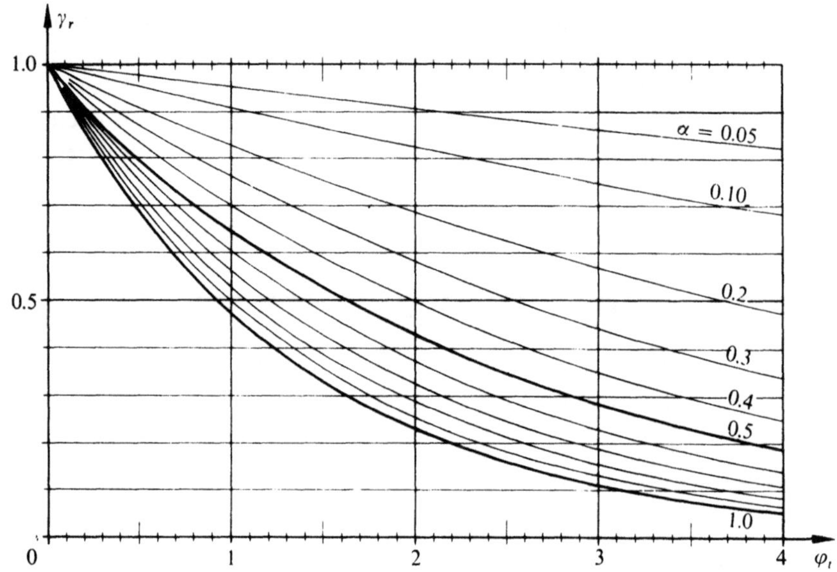

Figure 2.6. Improved coefficient γ_r to describe the effect of creep on sudden restraint.

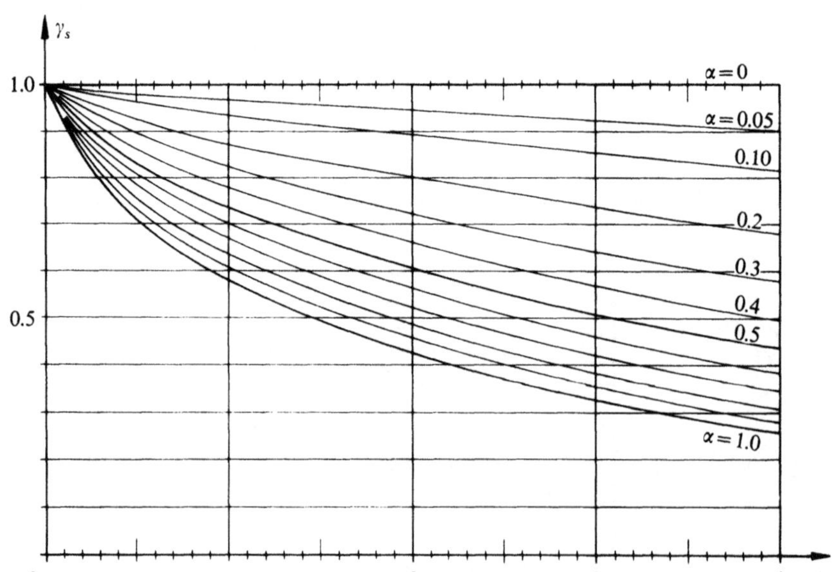

Figure 2.7. Improved coefficient γ_s to describe the effect of creep on shrinkage strain.

for the ratio φ_d/φ_f have been taken into account for the range of very small values of φ. The observed differences in γ values were so small that only the mean values had to be plotted.

In Section 9, the results thus achieved are compared to those obtained through other methods.

In Part C, the following Figs. 2.5–2.7 are given on a larger scale for practical operations.

The use of Eq. (2.22) and/or the γ values is not tied to the use of creep and shrinkage coefficients according to CEB/FIP or DIN 4227. We can also work with the diagrams of the γ values for the creep and shrinkage coefficients, determined in accordance with the ACI guidelines, because only the final values are used in these diagrams.

2.2.3. Continuous Coupling of a Purely Elastic Element with an Element Which Creeps

Analogous to the procedures in Section 2.1.3, we obtain the following equations as a result of the new assumptions:

$$f_{et} = f_{ewt} + \gamma_{sus} f_{e\,sus\,0} + \gamma_r f_{er0} + \gamma_s \varepsilon_{st} E_e(1 - \alpha), \tag{2.28}$$

$$f_{ecst} = (\gamma_{sus} - 1) f_{e\,sus\,0} + (\gamma_r - 1) f_{er0} + \gamma_s \varepsilon_{st} E_e(1 - \alpha). \tag{2.29}$$

For the γ coefficients in Eqs. (2.28) and (2.29), the data given in Section 2.2.2 and the diagrams in Figs. 2.5–2.7 still apply.

2.2.4. Local Coupling of Two Structural Elements with Varied Creep and Shrinkage Behavior (Expanded Method According to Dischinger and Kupfer)

The solution given in Section 2.1.4 does not take into account the fact that the creep coefficient φ also contains reversible parts. As in the original Dischinger equation, this causes noticeable shortcomings only if creep results in considerable stress redistribution. This, however, is not the case for most practical cases.

Thus, we have no problem replacing E_c and φ with improved values, as in Section 2.2.2 and 2.2.3, which interpret the delayed elastic strains to occur immediately.

We replace all φ values with

$$\frac{\varphi - 0.4}{1.4}$$

and all α values with

$$\alpha = \frac{1.4\varphi'_{UW}}{\varphi - 0.4},$$

whereby

$$\varphi'_{UW} = \frac{\delta_{U_{11}}(\varphi_U - 0.4) + \delta_{W_{11}}(\varphi_W - 0.4)}{1.4(\delta_{U_{11}} + \delta_{W_{11}})} = \frac{\varphi_{UW} - 0.4}{1.4}.$$

Analogous to Eq. (2.14), we then obtain

$$X_t = X_{Wt} + \frac{1 - [1 - 1.4\varphi'_{UW}/(\varphi_U - 0.4)]e^{-\varphi'_{UW}}}{1.4\varphi'_{UW}/(\varphi_U - 0.4)} X_{U0}$$

$$+ \frac{1 - [1 - 1.4\varphi'_{UW}/(\varphi_W - 0.4)]e^{-\varphi'_{UW}}}{1.4\varphi'_{UW}/\varphi_W - 0.4)} X_{W0}$$

$$+ \frac{1}{1.4} e^{\varphi'_{UW}} X_{t0} + \frac{1}{1.4} \frac{(1 - e^{\varphi'_{UW}})}{\varphi'_{UWt}} X_{st}^{(e)}. \tag{2.30}$$

Instead of evaluating Eq. (2.3), we can use γ coefficients, as in the preceding sections.

$$X_t = X_{Wt} + \gamma_{\text{sus}\,U} X_{U0} \frac{1 + 0.4\alpha_U}{1.4} + \gamma_{\text{sus}\,W} X_{W0} \frac{1 + 0.4\alpha_W}{1.4} + \gamma_r X_{r0} + \gamma_s X_{st}^{(e)}, \tag{2.31}$$

$$X_{cst} = \left(\gamma_{\text{sus}\,U} \frac{1 + 0.4\alpha_U}{1.4} - 1\right) X_{U0} + \left(\gamma_{\text{sus}\,W} \frac{1 + 0.4\alpha_W}{1.4} - 1\right) X_{W0}$$

$$+ (\gamma_r - 1) X_{r0} + \gamma_s X_{st}^{(e)}. \tag{2.32}$$

These γ coefficients can also be taken from Figs. 2.5–2.7, if we determine the values φ and α in accordance with the rules given in the subsequent table.

Coefficients	Parameters	
	α	φ
$\gamma_{\text{sus}\,U}$	$\alpha_U = \dfrac{\varphi_{UW} - 0.4}{1.4\varphi_U - 0.4\varphi_{UW} - 0.4}$	$\varphi = \varphi_U$
$\gamma_{\text{sus}\,W}$	$\alpha_W = \dfrac{\varphi_{UW} - 0.4}{1.4\varphi_W - 0.4\varphi_{UW} - 0.4}$	$\varphi = \varphi_W$
γ_r	$\alpha = 1$	$\varphi = \varphi_{UW}$
γ_s	$\alpha = 1$	$\varphi = \varphi_{UW}$

Since we can find values of $\alpha > 1$ and $\alpha < 0$ due to transformation, when coupling two creeping structural elements, Fig. 2.5 is expanded accordingly.

In accordance with Eq. (2.12), Section 2.1.4, the following applies here:

$$\varphi_{UW} = \frac{\delta_{U_1 1}\varphi_U + \delta_{W_1 1}\varphi_W}{\delta_{U_1 1} + \delta_{W_1 1}},$$

whereby the δ values can be inserted without change because the correction factors necessary, according to Eq. (2.16), cancel.

3. Finite Difference Methods

3.1. Step-by-Step Integration

If very complex conditions have to be considered for special cases, such as a time-dependent change in the sustained load, the interaction of concretes of different ages at loading, or a systems change during the creep and shrinkage process, the step-by-step method frequently offers the only, if rather laborious, way to solve the problem.

With this method, we separate the development of creep and shrinkage—which may also be different for the individual structural elements—into time intervals. Furthermore, we assume that the stress which causes creep remains constant in the pertinent time interval. The analysis is done in stages, whereby equilibrium and compatibility conditions are expressed in terms of finite differences. The changes ΔX in the action effects (or stress changes Δf) which occur at each interval are interpreted at each subsequent interval as a new additional constant load. Other methods are mentioned in Section 1.

If the time intervals chosen for this method are not short enough, then the action effect ΔX, which develops within an interval, cannot be neglected any longer. It has to be considered in the same calculation interval. We then assume as an approximation that the value ΔX and/or Δf within the interval Δt increase linearly with $\Delta\varphi$, and we calculate the elastic strain at the end of the interval with $\Delta f'/E_c'$ in accordance with Section 2.2, or approximately by using $\Delta f/E_c$. The creep strain is obtained from

$$\frac{\Delta f'}{2}\frac{\Delta\beta_f\varphi_{fo}}{E_c'} \quad \text{and/or} \quad \frac{\Delta f}{2}\frac{\Delta\varphi}{E_c}.$$

Here we have to take into account the dependence of the creep coefficient $\Delta\varphi$ on the respective load age and the creep interval. In the following, we will demonstrate the use of this method with a simple example of rigid constraint (see, also, Fig. 3.1).

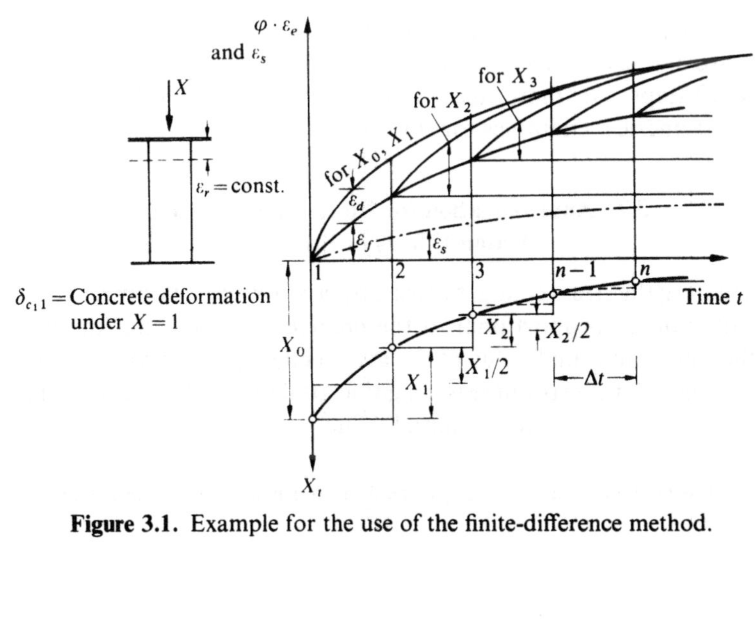

Figure 3.1. Example for the use of the finite-difference method.

According to Fig. 3.1, the following applies:

$$X_0\,\delta_{c_11} + \varepsilon_r = 0.$$

Step 1–2

$$X_0\delta_{c_11}(\varphi_{d2-1} - \varphi_{d1-1} + \varphi_{f2} - \varphi_{f1}) + X_1\delta_{c_11}\left(1 + \varphi_{d2-1} - \varphi_{d1-1} + \frac{\varphi_{f2} - \varphi_{f1}}{2}\right) + \varepsilon_{s2} - \varepsilon_{s1} = 0. \tag{3.1}$$

Step 2–3

$$X_0(\varphi_{d3-1} - \varphi_{d2-1} + \varphi_{f3} - \varphi_{f2}) + X_1(\varphi_{d3-1} - \varphi_{d2-1} + \varphi_{f3} - \varphi_{f2}) + X_2\left(1 + \varphi_{d3-2} - \varphi_{d2-2} + \frac{\varphi_{f3} - \varphi_{f2}}{2}\right) + \frac{\varepsilon_{s3} - \varepsilon_{s2}}{\delta_{c_11}}$$

$$= \left(X_0 + X_1 + \frac{X_2}{2}\right)(\varphi_{f3} - \varphi_{f2}) + (X_0 + X_1)(\varphi_{d3-1} - \varphi_{d2-1}) + X_2(\varphi_{d3-2} - \varphi_{d2-2}) + \frac{X_2(\varepsilon_{s3} - \varepsilon_{s2})}{\delta_{c_11}} = 0. \tag{3.2}$$

Step $(n-1) - n$:[1]

$$\left(\sum_{m=0}^{n-2} X_m + \frac{X_{n-1}}{2}\right)(\varphi_{fn} - \varphi_{fn-1}) + X_0(\varphi_{dn-1} - \varphi_{d(n-1)-1}) + \sum_{m=1}^{n-2} X_m(\varphi_{dn-m} - \varphi_{d(n-1)-m}) + X_{n-1}(\varphi_{dn-(n-1)} - \varphi_{d(n-1)-(n-1)}) + X_{n-1} + \frac{\varepsilon_{sn} - \varepsilon_{sn-1}}{\delta_{c_11}} = 0, \tag{3.3}$$

$$X_{tn} = X_0 + X_1 + X_2 + \cdots + X_{n-1}. \tag{3.4}$$

It is unnecessary to choose a large number of intervals. Even in the most unfavorable case of relaxation, the final value changes by only ~5% with an increase from six to 10 steps. For more than 10 steps, no noticeable difference is obvious. We can find numerical examples for this in Part C.

3.2. Approximation for Mean Creep-Inducing Action Effects or Stresses

When calculating the stress changes caused by creep, we can assume, as a simplification, that they occur proportional to the change in φ and that the mean value between the initial and final value of the action effect (or stress) is creep inducing. As shown in Fig. 3.2, the error here is insignificant if the final value to be determined does not differ too much from the initial value.

[1] φ_{fn} is the flow between t_1 and t_n; φ_{dn-1} is the delayed elastic strain which commences at t_1 and continues up to t_n.

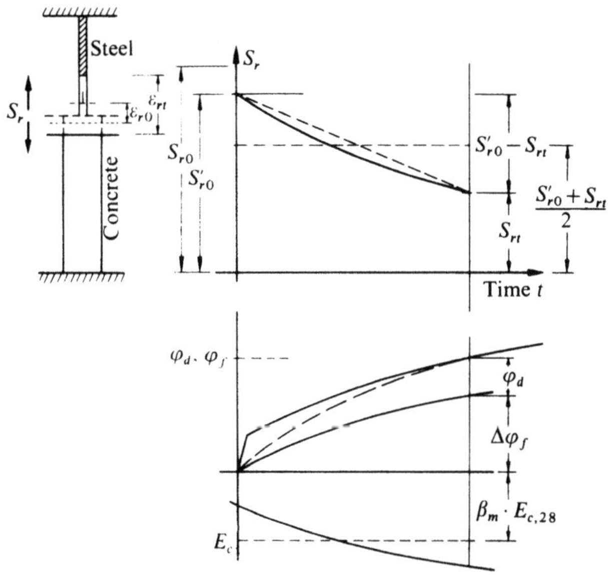

Figure 3.2. The mean creep inducing stress.

The error can be reduced further if we divide the creep process into several intervals, in accordance with Fig. 3.3, whereby the correct curve is replaced by a polygon. This method is successful especially for complicated cases which are hard to grasp analytically (see examples in Sections 9.3.5 and 10.5.3.2, Part C).

This method is not new. For more than 20 years it has been used to determine the stress loss in prestressed concrete, as is shown in the approximation described in Section 7.4.6.1, for stress loss in single-strand prestressing [15]. In [26] as well, some interesting applications are described. In the following, the method is described in a more general form by taking into account the influence of the effects of delayed elasticity and the age dependence of the modulus of elasticity E_c.

3.2.1. General Solution

In accordance with Fig. 3.2, it is assumed that in the time interval to be analized, the imposed action changes from ε_{r0} to ε_{rt} and the pertinent action effect from

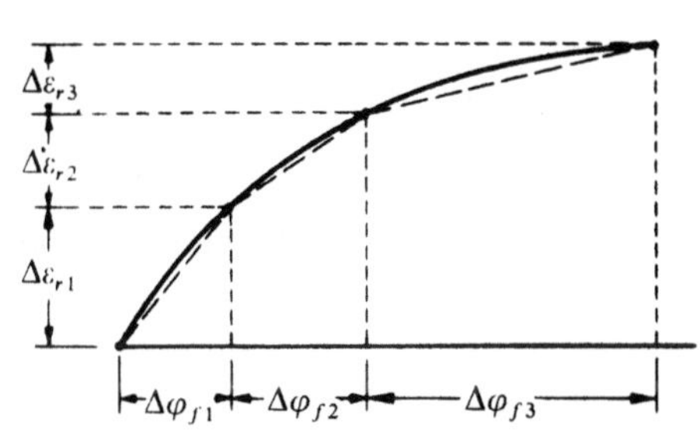

Figure 3.3.

S_{r0} to S_{rt}. The following strains have to be considered when taking into account the effects of concrete age on the modulus E_c and the separation of creep into φ_f for flow and φ_d for delayed elasticity.[1]

The change of the imposed action is

$$\varepsilon_{rt} - \varepsilon_{r0} = \Delta\varepsilon_r.$$

The flow is

$$\frac{S'_{r0} + S_{rt}}{2}\,\delta_{c_11}^{28}\Delta\varphi_f.$$

The delayed elastic and elastic recovery are

$$(S'_{r0} - S_{rt})\,\frac{1/\beta_{em} + \varphi_d}{\alpha'_m}\,\delta_{c_11}^{28}.$$

If we set $1/\beta_{em} + \varphi_d = 0.5(1/\beta_{e0} + 1/\beta_{et}) + \varphi_d = A_m$ and

$$\frac{1}{\alpha'_m} = \frac{A_m\delta_{c_11}^{28} + \delta_{e_11}}{A_m\delta_{c_11}^{28}} = 1 + \frac{\delta_{e_11}}{A_m\delta^{28}{}_{c_11}} = 1 + \frac{1-\alpha}{\alpha A_m},$$

as well as $\varphi_d = 0.4$, we then obtain

$$\frac{1/\beta_{em} + 0.4}{\alpha'_m} = C_m = \frac{1}{\alpha} + 0.5\left(\frac{1}{\beta_{e0}} + \frac{1}{\beta_{et}}\right) - 0.6, \tag{3.5}$$

and also

$$C_0 = \frac{1}{\alpha} + \frac{1}{\beta_{e0}} - 0.6. \tag{3.6}$$

The share of elastic and delayed elastic recovery is

$$(S'_{r0} - S_{rt})C_m\delta_{c_11}^{28}.$$

The general compatibility condition is

$$\frac{S'_{r0} + S_{rt}}{2}\Delta\varphi_f - (S'_{r0} - S_{rt})C_m - \frac{\Delta\varepsilon_r}{\delta_{c_11}^{28}} = 0, \tag{3.7}$$

and the solution is

$$S_{rt} = \frac{(C_m - 0.5\Delta\varphi_f)S'_{r0} + \Delta\varepsilon_r/\delta_{c_11}^{28}}{C_m + 0.5\Delta\varphi_f}$$

$$= \frac{(C_m - 0.5\Delta\varphi_f)\varepsilon_{r0}/C_0}{(C_m + 0.5\Delta\varphi_f)\delta_{11}^{28}\alpha} + \frac{\Delta\varepsilon_r}{(C_m + 0.5\Delta\varphi_f)\delta_{11}^{28}\alpha},$$

$$\delta_{11}^{28} = \delta_{c_11}^{28} + \delta_{e_11}. \tag{3.8}$$

[1] All of the delayed elastic strain is assumed to occur immediately. An improvement on this method is shown in Part C, Section 10.4, whereby the true value of the delayed elastic strain is included each time.

Hereby is

$$S'_{r0} = \frac{\alpha'_0 \varepsilon_{r0}}{\delta'_{c_1 1}} = \frac{\alpha'_0 \varepsilon_{r0}}{A_0 \delta^{28}_{c_1 1}} = \frac{\varepsilon_{r0}}{C_0 \delta^{28}_{c_1 1}} = \frac{\varepsilon_{r0}}{C_0 \alpha \delta^{28}_{11}}. \tag{3.9}$$

The value S_{r0} is frequently known from a previous elastic analysis, particularly in the case of step-by-step calculations. S'_{r0} then has to be replaced by S_{r0} for use in Eq. (3.8) which presupposes that the effects of delayed elastic strain occur immediately. Since $S'_{r0} = \varepsilon_{r0}/(C_0 \delta^{28}_{11}\alpha)$ and $S_{r0} = \varepsilon_{r0}/\delta^0_{11}$ in accordance with Eq. (3.9), we have to replace, in such cases, S'_{r0} in Eq. (3.8) with

$$S'_{r0} = \frac{\delta^0_{11}}{C_0 \alpha \delta^{28}_{11}} S_{r0} = \frac{S_{r0}}{C_0 \alpha_0 \beta_{e0}} = \frac{S_{r0}}{C_0}\left(\frac{1}{\beta_{e0}} + \frac{1-\alpha}{\alpha}\right). \tag{3.10}$$

Frequently, however, the effects of age on the modulus of elasticity E_c are neglected when determining S_{r0}. Equation (3.10) then reads

$$S'_{r0} = \frac{S_{r0}}{C_0 \alpha} = \frac{S_{r0}}{1 + 0.4\alpha}. \tag{3.10a}$$

3.2.2. Special Case of a Constant Imposed Action: $\varepsilon_{r0} = \varepsilon_{rt} = \varepsilon_r$; $\Delta\varepsilon_r = 0$

$$S_{rt} = \frac{C_m - 0.5\Delta\varphi_f}{C_m + 0.5\Delta\varphi_f} \frac{\varepsilon_r}{C_0 \delta^{28}_{11}\alpha} = \frac{C_m - 0.5\Delta\varphi_f}{C_m + 0.5\Delta\varphi_f} \frac{S_{r0}}{C_0}\left(\frac{1}{\beta_{e0}} + \frac{1-\alpha}{\alpha}\right). \tag{3.11}$$

If the concrete is loaded at a later date, it usually does not pay to consider the effects of age on the modulus of elasticity E_c which, itself, is subject to uncertainty.

With $\beta_{e0} = \beta_{et} = 1$, Eqs. (3.5) and (3.6) read as follows:

$$C_m = C_0 = 0.4 + \frac{1}{\alpha}. \tag{3.12}$$

After introducing $\Delta\varphi_t = \varphi_t = \varphi - 0.4$ and $S_{r0} = \varepsilon_{r0}/\delta^0_{11} = \varepsilon_{r0}/\delta^{28}_{11}$, the following applies:

$$S_{rt} = \frac{1.0 + 0.6\alpha - 0.5\alpha\varphi}{1.0 + 0.6\alpha + 0.08\alpha^2 + \alpha\varphi(0.5 + 0.2\alpha)} S_{r0}, \tag{3.13}$$

and for stresses,

$$f_{rt} = \frac{1 + 0.6\alpha - 0.5\alpha\varphi}{1.0 + 0.6\alpha + 0.08\alpha^2 + \alpha\varphi(0.5 + 0.2\alpha)} f_{r0}. \tag{3.14}$$

For imposed action stresses, α frequently equals 1. This leads to

$$S_{rt} = \frac{1.6 - 0.5\varphi}{1.68 + 0.7\varphi} S_{r0} \text{ and } f_{rt} = \frac{1.6 - 0.5\varphi}{1.68 + 0.7\varphi} f_{r0}. \tag{3.15}$$

This approach results in an error of less than 1% to 3%, even at point S_{rt}/S_{r0} and/or $f_{rt}/f_{r0} = 0.3$. The same is true for Eqs. (3.13) and (3.14). If f_{rt} does not decrease considerably compared to f_{r0}, we can even neglect the effects of delayed elasticity. We then have $C_m = C_0 = 1/\alpha$, and Eq. (3.15) is simplified to read

$$\frac{S_{rt}}{S_{r0}} = \frac{f_{rt}}{f_{r0}} = \frac{1 - \varphi/2}{1 + \varphi/2}. \tag{3.16}$$

If we use this formula up to $S_{rt}/S_{r0} \geq 0.3$, then the result deviates from that of the more exact method by up to approximately 50%. Its use should therefore be limited to $\varphi \leq 0.5$, thus limiting the error to approximately 10%. The reliability of this formula can be increased by applying φ in n stages. We then have the following formula which coincides with the Dischinger equation (2.2) for $n \to \infty$:

$$\frac{S_{rt}}{S_{r0}} = \frac{f_{rt}}{f_{r0}} = \left(\frac{1 - \varphi/2n}{1 + \varphi/2n}\right)^n. \tag{3.17}$$

Even this formula should only be used up to $\varphi = 0.5$, due to the fact that the effects of delayed elasticity are neglected, especially since Eq. (3.15) is much more reliable and hardly more complicated.

3.2.3. Special Case of Imposed Action Which Increases from $\varepsilon_{r0} = 0$ to ε_{rt}

For $\varepsilon_{r0} = 0$, $S_{r0} = 0$ and $\Delta\varepsilon_r = \varepsilon_{rt} = \varepsilon_r$. We then have

$$S_{rt} = \frac{\Delta\varepsilon_r}{C_m + 0.5\Delta\varphi_f} \frac{1}{\delta_{11}^{28}\alpha}. \tag{3.18}$$

When neglecting the effects of concrete age on E_{ct}, $C_m = 0.4 + 1/\alpha$ and, in accordance with Eq. (3.13), we then have

$$S_{rt} = \frac{\varepsilon_r/\delta_{11}^{28}}{1 + 0.2\alpha + 0.5\alpha\varphi} \quad \text{and} \quad f_{rt} = \frac{f_{r0}^{(e)}}{1 + 0.2\alpha + 0.5\alpha\varphi}. \tag{3.19}$$

These equations apply also for the effects of shrinkage.

3.2.4. Step-by-Step Calculation

If, in accordance with Fig. 3.3, a large constraint is applied in several steps to increase accuracy, at stage i, two effects are superimposed: the results of the constant constraint from the previous stages, which leads to action effect $S_{(i-1)}$ and the increase of constraint, $\Delta\varepsilon_i$. The first effect is to be dealt with in accordance with Section 3.2.2 and the second in accordance with Section 3.2.3. This brings us to the following formulas.

In the first stage, where because $\varepsilon_{r0} = 0$, only the increase $\Delta\varepsilon_{rt}$ is effective,

$$S_{r1} = \frac{\Delta\varepsilon_{r1}}{C_{mi} + 0.5\Delta\varphi_{f1}} \frac{1}{\delta_{11}^{28}}. \tag{3.20}$$

At stage i,

$$S_{ri} = \frac{(C_{mi} - 0.5\Delta\varphi_{fi})S_{ri-1} + \Delta\varepsilon_{ri}/(\delta_{11}^{28}\alpha)}{C_{mi} + 0.5\Delta\varphi_{fi}}. \tag{3.21}$$

If a constraint exists at the beginning of the first stage, then Eq. (3.21) has to be used for that stage as well, whereby the initial action effect S'_{r0} has to replace S_{ri-1} in accordance with Eq. (3.10). If $\alpha = 1$, we obtain, in stage 1,

$$S_{r1} = \frac{\Delta\varepsilon_{r1}}{C_{m1} + 0.5\Delta\varphi_{f1}} \frac{1}{\delta_{11}^{28}}, \tag{3.22}$$

and in stage i,

$$S_{ri} = \frac{(C_{mi} - 0.5\Delta\varphi_{fi})S_{ri-1} + \Delta\varepsilon_{ri}/\delta_{11}^{28}}{C_{mi} + 0.5\Delta\varphi_{fi}}. \tag{3.23}$$

If we can disregard the effects of the concrete age on E_c, then Eqs. (3.20) and (3.21) are as follows:

$$S_{r1} = \frac{\Delta\varepsilon_{r1}/\delta_{11}^{28}\alpha}{0.4 + 1/\alpha + 0.5\varphi_{f1}}, \tag{3.24}$$

$$S_{ri} = \frac{(0.4 + 1/\alpha - 0.5\Delta\varphi_{fi})S_{ri-1} + \Delta\varepsilon_{ri}\delta_{11}^{28}\alpha}{0.4 + 1/\alpha + 0.5\Delta\varphi_{fi}}. \tag{3.25}$$

We can simplify the calculation by choosing the n steps such that $\Delta\varphi_f = \varphi_{ft}/n =$ constant and by adjusting the $\Delta\varepsilon_{ri}$ values accordingly.
For each step, we then obtain,

$$S_{ri} = \frac{a}{b} S_{ri-1} + \frac{\Delta\varepsilon_{ri}}{b\delta_{11}^{28}\alpha},$$

whereby the constants a and b assume the following values:

$$a = 0.4 + 1/\alpha - 0.5\varphi_f/n \qquad \text{and/or for } \alpha = 1\text{: } 1.4 - 0.5\varphi_f/n;$$

$$b = 0.4 + 1/\alpha + 0.5\varphi_f/n \qquad \text{and/or for } \alpha = 1\text{: } 1.4 + 0.5\,\varphi_f/n.$$

Usually, we can obtain sufficiently accurate results with three to five steps. More steps are needed only if we want to determine the maximum stress for a quickly developing imposed action (see Sections 9 and 10 in Part C).

3.2.5. *Final Remarks*

In Fig. 8.1, the results calculated for $\alpha = 1$ [Eqs. (3.15) and (3.16)] are compared with the values obtained by other methods. In Eqs. (3.24) and (3.25), analized in four steps, we already obtain close agreement with the more exact solutions according to Sections 2.2 and 4.

The principle of the mean creep-inducing stress can also be employed successfully with multiple statically indeterminate problems. As will be shown in Section 7.2, in detail, this is a simple way to expand the deformation matrix by a plastic strain component.

A multitude of examples for the use of this method can be found in Part C.

4. The Trost Method

4.1. General Approach

Trost [6], [10] suggested to express the time-dependent strain ε_t, which develops under a variable stress, with the aid of a simple algebraic equation:

$$\varepsilon_t = \frac{1}{E_c}[f_0(1 + \varphi_t) + (f_t - f_0)(1 + \rho\varphi_t)] + \varepsilon_{st}. \tag{4.1}$$

In this equation, the first term within brackets describes the effects of the initial stress f_0, and the second term, the stress changes caused by creep. Trost considered the fact of a gradual stress change with the aid of a mean creep coefficient and not, as in Section 3.2, with the aid of mean stress reduction. Trost calls the reduction factor ρ *relaxation coefficient*.

This method has the advantage that ρ can be adjusted to the theoretical requirements. The better the adjustment, e.g., in the form of tables, the better the agreement with the more exact solution and thus with reality.

The determination of the relaxation coefficients described in [10] is still based on the formulation of creep coefficients φ_t, as given in the CEB/FIP recommendations, 1970. This method considers the effects of delayed elasticity only in approximation, as shown in Sections 3 and 5 of Part B.

Because of the different basis, we can not directly compare the new results with the γ values which had been calculated from Trost's relaxation coefficients. As for any differences which might appear, they will certainly be largest for the case of imposed action. Although they have a different basis, to some extent, such a comparison is of interest. For this purpose, the γ values corresponding to Eqs. (2.22)–(2.25) were determined according to Trost. The derivation is shown in the following for the case of imposed action.

The strain at time $t = 0$: $\varepsilon = f_0/E_c$.

The strain at time t:

$$\varepsilon = \frac{f_0}{E_c} = \text{constant} = \frac{f_0}{E_c}(1 + \varphi_t) + \frac{f_t - f_0}{E_c}(1 + \rho\varphi_t),$$

$$\beta_t/\beta_0 = 1 - \frac{\varphi_t}{1 + \rho\varphi_t} = \gamma_r. \tag{4.2}$$

After introducing the stiffness ratio α, we obtain for the sustained load:

$$\gamma_{\text{sus}} = 1 + \frac{1 - \alpha}{\alpha}\frac{\alpha\varphi_t}{1 + \alpha\rho\varphi_t}; \tag{4.3}$$

for suddenly appearing imposed action:

$$\gamma_r = 1 - \frac{\alpha\varphi_t}{1 + \alpha\rho\varphi_t}; \tag{4.4}$$

for shrinkage:

$$\gamma_s = \frac{1}{1 + \alpha\rho\varphi_t}. \tag{4.5}$$

The γ values obtained in this way are compared to the results from other methods in Figs. 8.2–8.4. The observed differences can almost exclusively be traced back to differences in separating creep into ε_d and ε_f.

Also, the approximation mentioned at the beginning of Section 3.2 for calculation of the stress loss in prestressed concrete structures could be improved according to Trost by changing $(1 + \varphi/2)$ in the denominator to $(1 + \rho\varphi)$.

The approach of Trost permits the use of a simple calculation process, even for statically indeterminate systems with varied creep behavior (see Sections 7.1 and 7.2).

4.2. Adaptation of Characteristic Relaxation Coefficients to the Creep Prediction as Given in Part B, Section 4.3

Besides the work done by Trost, a number of other methods and computer programs are based on a product assumption approach to predict the magnitude and development, over time, of creep coefficients as given in the 1970 CEB/FIP recommendations. If, for such cases, no satisfactory way can be found to convert the analytical methods to the prediction of Rüsch–Jungwirth, as given in Part B, Section 4.3, it is possible to express approximately, the data of the Rüsch–Jungwirth method through a product formulation [16]. This concerns, above all, the two factors ξ_d and ξ_t which, in the 1970 CEP/FIP recommendations, are used to express the influence of the load age and duration of loading on φ_t.

These coefficients are to be selected such that both methods will lead to the same results.

$$\bar{\varphi}\xi_{dt_1}\xi_{t(t-t_1)} = 0.4\beta_{d(t-t_1)} + \varphi_{f0}(\beta_{ft} - \beta_{ft_1}). \tag{4.6}$$

$\bar{\varphi}$ is to include the influence of environmental conditions, thickness of section, and concrete composition. For $t = \infty$, $\xi_t = 1$ and $\beta_d = 1$, and we then have

$$\xi_{dt_1} = \frac{0.4 + \varphi_{f0}(\beta_{ft} - \beta_{ft_1})}{\bar{\varphi}}. \tag{4.7}$$

An evaluation of Eq. (4.7) is given in Fig. 4.1. It shows that the coefficient ξ_{dt_1} depends on load age as well as effective thickness.

The ξ_t coefficient is obtained

$$\xi_{t(t-t_1)} = \frac{0.4\beta_{d(t-t_1)} + \varphi_{f0}(\beta_{ft} - \beta_{ft_1})}{\bar{\varphi}\xi_{dt_1}}. \tag{4.8}$$

The time development therefore depends on the basic flow coefficient φ_{f0}, the effective thickness, and the load age. The ratio of flow to delayed elastic strain depends on the load age. However, it can be assumed to be constant in view of the degree of accuracy which may be achieved in the prediction of creep coefficients.

We now can calculate improved ρ values, either through the method indicated here, or with the aid of a computer program (e.g., [17]) which results in characteristic relaxation coefficients for arbitrary creep curves. The results can be seen in [18]. It is much simpler to obtain the ρ values, in accordance with a suggestion by Trost, by equating Eqs. (2.26) and (4.4) with $\alpha = 1$ (see Part C, Section 2.2).

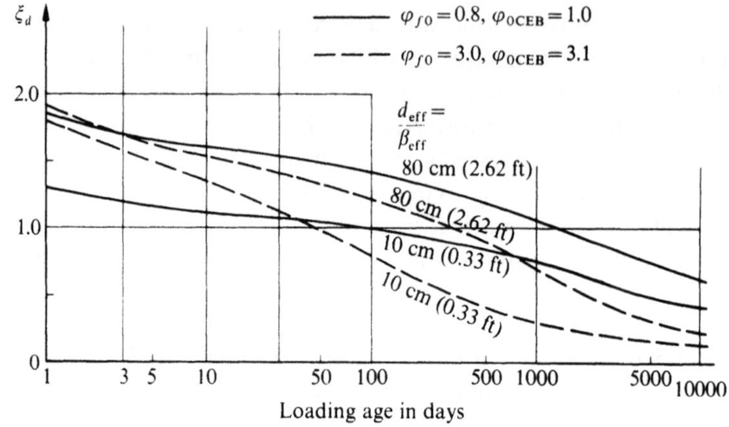

Figure 4.1. Adaptation of the coefficient ξ_d to the Rüsch–Jungwirth method. ξ_d describes the effect of concrete age on creep.

5. The Bažant Method Based on the Effective Modulus

Sometimes, it is still common practice to consider the effects of creep by using a reduced modulus $E_{eff} = E_c/(1 + \varphi)$ to take into account creep effects. Since in this way the permanent deformations are rated equal to the elastic ones, the results were unsatisfactory initially. Bažant [12] suggested an improvement, analogous to Trost, by adapting, through correction factors, the value of the effective modulus E'' to the respective requirements of the relevant differential and/or integral equations.

Some excerpts from the work of Bažant [12] are included in the following.

"... Then $f(t)$ varies linearly with $E_R(t, t_0)$ (relaxation function) and the stress-strain relations may be written in the form of an incremental elastic law:

$$\Delta f(t) = E''(t, t_0)[\Delta\varepsilon(t) - \Delta\varepsilon''(t)] \tag{5.1}$$

in which

$$\Delta\varepsilon(t) = \varepsilon(t) - \varepsilon(t_0), \tag{5.2}$$

$$\Delta f(t) = f(t) - f(t_0), \tag{5.3}$$

$$\Delta\varepsilon''(t) = \frac{f(t_0)}{E(t_0)}\varphi(t, t_0) + \varepsilon^0(t) - \varepsilon^0(t_0), \tag{5.4}$$

$$E''(t, t_0) = \frac{E(t_0)}{1 + \chi(t, t_0)\varphi(t, t_0)}, \tag{5.5}$$

$$\chi(t, t_0) = \left[1 - \frac{E_R(t, t_0)}{E(t_0)}\right]^{-1} - \frac{1}{\varphi(t, t_0)}, \tag{5.6}$$

where $\chi(t, t_0)$, $E''(t, t_0)$, and $\Delta\varepsilon''(t)$ will be termed aging coefficient, age-adjusted effective modulus, and fictitious inelastic strain increment.

Determination of χ requires the knowledge of the relaxation function, which can be obtained from the creep function $J_c(t, t')$ with the help of a computer. Table 5.1 shows the values of χ which have been found for the following material properties:

$$\varphi(t, t') = \varphi_u(t') \frac{(t - t')^{0.6}}{10 + (t - t')^{0.6}} \tag{5.7}$$

or

$$\varphi(t, t') = \varphi_u(t')0.113 \ln(1 + t - t'), \tag{5.8}$$

where

$$\left.\begin{aligned} E(t') &= E(28)\left(\frac{t'}{4 + 0.85t'}\right)^{1/2}, \\ \varphi_u(t') &= \varphi(\infty, 7)1.25t^{-0.118}, \end{aligned}\right\} \tag{5.9}$$

where t, t' is the time from casting of concrete being given in days and t_0 is the time at first load application.

Equation (5.7) through Eq. (5.9) have been recently recommended by ACI Committee 209, along with a method of determination of the constant $\varphi(\infty, 7)$. Equation (5.7) is acceptable for structures of normal dimensions that are exposed to a mild climate and allowed to dry. Equation (5.8) is suitable for mass concrete.

For the purpose of comparison, the χ values have also been computed for creep functions Eq. (5.7) and (5.8) with a constant modulus E_c (Table 5.1)."

We can see that for the normal case (normal creep coefficient, normal load age, and duration of loading), the aging coefficients χ vary little, even for a variable modulus of elasticity. This also follows from the works of Trost [10] and Schade [18]. The mean value is approximately 0.8.

In conclusion, it has to be said that the relaxation coefficient ρ according to Trost is of the same importance as the aging coefficient χ according to Bažant. The numerical value of χ and ρ should be the same. Equation (4.1), as written with the age-adjusted effective modulus, will read as follows:

$$\varepsilon_t = \frac{1}{E_c} f_0(1 + \varphi_t) + \frac{1}{E''}(f_t - f_0) + \varepsilon_{st},$$

where

$$E'' = \frac{E_c}{1 + \rho\varphi_t}.$$

Table 5.1. Aging coefficient χ for two different creep laws, with and without consideration of variation of elastic modulus.

Creep law	$t-t_0$ days	$\varphi(\infty,7)$	Variable E, Eq. (5.9) t_0, days 10^1	10^2	10^3	10^4	Constant E t_0, days 10^1	10^2	10^3	10^4	$\frac{\varphi(t,t_0)}{\varphi_u(t_0)}$
		0.5	0.525	0.804	0.811	0.809	0.798	0.811	0.811	0.809	
	10^1	1.5	0.720	0.826	0.825	0.820	0.820	0.829	0.825	0.820	0.273
		2.5	0.774	0.842	0.837	0.830	0.839	0.844	0.837	0.830	
		3.5	0.806	0.856	0.848	0.839	0.855	0.857	0.848	0.839	
Eq.		0.5	0.505	0.888	0.916	0.915	0.848	0.905	0.916	0.915	
(5.7)	10^2	1.5	0.739	0.919	0.932	0.928	0.878	0.926	0.932	0.928	0.608
and		2.5	0.804	0.935	0.943	0.938	0.899	0.939	0.943	0.938	
(5.9)		3.5	0.839	0.946	0.951	0.946	0.914	0.949	0.951	0.946	
		0.5	0.511	0.912	0.973	0.981	0.846	0.937	0.974	0.981	
	10^3	1.5	0.732	0.943	0.981	0.985	0.878	0.953	0.981	0.985	0.857
		2.5	0.795	0.956	0.985	0.988	0.899	0.963	0.985	0.988	
		3.5	0.830	0.964	0.987	0.990	0.914	0.969	0.987	0.990	
		0.5	0.501	0.899	0.976	0.994	0.828	0.927	0.977	0.994	
	10^4	1.5	0.717	0.934	0.983	0.995	0.863	0.945	0.983	0.995	0.954
		2.5	0.781	0.949	0.986	0.996	0.887	0.956	0.987	0.996	
		3.5	0.818	0.958	0.989	0.997	0.903	0.963	0.989	0.997	
		0.5	0.522	0.815	0.822	0.821	0.809	0.823	0.822	0.821	
	10^1	1.5	0.727	0.838	0.836	0.832	0.831	0.840	0.836	0.832	0.269
		2.5	0.783	0.854	0.849	0.842	0.850	0.855	0.849	0.842	
		3.5	0.815	0.867	0.860	0.851	0.865	0.868	0.860	0.851	
		0.5	0.493	0.901	0.929	0.929	0.864	0.919	0.930	0.929	
Eq.	10^2	1.5	0.742	0.928	0.941	0.939	0.889	0.935	0.941	0.939	0.518
(5.8)		2.5	0.807	0.941	0.950	0.947	0.906	0.945	0.950	0.947	
and		3.5	0.842	0.950	0.956	0.952	0.919	0.953	0.956	0.952	
(5.9)		0.5	0.461	0.887	0.956	0.965	0.826	0.917	0.957	0.965	
	10^3	1.5	0.702	0.924	0.966	0.972	0.859	0.935	0.966	0.972	0.775
		2.5	0.770	0.940	0.972	0.976	0.882	0.947	0.973	0.976	
		3.5	0.808	0.950	0.977	0.980	0.898	0.955	0.977	0.980	
		0.5	0.434	0.838	0.940	0.972	0.767	0.873	0.942	0.972	
	10^4	1.5	0.657	0.887	0.955	0.979	0.811	0.901	0.956	0.979	1.034
		2.5	0.727	0.909	0.964	0.983	0.841	0.919	0.964	0.983	
		3.5	0.768	0.924	0.970	0.985	0.863	0.932	0.970	0.985	
$\varphi_u(t_0)/\varphi_u(7)$			0.960	0.731	0.558	0.425	0.960	0.731	0.558	0.425	
$E(t_0)/E(28)$			0.895	1.060	1.083	1.089	1.000	1.000	1.000	1.000	

6. The Busemann Method for Internally and Multiple Statically Indeterminate Systems

For multiple statically indeterminate composite problems, the Dischinger approach leads to coupled linear differential equations which can be solved, but only with an extensive effort as Knittel demonstrated with the example of two-strand prestressing [19]. This problem can be avoided by using either an iteration method (see Part C, Section 6.2) or the method of the mean creep-inducing stress shown in Section 3.2.

For such cases, the creep fiber method of Busemann [21], which has almost fallen into oblivion, also proves to be particularly practical. With this method, a decoupling of both differential equations is reached by selecting two particular fibers within the cross section. A normal force acting in one fiber does not affect the other fiber. We then assume the cross-section characteristics and action effects to be concentrated in those so-called creep fibers. Thus we can reduce the complicated problem to a simple analysis of two concentrically loaded composite members which are simply and internally statically indeterminate.

This method is suited not only for two-strand prestressing, but also for composite structures. It can be used successfully even if a cross section consists of two different types of concrete (e.g., prefabricated member and cast in situ concrete).

In order to determine the position of the creep fibers, Busemann suggested a graphic method. With this method, we plot the center of gravity S_c of the concrete cross section and S_s of the steel cross section, their radii of gyration $r_c = \sqrt{I_c/A_c}$ and $r_s = \sqrt{I_s/A_s}$ (see an example of a composite structure in Fig. 6.1) and we draw a semicircle through extreme points A and B, the center point M of which is on the baseline. The points of intersection of the circle and baseline determine the creep fibers CI and CII.

Most of the time, however, the position of the creep fibers can be determined just as quickly and more reliably as follows. We then have

$$cb = r_s^2, \qquad a'b' = r_c^2 \tag{6.1}$$

and

$$a + b = a' + b' = c = m + a + b'. \tag{6.2}$$

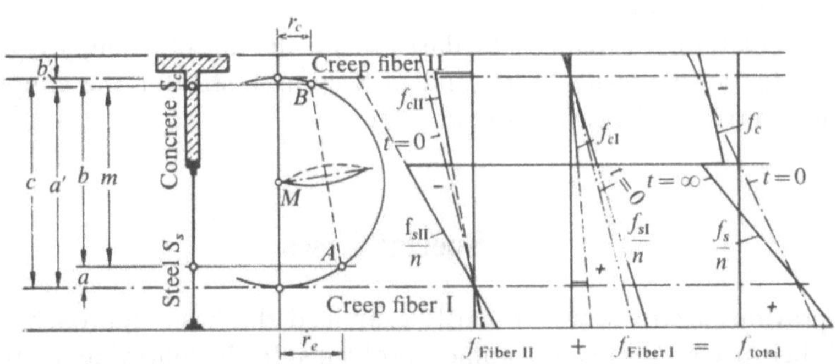

Figure 6.1. Example of the creep fiber method according to Busemann.

By using the auxiliary value x,

$$\varkappa = r_s^2 + m^2 - r_c^2, \tag{6.3}$$

we obtain,

$$b' = \frac{\sqrt{\varkappa^2 + 4m^2 r_c^2} - x}{2m}, \tag{6.4}$$

as well as

$$a' = r_c^2/b', \tag{6.5}$$

$$a = a' - m, \tag{6.6}$$

$$c = a' + b', \tag{6.7}$$

$$b = c - a. \tag{6.8}$$

If we divide both cross sections into C_{I} and C_{II}, then each creep fiber will be assigned two equivalent cross sections.

$$A_{s\mathrm{I}} = A_s \frac{b}{c}, \qquad A_{c\mathrm{I}} = A_c \frac{b'}{c}, \tag{6.9}$$

$$A_{s\mathrm{II}} = A_s \frac{a}{c}, \qquad A_{c\mathrm{II}} = A_c \frac{a'}{c}. \tag{6.10}$$

The moment M, which acts on the total cross section and the normal force N are also replaced by equivalent normal forces N_{I} and N_{II}, which act in C_{I} and C_{II}. If the action effects are related to the point M, we obtain

$$N_{\mathrm{I}} = \frac{N}{2} + \frac{M}{c} \quad \text{and} \quad N_{\mathrm{II}} = \frac{N}{2} - \frac{M}{c}. \tag{6.11}$$

We now have to calculate separately the creep redistribution for each of the columns thus formed. Subsequently, we have to connect the concrete and steel stresses calculated for C_{I} and C_{II} through a straight line with C_{II} and C_{I}. Through superposition (compare Fig. 6), we then obtain the stress distribution in the composite cross section.

This calculation is explained in Part C with the aid of examples; that is, in Section 5, for a composite beam, and in Section 11.2.2, for the coupling of a prefabricated part and a cast in situ concrete.

7. Special Cases

In the previous sections, we primarily examined the basic approaches and explained their use by means of common problems. In the following, more complicated cases are described.

7.1. Multiple Externally Statically Indeterminate Structures

The matrix for the elastic deformations has to be expanded by the time-dependent plastic deformations. We then obtain a system of coupled linear differential equations which can be solved only with considerable calculation effort. In many cases, it is preferable and sufficient to use approximations.

Iteration and finite differences are not the only methods which can be used for this. The methods of the mean creep-inducing stress, in accordance with Section 3.2, and the creep approaches by Trost (see Section 4) and/or Bažant (Section 5) have proven particularly practical and, above all, very illustrative for such cases. This is demonstrated in the example in Part C, Section 8.3, where different methods were employed.

7.2. Varied Creep Behavior in a Statically Indeterminate Structure which Has Been Constructed in Sections

As in Section 7.1, the approaches of Trost [20] and/or Bažant [12], as well as the method for the mean creep-inducing stress according to Section 3.2, are particularly suited for solving this problem. It is explained in the following with the example of a frame, the column of which is older than the girder (see Fig. 7.1). We are looking for the fixed end moment X_1 of the girder. In the example, the superscript S stands for the column and R for the girder.

The compatibility conditions at time $t = 0$ are

$$X_{1t=0}\delta_{11} = X_0\delta_{11} = -\delta_{10}, \tag{7.1}$$

$$X_0 = -\frac{\delta_{10}}{\delta_{11}} = -\frac{\delta_{10}^S + \delta_{10}^R}{\delta_{11}^S + \delta_{11}^R} = \frac{dh/2E_c'A + dl^3/24E_cI}{h/l^2E_c'A + l/3E_cI}. \tag{7.2}$$

To solve the compatibility conditions at time t, the following methods of analysis are available:

(a) According to Trost (see Section 4); or Bažant (see Section 5)

$$\varepsilon_t = \frac{f_0}{E_c}(1 + \varphi) + \frac{f_t - f_0}{E_c}(1 + \rho\varphi); \tag{4.1}$$

when used with Eqs. (7.1) and (7.2), we obtain for X_t:

$$X_0[\delta_{11}^S(1 + \varphi_1) + \delta_{11}^R(1 + \varphi_2)] + (X_t - X_0)$$
$$[\delta_{11}^S(1 + \rho_1\varphi_1) + \delta_{11}^R(1 + \rho_2\varphi_2)] = -\delta_{10}^S(1 + \varphi_1) - \delta_{10}^R(1 + \varphi_2). \tag{7.3}$$

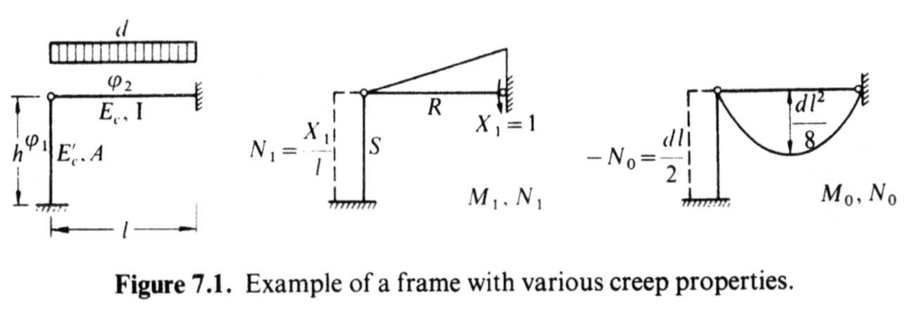

Figure 7.1. Example of a frame with various creep properties.

(b) According to Section 3.2:

$$\varepsilon_t = \frac{f_0}{E_c} - \frac{f_0 - f_t}{E_c} + \frac{f_0 + f_t}{2E_c}\varphi = \frac{f_t}{E_c} + \frac{f_0 + f_t}{2E_c}\varphi; \qquad (3.5)$$

when used with Eq. (7.2), we obtain, for X_t,

$$X_t(\delta_{11}^S + \delta_{11}^R) + (X_0 + X_t)\left(\delta_{11}^S \frac{\varphi_1}{2} + \delta_{11}^R \frac{\varphi^2}{2}\right)$$

$$= -\delta_{10}^S(1 + \varphi_1) - \delta_{10}^R(1 + \varphi_2). \qquad (7.4)$$

(c) According to the Dischinger–Kupfer method (see Section 2.2.4): for the statically indeterminate system, shown in Fig. 7.1, we can fall back on the expanded approaches by Dischinger, described in Section 2.2.4. We then obtain

$$X_t = \frac{\delta_{10}^S \gamma_{\text{sus}\,S}(1 + 0.4\alpha_S)/1.4 + \delta_{10}^R \gamma_{\text{sus}\,R}(1 + 0.4\alpha_R)/1.4}{\delta_{11}^S + \delta_{11}^R}. \qquad (7.5)$$

The coefficients $\gamma_{\text{sus}\,S}$ and $\gamma_{\text{sus}\,R}$ can be taken from Fig. 2.5 by using the following values.

For $\gamma_{\text{sus}\,S}$,

$$\alpha_S = \frac{\varphi_{SR} - 0.4}{1.4\varphi_1 - 0.4\varphi_{SR} - 0.4} \text{ and } \varphi = \varphi_1;$$

for $\gamma_{\text{sus}\,R}$,

$$\alpha_R = \frac{\varphi_{SR} - 0.4}{1.4\varphi_2 - 0.4\varphi_{SR} - 0.4} \text{ and } \varphi = \varphi_2,$$

where

$$\varphi_{SR} = \frac{\delta_{11}^S\varphi_1 + \delta_{11}^R\varphi_2}{\delta_{11}^S + \delta_{11}^R}. \qquad (7.6)$$

All three methods of analysis lead to satisfactory solutions, as shown in Table 7.1.

Table 7.1. Comparison of solutions for $E_c'A/E_cI = 1$ and $l/h = 1$, $l = 7$ m (23 ft).

$X_t = K\dfrac{dl^2}{8}$	K for $t = 0$ and/or	K for $t = \infty$	
	$\varphi_1 = \varphi_2$	$\varphi_1 = 0; \varphi_2 = 2$	$\varphi_1 = 2; \varphi_2 = 0$
According to Trost, Bažant		1.03	1.46
According to Section 3.2	1.17	1.01	1.48
According to Dischinger–Kupfer (systems creep factor)		1.02	1.47

7.3. Restraint Develops Gradually, Though Not Similar to Creep

7.3.1. *According to Dimel*

The case of an imposed action which does not develop similar to creep is particularly important for calculating the effects of foundation settlement in cohesive soil. Dimel [22] demonstrated a method for solving this which is based upon Dischinger's considerations. He set up the following simple relationship between time development of the imposed action and creep:

$$\varphi_t = (1 - e^{-ct})\varphi_\infty \tag{7.7}$$

describes the time development of creep. (For the evaluation in [22], $c = 0.0289$, had been chosen whereby t designates the time in months).

The settlement function is described by

$$a_t = (1 - e^{-\lambda ct})a_\infty. \tag{7.8}$$

The general solution for the differential equation is, then,

$$X_t = e^{\varphi_\infty e^{-ct}}\left(C + \lambda c X^{(e)} \int_0^t e^{-\lambda ct} e^{-\varphi_\infty e^{-ct}}\, dt\right). \tag{7.9}$$

If, with $\lambda < 1$, the imposed action develops more slowly than creep, we have the maximum action effects at the end of the process. If it develops faster than creep ($\lambda > 1$), we have to pay attention to the high intermittent values. In Fig. 7.2, the

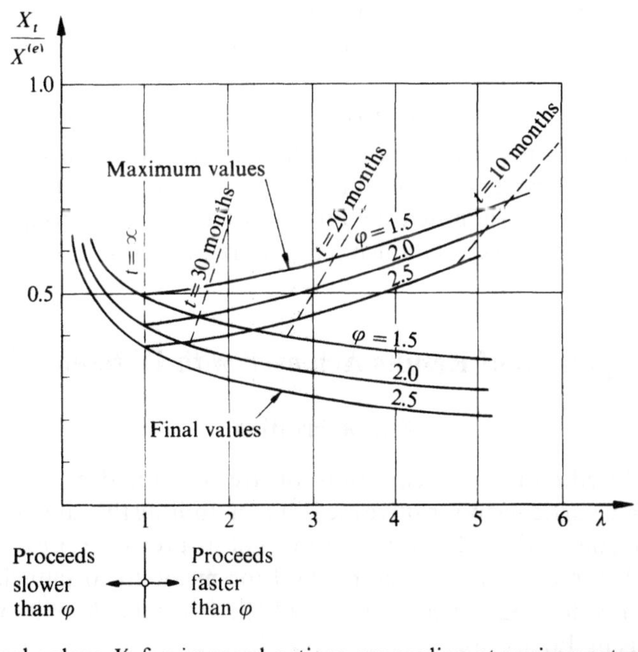

Figure 7.2. Final values X_t for imposed actions proceeding at various rates related to $X^{(e)}$ without creep effects. The dashed lines provide a reference point for the point in time when maximum values occur.

ratio of the maximum values X_t and the values $X^{(e)}$, calculated without creep, are plotted for some creep coefficients as a function of λ. The expansion of the original Dischinger solution, developed in Section 2.2, had been taken into account. The evaluation was done for $c = 0.0289$. According to the Rüsch–Jungwirth method (Part B, Section 4.3), this corresponds to the development of creep in time for $d_{\text{eff}} = 80$ cm.

7.3.2. *According to Trost and Mainz* [23]

Similar to Dimel, Trost and Mainz established a simple relation between the development of creep $\varepsilon(t)$ and of the imposed action $S_r(t)$ (e.g., settlement):

$$S_r(t) = \varepsilon(t)^{\bar{\lambda}} S_{\infty}. \tag{7.10}$$

As can be seen in the works of Dimel and Trost, λ Eq. (7.8) and $\bar{\lambda}$ Eq. (7.10) have a different meaning. In Trost's work the assessed time development of concrete creep holds true only for very thin bodies, $d_{\text{eff}} = 10$ cm. $\bar{\lambda} = 0$ corresponds to a suddenly occurring imposed action, and $\bar{\lambda} = 1$ applies to that imposed action which develops in time similar to creep. In accordance with Section 4.1, we then have

$$f_t = f_0\left(1 + \frac{\varphi_t}{1 + \rho\varphi_t}\right) + \frac{\varepsilon_t E_c}{1 + \rho\varphi_t} = f_0\left(1 - \frac{\varphi_t}{1 + \sigma\varphi_t}\right) + \frac{f_{rt}^{(e)}}{1 + \rho\varphi_t}. \tag{7.11}$$

The first term in Eq. (7.11) covers the suddenly occurring restraint and the second term covers the gradually developing restraint.

The magnitude of ρ has been obtained step-by-step in [23] and was tabulated as a function of the creep coefficient, the load age, and the parameter $\bar{\lambda}$.

7.3.3. *With Mean Creep-Inducing Action Effects*

The method for mean creep-inducing action effects, mentioned in Section 3.2, provides a good approximation in three to four steps. It has the additional advantage that creep and imposed action do not have to be coupled through a mathematically formulated relationship (see Part C, Section 9.3.6).

8. Comparison of Results Achieved with Different Methods

8.1. Relaxation

The creep-induced reduction (relaxation) of stresses caused by imposed constant deformation is a particularly sensitive measure for limiting the range of applicability of approximate methods. In Fig. 8.1, the results of six previously described methods are compared. We can see that the method for mean creep inducing stresses is reliable over a wide range when using Eq. (3.15). It is reliable only over a narrow range when using Eq. (3.16).

It had to be expected that the original Dischinger equation would predict too great a stress loss. The values determined in accordance with Trost and/or Bažant

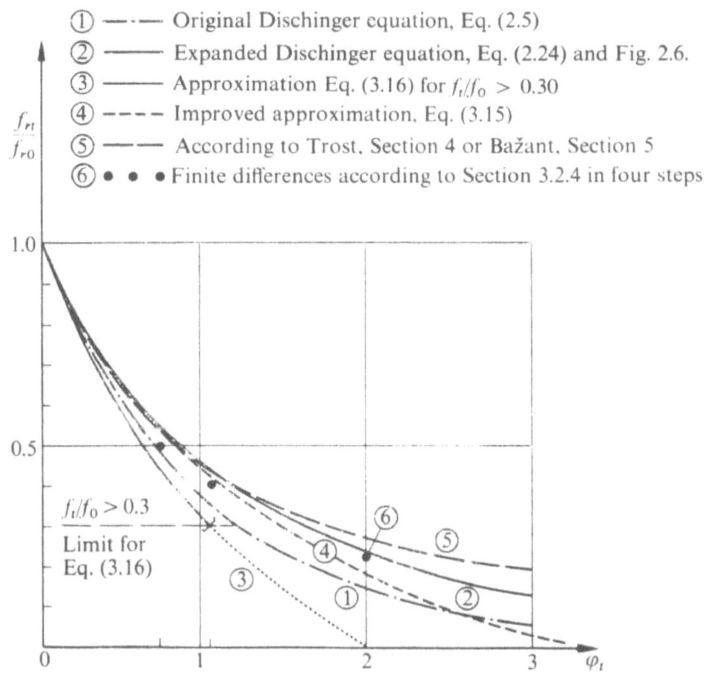

Figure 8.1. Relaxation of immediately occurring restraint $\alpha = 1$; comparison of results of different methods of analysis.

are somewhat higher than those predicted by the improved Dischinger equation. The observed differences are not the result of the different methods, but stem from the different assumptions regarding the magnitude of delayed elasticity.

The range for which we obtain acceptable results with the simple equations (3.24) and (3.25), which are based on the mean creep-inducing stress, is surprising (see Section 3.24).

Table 8.1 attempts to obtain an objective evaluation by comparing analytical results to experimental data. An attempt has been made to minimize the inevitable high scatter of experimental data, particularly obvious in φ_t, as much as possible by averaging the experimental data. As far as the limited number of experiments allows conclusions the values from the improved Dischinger equation are somewhat too low.

8.2. General Behavior

The above described differences are considerably smaller for normal cases of application. This is particularly obvious in Fig. 8.2 which shows the change in the small coupling force which is caused by sustained loads. For imposed action stresses (Fig. 8.3) as well, the differences are normally smaller than in Fig. 8.1, which shows only the special case $\alpha = 1$. The same is true for shrinkage effects (Fig. 8.4).

Table 8.1. Comparison of the results of relaxation experiments and analytical methods.

Reference	Experimental series	f_t/f_0 N/mm² (psi)	Observation period (days)	φ_t	f_t/f_0 Dischinger (Section 2.1)	f_t/f_0 Dischinger (Section 2.2)	Trost (Section 4) or Bažant (Section 5)
[24]	O	3.2/10.32 (459/1480) = 0.310		1.866	0.150	0.250	0.285
	L	3.5/ 9.58 (502/1374) = 0.360	1315	1.60	0.204	0.305	0.320
	LS	3.7/10.81 (531/1550) = 0.345		1.37	0.255	0.356	0.360
	LF	3.3/ 9.47 (473/1358) = 0.345		1.25	0.285	0.390	0.390
Technical University of Munich	KL–R 1	0.400		1.40	0.250	0.350	0.350
			90				
(unpublished)	LM–R 160	0.290		1.74	0.180	0.280	0.310
Sum:		2.050			1.324	1.931	2.015
Deviation of average values from measured values in %					$\frac{2.050 - 1.324}{2.050}100 = 35\%$	$\frac{2.050 - 1.931}{2.050}100 = 6\%$	$\frac{2.050 - 2.015}{2.050}100 = 2\%$

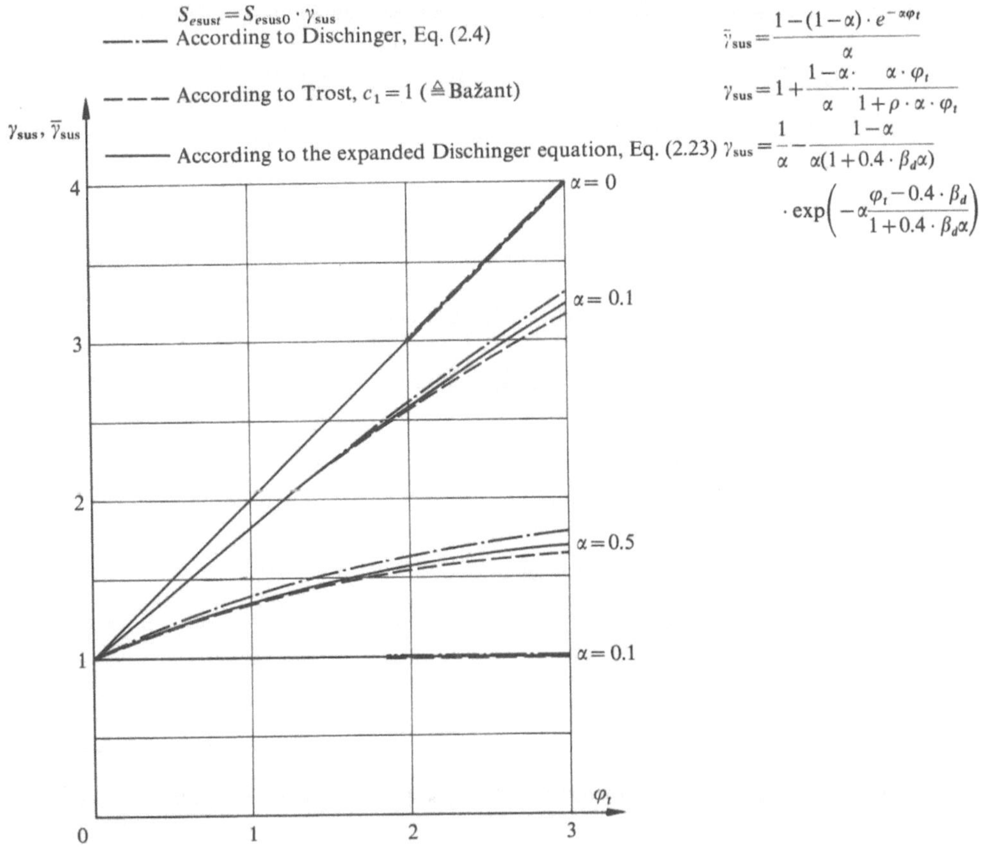

Figure 8.2. Comparison of γ_{sus} for sustained loads as obtained from different methods.

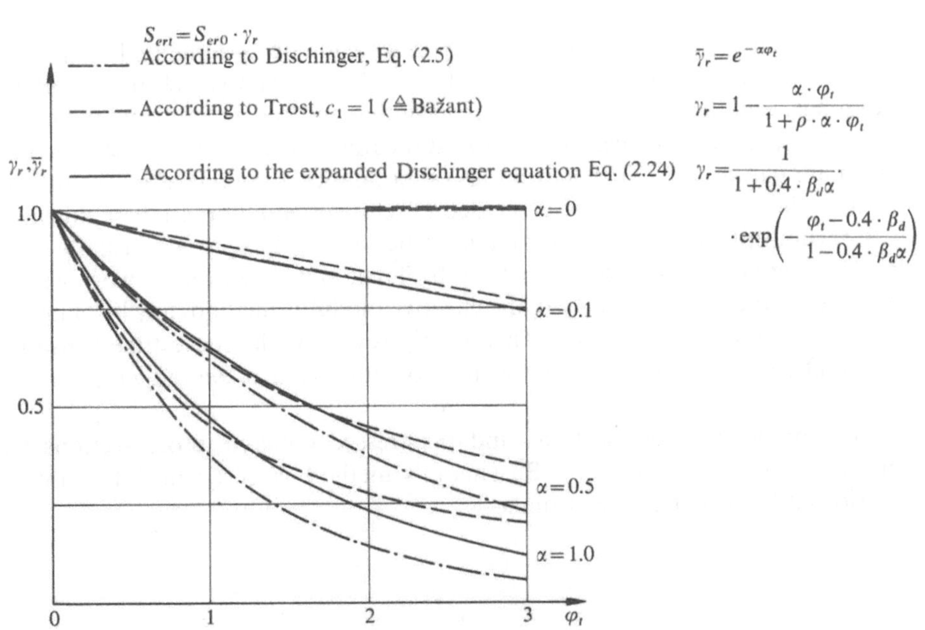

Figure 8.3. Comparison of γ_r for immediately occurring restraint as obtained from different methods.

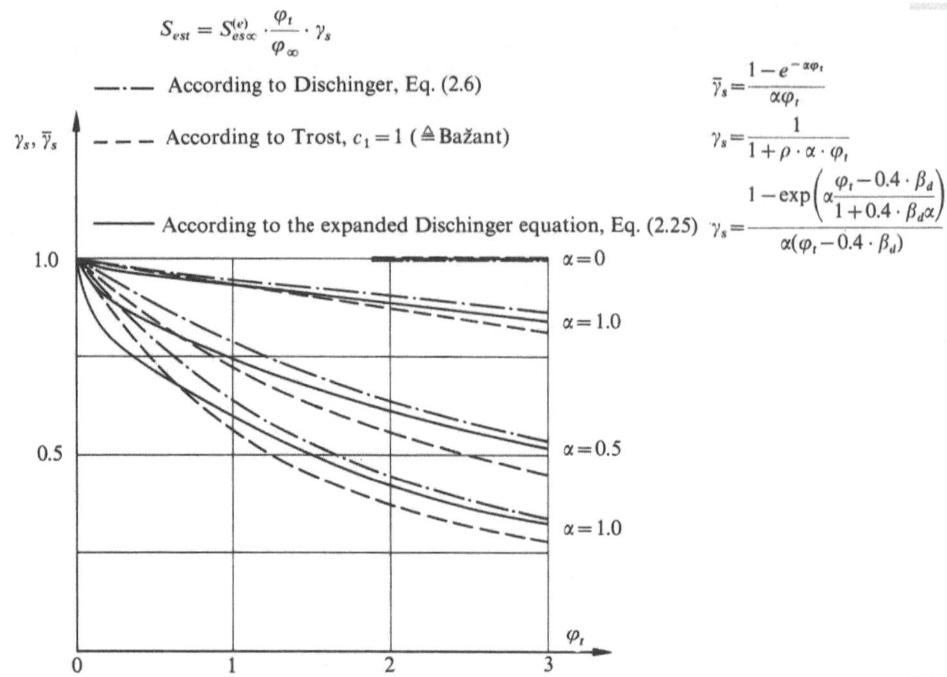

Figure 8.4. Comparison of γ_s for restrained shrinkage as obtained from different methods.

9. Outlook

It is apparent that for simply statically indeterminate systems and small stress changes, all discussed methods can be used. The simple, clear formulas of Dischinger and the method for mean creep-inducing stress recommend themselves.

For large stress changes and simply statically indeterminate systems or those which are based on it, Trost's formulas (after adapting the relaxation values) and those of Bažant, as well as the expanded Dischinger equation are best suited.

For multiple indeterminate systems, the stress changes are usually small. We can therefore either use the method by Trost and/or Bažant or work with the mean creep-inducing stress. Also, the effective modulus method may be of interest. In very complicated cases, we can possibly revert to the more time-consuming finite-difference method and/or the method for mean creep-inducing stress or action effects.

For composite cross sections and prestressed concrete cross sections with multiple-strand reinforcement, Busemann's method is best suited to solve the interior multiple static indeterminacy.

References for Appendix II

[1] Dischinger, F.: "Elastische und plastische Verformungen der Eisenbetontragwerke und insbesondere der Bogenbrücken" (Elastic and plastic deformations of reinforced concrete structures and particularly arched bridges), *Der Bauingenieur*, **20**, No. 5/6, 1939; and "Untersuchungen über die Knicksicherheit, die elastische Verformung und das Kriechen des Betons bei Bogenbrücken" (Studies on the buckling safety, the elastic deformation and concrete creep in arched bridges), *Der Bauingenieur*, **18**, No. 33/34, 1937.

[2] Whitney, C. S.: "Plain and reinforced concrete arches. Progress report on the limitations of the theory of elasticity and the effect of plastic flow, temperature variations and the Freyssinet-method of adjustment," *J. Amer. Concr. Inst.* **3**, 7, 1932.

[3] Boltzmann, L.: "Zur Theorie der elastischen Nachwirkung" (Concerning the theory of elastic aftereffects), Wiener Bericht, 10, 1874.

[4] Arutjunjan, N. Kh.: "Some Problems in the Theory of Creep" (in Russian), Gostechisdat, 1962; and "Some Problems in the Theory of Creep in Concrete Structures," Pergamon Press, Oxford, 1966 (in Russian, Moscow, 1952).

[5] Alexandrowski, S. W.: "Analysis of reinforced concrete and concrete structures in view of the effects of temperature and humidity, taking into consideration creep" (in Russian), Moscow, 1966.

[6] Trost, H.: "Spannungs-Dehnungs-Gesetz eines viskoelastischen Festkörpers wie Beton und Folgerungen für Stabtragwerke aus Stahlbeton und Spannbeton" (Stress-strain-law for a viscoelastic body such as concrete and the consequences concerning the behavior of reinforced and prestressed concrete structures), *Beton*, **16**, No. 6, 1966.

[7] Zerna, W. and Trost, H.: "Rheologische Beschreibungen des Werkstoffes Beton" (Rheological description of the structural material concrete), *Beton- und Stahlbetonbau*, **62**, No. 7, 1967.

[8] Schade, "Einige eindimensionale Ansätze zur Berechnung des Kriechens und der Relaxation von Betontragwerken" (Some one-dimensional approaches to creep and relaxation in concrete structures), *Beton- und Stahlbetonbau*, **67**, No. 3, 1972.

[9] Comité Européen du Béton: "Structural effects of time-dependent behaviour of concrete," CEB Bulletin d'Information No. 90, Paris, April 1973 and No. 94, Paris, October 1973.

[10] Trost, H.: "Auswirkungen des Superpositionsprinzips auf Kriech- und Relaxationsprobleme bei Beton und Spannbeton" (Effects of the superposition principle on creep and relaxation problems in concrete and prestressed concrete), *Beton- und Stahlbetonbau*, **62**, No. 10, 1967.

[11] Zerna, W.: "Spannungs-Dehnungs-Beziehung für Beton bei einachsiger Beanspruchung" (Stress-strain-relation for concrete under uniaxial stresses), taken from "Aus Theorie und Praxis des Stahlbetonbaues," Verlag Wilhelm Ernst and Sohn, Berlin, 1969.

[12] Bažant, Z. P.: "Prediction of concrete creep effects using age-adjusted effective modulus method," *J. Amer. Concr. Inst.* **69**, 1972.

[13] Grasser, E.: "Der Einfloß der verzögert elastischen Verformung auf das Verhalten von Betontragwerken" (The influence of delayed elastic strain on the behavior of concrete structures), Internal Report of the Institute for Concrete Construction, Technical University of Munich, July, 1964.

[14] Fuglsang Nielsen, L.: "Kriechen und Relaxation des Betons" (Creep and relaxation of concrete), *Beton- und Stahlbetonbau*, **65**, No. 11, 1970.

[15] Rüsch, H. and Kupfer, H.: "Bemessung von Spannbetonbauteilen" (Dimensioning of reinforced concrete elements), Beton-Kalender, Verlag Wilhelm Ernst and Sohn, Berlin, 1980.

[16] Schade, D. and Haas, W.: "Elektronische Berechnung der Auswirkungen von Kriechen und Schwinden bei abschnittsweise hergestellten Verbundstabwerken" (Computer calculation of the effects of creep and shrinkage on composite frames with rigid joints which have been constructed in sections), Proceedings Deutscher Ausschuß für Stahlbeton, No. 244, Berlin, 1975.

[17] Wolff, H. J. and Mainz, B.: "Einfluß des Betonzeitverhaltens" (The influence of concrete time behavior), Werner-Verlag, Düsseldorf, 1972.

[18] Schade, D.: "Alterungsbeiwerte für das Kriechen von Beton nach den Spannbetonrichtlinien" (Aging coefficients for concrete creep in accordance with the German code for prestressed concrete), *Beton- und Stahlbetonbau*, **72**, No. 5, 1977.

[19] Knittel, G.: "Der Einfluß des Kriechens und Schwindens auf den Spannungszustand in Tragwerken aus vorgespanntem Beton" (The influence of creep and shrinkage on the state of stress in prestressed concrete structures), *Der Bauingenieur*, **29**, No. 1, 1954.

[20] Trost, H. and Wolff, H. J.: "Zur wirklichkeitsnahen Ermittlung der Beanspruchungen in abschnittsweise hergestellten Spannbetontragwerken" (Realistic determination of loads in prestressed concrete structures which have been erected in sections), *Der Bauingenieur*, **45**, No. 5, 1970.

[21] Busemann, R.: "Kriechberechnung von Verbundträgern unter Benutzung von zwei Kriechfasern" (Creep calculation of composite girders using two creep fibers), *Der Bauingenieur*, **25**, No. 11, 1950.

[22] Dimel, E.: "Der Einfluß des Betonkriechens auf Setzungszwängungen" (The influence of concrete creep on settlement restraint), *Beton- und Stahlbetonbau*, **60**, No. 6, 1965.

[23] Trost, H. and Mainz, B.: "Zur Auswirkung von Zwängungen in Spannbetontragwerken" (On the effects of restraint in prestressed concrete structures), *Beton- und Stahlbetonbau*, **65**, No. 8, 1970.

[24] Rostasy, F. S., Teichen, K. Th., and Engelke, H.: "Beitrag zur Klärung des Zusammenhanges von Kriechen und Relaxation bei Normalbeton" (Correlation of creep and relaxation of normal weight concrete), Report No. 57 of the Otto-Graf-Institute, University of Stuttgart, Germany, 1971.

Index

FSC
www.fsc.org
MIX
Papier aus verantwortungsvollen Quellen
Paper from responsible sources
FSC® C105338

If you have any concerns about our products,
you can contact us on
ProductSafety@springernature.com

In case Publisher is established outside the EU,
the EU authorized representative is:
Springer Nature Customer Service Center GmbH
Europaplatz 3, 69115 Heidelberg, Germany

Printed by Libri Plureos GmbH
in Hamburg, Germany